Variational Principles
for Nonpotential Operators

TRANSLATIONS OF MATHEMATICAL MONOGRAPHS

VOLUME **77**

Variational Principles
for Nonpotential Operators

V. M. FILIPPOV

American Mathematical Society · Providence · Rhode Island

В. М. ФИЛИППОВ

ВАРИАЦИОННЫЕ ПРИНЦИПЫ ДЛЯ НЕПОТЕНЦИАЛЬНЫХ ОПЕРАТОРОВ

«НАУКА», МОСКВА, 1985

Translated from the Russian by J. R. Schulenberger
Translation edited by Ben Silver

1980 *Mathematics Subject Classification* (1985 *Revision*). Primary 35A15, 47A50, 47H17; Secondary 46E35.

ABSTRACT. A variational method for solving linear equations with B-symmetric and B-positive operators and its generalization to nonlinear equations with nonpotential operators are developed. A constructive extension of the variational method to "nonvariational" equations (including parabolic equations) is carried out in classes of functionals which differ from the Euler-Lagrange functionals; in connection with this some new function spaces are considered.

The book is intended for mathematicians working in the areas of functional analysis and differential equations and also for specialists, graduate students, and students in advanced courses who use variational methods in solving linear and nonlinear boundary value problems in continuum mechanics and theoretical physics.

Bibliography: 356 titles.

Library of Congress Cataloging-in-Publication Data

Filippov, V. M.
    [Variatsionnye printsipy dlia nepotentsial′nykh operatorov. English]
    Variational principles for nonpotential operators/V.M. Filippov.
    p. cm. – (Translations of mathematical monographs; v. 77)
    Title on verso t.p.: Variatsionnye printsipy dlia nepotentsial′nykh operatorov.
    Translation of: Variatsionnye printsipy dlia nepotentsial′nykh operatorov.
    Bibliography: p.
    ISBN 0-8218-4529-2 (alk. paper)
    1. Nonlinear operators. 2. Variational principles. 3. Differential equations, Partial.
I. Title. II. Title: Variatsionnye printsipy dlia nepotentsial′nykh operatorov. III. Series.
QA329.8F5513 1989   89-6904
515′.7248–dc20   CIP

Information on Copying and Reprinting can be found at the back of this volume.
The paper used in this book is acid-free and falls within the guidelines
established to ensure permanence and durability. ⊗
This publication was typeset using $\mathcal{A}_{\mathcal{M}}\mathcal{S}$-TEX,
the American Mathematical Society's TEX macro system.

Contents

Preface

In the domestic and foreign literature there are remarkable books devoted to the variational method of investigating equations with selfadjoint operators (in the nonlinear case with potential operators). At a mathematical level these are primarily the monographs of S. L. Sobolev, *Applications of Functional Analysis in Mathematical Physics*, L. D. Kudryavtsev, *Direct and Inverse Imbedding Theorems. Applications to the Solution of Elliptic Equations by a Variational Method*, and M. M. Vaĭnberg, *Variational Methods of Investigating Nonlinear Equations*. Other books by S. G. Mikhlin, K. O. Friedrichs, and K. Rektorys were of major importance for the popularization of variational methods in applications.

The development of computational techniques and the possibility of automating variational methods aroused the current interest in questions of further improving them and constructive extension to new classes of equations. The books of W. Velte [337], G. I. Marchuk and V. I. Agoshkov [207], and P. Blanchard and E. Brüning [39] which have appeared in recent years are partially devoted to these problems. However, in the mathematical literature there is no single monograph specially devoted to variational principles for equations with nonsymmetric (nonpotential) operators, although the basic mathematical aspects of this direction were worked out by A. E. Martynyuk, W. V. Petryshyn, and V. M. Shalov back in 1957–1965. The present book is intended to fill this gap to some extent.

In applications most people know the practical value of the fact that for a particular boundary value problem (generally speaking, with an unbounded nonpotential operator) it is possible to construct a functional analogous in variational properties to the Dirichlet functional for the Laplace equation; but variational principles for nonsymmetric nonpotential operators have not been widely used either by mathematicians or in applications. One of the basic reasons for this is, apparently, the complexity of a constructive approach to the necessary "symmetrizing" operators. In Marchuk and Agoshkov's book [207] many results are presented for equations with B-symmetric and B-positive-definite operators (in the sense of W. V. Petryshyn), but the symmetrization of the neutron-transport equation presented as a substantial example (a result obtained by V. S. Vladimirov [340] in his doctoral dissertation) is one of the few cases of constructive execution of symmetrization of an operator of a rather complex boundary value problem.

The theoretical foundations of a variational method for investigating linear equations with, generally speaking, nonsymmetric and nonpositive operators are presented in the first part of the book (Chapters 1 and 2). Chapter 3 is devoted to a constructive approach to variational principles for various boundary value problems for partial differential equations. Generalizations of the variational method of investigating nonlinear equations with nonpotential operators proposed by M. Z. Nashed [235], A. D. Lyashko [202], and E. Tonti [324] are developed in Chapter 4.

The development of a variational method for investigating a differential equation $N(u) = f$ is closely connected with the inverse problem of the calculus of variations, and throughout the entire book a quasiclassical solution of this inverse problem is investigated in the sense of seeking functionals $F[u]$, bounded below in some Hilbert space, which contain derivatives of the unknown function u of lower order than in the equation $N(u) - f = 0$ and are such that the set of solutions of the equation coincides with the set of critical points of the functional.

Many of the results presented here were reported in 1961–1983 by V. M. Shalov and by the author to the Steklov Mathematical Institute of the Academy of Sciences of the USSR in the seminar of Academician S. M. Nikol'skiĭ and Corresponding Member of the Academy of Sciences of the USSR L. D. Kudryavtsev; the author is deeply grateful to all participants of the seminar for their attention. During one report of the author Professor Kudryavtsev pointed out the expediency of investigating nonclassical boundary value problems for partial differential equations (PDE) by the variational method developed here; in Chapter 3 we carry out constructive approaches to variational principles and their investigation, mainly for known equations of mathematical physics with nonclassical boundary conditions or for nonclassical PDE.

Also during one of the author's reports Academician Nikol'skiĭ noted that if a linear boundary value problem can be rigorously investigated by the Bubnov-Galerkin method, then it is possible to select a corresponding quadratic functional of the variational method, while in the nonlinear case with a nonnegative operator the selection is, of course, almost impossible; in Chapter 4 we prove the existence of solutions of the inverse problem of the calculus of variations for broad classes of nonlinear equations and develop a variational method for investigating nonlinear equations with nonpotential operators.

I am especially grateful to Corresponding Member of the Academy of Sciences of the USSR L. D. Kudryavtsev, whose ideas regarding the development and application of variational methods constantly aided me in the work.

A large part of the book was written during my stay in 1983–1984 at the Free University of Brussels, and I should thank Professors J. P. Gosset, I. R. Prigogine, and P. Glansdorf (Belgium) for the assistance and support they showed me. I am grateful also to Professor V. I. Burenkov, Professor A. N. Skorokhodov, and A. Yu. Rodionov, who examined individual sections and made a number of valuable remarks. Especially valuable were the remarks and discussion of the results obtained with one of the founders of the theory developed here—Professor W. V. Petryshyn (USA)—to whom I am deeply grateful.

The Author

Notation and Terminology

1. R_n is the n-dimensional Euclidean, arithmetic space of points $x = (x_1, \ldots, x_n)$, $|x| = (\sum_1^n x_i^2)^{1/2}$, and $x\xi = \sum_1^n x_i \xi_i$.

2. Ω is a domain, an open connected set in R_n (which is bounded everywhere in this book), with boundary $\partial\Omega$ which is piecewise smooth unless higher smoothness is specified; and $\bar{\Omega}$ is the closure of Ω in R_n ($\bar{\Omega} = \Omega \cup \partial\Omega$; $\bar{\Omega}$ is compact).

3. $C^k(\Omega)$ $(C^k(\bar{\Omega}))$ is the set of functions having uniformly continuous derivatives in Ω $(\bar{\Omega})$ through kth order, $k \geq 0$, and the norm is

$$\|u|C^k(\bar{\Omega})\| = \sum_{|\bar{\alpha}| \leq k} \max_{\Omega} |D^{\bar{\alpha}} u(x)|.$$

$\overset{\circ}{C}{}^l(\Omega) \equiv \overset{\circ}{C}{}^l(\bar{\Omega})$ is the set of functions in $C^l(\bar{\Omega})$ which vanish on $\partial\Omega$ together with all partial derivatives through order l.

$C_0^l(\Omega)$ is the set of functions in $C^l(\bar{\Omega})$ with compact support in Ω.

$\overset{\circ}{C}{}^\infty(\Omega)$ is the set of functions for which all partial derivatives of any order in Ω exist and vanish on $\partial\Omega$.

4. $D^{\bar{\alpha}} = \partial^{|\bar{\alpha}|}/\partial x_1^{\alpha_1} \cdots \partial x_n^{\alpha_n}$, where $\bar{\alpha} = (\alpha_1, \ldots, \alpha_n)$ and $|\bar{\alpha}| = \sum_1^n \alpha_j$; $\alpha_j \geq 0$, $j = 1, \ldots, n$.

$\Delta = \sum_1^n \partial^2/\partial x_k^2$ is the Laplacian.

$\nabla u \equiv \operatorname{grad} u = (u_{x_1}, \ldots, u_{x_n})$, where $u_{x_i} = \partial u(x)/\partial x_i$, $i = 1, \ldots, n$,

Finally, $|\nabla u|^2 = \sum_1^n (u_{x_i})^2$.

5. $L_p(\Omega)$ is the (linear, normed, complete) real Banach space of measurable functions for which the norm

$$\|u|L_P(\Omega)\| = \left(\int_\Omega |u(x)|^p \, dx \right)^{1/p}, \qquad \infty > p \geq 1,$$

exists and is finite. An equivalent definition of $L_p(\Omega)$ is the completion in the norm $\| \cdot |L_p(\Omega)\|$ of the sets of functions $C(\Omega)$, $\overset{\circ}{C}{}^{\bar{\alpha}}(\Omega)$, $C_0^{\bar{\alpha}}(\Omega)$

$(\bar{\alpha} = (\alpha_1,\ldots,\alpha_n),\ \alpha_j \geq 0,\ j = 1,\ldots,n)$, and $C_0^\infty(\Omega)$. The Sobolev spaces $W_p^m(\Omega)$ are defined as follows:

$$W_p^m(\Omega) = \{u(x)\colon D^\Omega u \in L_p(\Omega),\ |\alpha| \leq m\}, \qquad \infty > m \geq 1,$$

$$\|u|W_p^m(\Omega)\| = \left[\int_\Omega \left\{|u|^p + \sum_{|\bar{k}|\leq m} |D^{\bar{k}}u(x)|^p\right\} dx\right]^{1/p},$$

where the $D^{\bar{k}}$ are generalized partial derivatives in the Sobolev sense, $\bar{k} = (k_1,\ldots,k_n)$; $\overset{\circ}{W}_p^m(\Omega)$ is the closure of $\overset{\circ}{C}^\infty(\Omega)$ in the norm of $W_p^m(\Omega)$; for the domains Ω considered here the functions $u(x) \in \overset{\circ}{W}_p^m(\Omega)$ vanish on $\partial\Omega$ in the sense of $L_p(\partial\Omega)$ together with all j generalized derivatives to order $m - 1$.

Note that we used both $\|\cdot|W\|$ and $\|\cdot\|_W$ as the norm for a function space W. Moreover, in the definition of the norm of a negative space

$$\|u|W_2^-\| = \sup_{v\neq 0} \frac{|(u,v)|}{\|v|W_2^+\|}$$

to shorten notation we shall not always indicate that the sup is taken over all functions v in W_2^+.

6. The symbol $\forall$ denotes "for every" or "for any"; $\exists\alpha \in M$ means "there exists an element α in the set M".

$\exists!\alpha \in M$ means "in the set M there exists a unique element α".

$M \Leftrightarrow N$ means that assertion M holds if and only if assertion N holds or, in other words, for the validity of M it is necessary and sufficient that N be satisfied.

$A := B$ means "A is set equal to B". $l = 1,\ldots,n$ indicates that the quantity l takes integral values from 1 to n.

$C_0, C_1, C_2,\ldots$ denote positive constants not depending on the functions explicitly written in the formula in question. The notation $C_n(g)$ for some n means that this coefficients $C_n > 0$ depends on the function g.

The enumeration of the constants $C_0, C_1, C_2,\ldots$ is done independently in each section.

7. H and H_i everywhere denote Hilbert spaces; $\|u'|H_1\| \sim \|u'|H_2\|$ means the two norms are equivalent, i.e., there exist $C_1, C_2 > 0$ such that $\|u'|H_1\| \leq C_1\|u'|H_2\| \leq C_2\|u'|H_1\|$ for all $u' \in H_1$. A space H_1 is *equivalent* to a space H_2, $H_1 \sim H_2$, if $H_1 \subseteq H_2 \subseteq H_1$ and $\|u'|H_1\| \sim \|u'|H_2\|$.

$\bar{Q} = H$ means that the completion of the set Q in the norm of the space H is the entire space (set) H.

8. We shall sometimes write "a sequence $\{u_n\} \subset H$", having in mind that this sequence is formed from elements of the space H.

Strong convergence "$u_n \to u$ $(n \to \infty)$ in H" means convergence in the norm of this space: $\|u_n - u|H\| \to 0$ $(n \to \infty)$, u, $u_n \in H$ for all n.

Weak convergence "$u_n \to u$ $(n \to \infty)$ in H" means that $|(u_n - u, v)| \to 0$ $(n \to \infty)$ for all $v \in H$ and u_n, $u \in H$, where $(\cdot, \cdot)$ is the inner product in H.

9. $W_1 \to W_2$, i.e., "the space W_1 is imbedded in W_2", means that $W_1 \subseteq W_2$ and $\|u|W_2\| \leq C_n\|u'|W_1\|$ for all $u \in W_1$.

10. A function u is some function space $W_1(\Omega)$ has stable boundary values in the sense of a space $W_2(\Gamma)$ if the norm $\|u'|W_2(\Gamma)\|$ exists and the estimate

$$\|u|W_2(\Gamma)\| \leq C_n\|u|W_1(\Omega)\| \quad \forall u \in W_1(\Omega), \ \Gamma = \partial\Omega,$$

holds with a constant $C_n > 0$ not depending on u.

11. $D(A)$ is the domain and $R(A)$ the range of an operator A. Linearity of an operator A is understood only in the following sense:

$$A(\alpha_1 u_1 + \alpha_2 u_2) = \alpha_i A u_1 + \alpha_2 A u \quad \forall \alpha_1, \ \alpha_2 \in R_1, \ \forall u_1, u_2 \in D(A); \tag{1}$$
$$D(A, B) = \{v: v \in D(A) \cap D(B)\};$$
$$R_A(B) = \{Bv: v \in D(A)\}; \tag{1}$$

A^* is the adjoint operator, A^{-1} is the inverse operator, and I is the identity operator.

The notation $A \supseteq B$ for two operators A and B means that A is an extension of B, i.e., $D(A) \supseteq D(B)$ and $Au = Bu$ for all $u \in D(B)$.

12. The convolution of functions $(f * g)(x) = \int_{R_n} f(x - y)g(y)\,dy$.

13. Two equations are considered equivalent if any solution (in a particular sense) of one of them is a solution of the other as well.

14. Well-posedness of some problem $Au = f$ in a pair of spaces W, G is understood in the Hadamard sense: for any element $f \in G$ there exists a unique element u_0 in W—the solution of the equation—and this solution depends continuously on f: $\|u_0|W\| \leq C_0\|f|G\|$.

15. The book is broken into chapters, Chapters into sections, and some of the sections into subsections.

In a number $(m.n)$ of a formula the first number (m) denotes the number of the section, while the second (n) denotes the order number of this formula in §m. Similarly, "Theorem M.N" (or "Corollary M.N", "Lemma M.N", "Remark M.N", "Assertion M.N") means that this is the theorem (corollary, etc.) with order number N in §M.

_____

[1] Operators possessing this property will sometimes be called *distributive* below.

Introduction

In mathematical physics variational methods are understood to be methods making it possible to reduce a problem of integrating a differential equation to an equivalent variational problem, i.e., to finding a function for which some integral has an extremal value.

After Hilbert's paper *Über das Dirichlet'sche Princip* the variational method of investigating boundary value problems for partial differential equations was developed and received theoretical justification in works of both Soviet and foreign mathematicians (see the surveys [171] and [224]). However, the method was extended mainly to equations

$$Au = f \tag{1}$$

considered in some Hilbert space H (for simplicity here H is real), $\|u\|_H = \|u\| = (u, u)^{1/2}$, with a symmetric and positive operator A, $\overline{D(A)} = H$:

$$(Au, v) = (u, Av) \quad \forall u, v \in D(A), \tag{2}$$

$$(Au, u) > 0 \quad \forall u \in D(A), \ u \neq 0. \tag{3}$$

In 1920–1926 Academicians M. F. Kravchuk and N. M. Krylov indicated the possibility of applying direct methods for the construction of an approximate solutions of equation (1) if a condition generalizing (3) is satisfied:

$$(Au, Bu) > 0 \quad \forall u \in (A, B), \ u \neq 0, \tag{4}$$

for some auxiliary linear operator B. In 1957–1965 A. E. Martynyuk [209], W. V. Petryshyn [250], and V. M. Shalov [293]([2]) purposefully used condition (4) to develop a variational method of investigating and solving equation (1), generally speaking, with a nonsymmetric operator A of alternating sign. It was established in [293] that for any uniquely solvable linear equation in a Hilbert space one can formulate an equivalent variational

---

([2])Regarding some other works using a condition of the form (4), see the commentary to Chapter 1.

problem of finding a minimum point of some quadratic functional, so that the original equation together with the boundary conditions is equivalent to the Euler equation of this functional.

Since, on the one hand, for example, in mechanics [28], [133] and thermodynamics [131], [34] a number of variational principles for the equations of mathematical physics are used, while, on the other hand, various "semivariational" and "quasivariational" methods have been developed for solving practical problems in applications (see [3], [129], [233], [245], [276], and [288]), we emphasize that by a variational problem for equation (1) we shall understand the problem of minimizing a quadratic functional so that

A) in the case of a differential operator A the functional contains derivatives of the unknown function of lower order than in (1);

B) the functional is bounded below in some Hilbert space; and

C) the set of critical points of the functional coincides with the set of solutions of the equation being studied.

As a rule, in this book we present a construction of functionals having a unique minimum point, so that the minimizing element coincides with the (unique) generalized solution of the original equation.

We mention the theoretical and practical significance of conditions A)–C). Condition A) at a theoretical level makes it possible to investigate various generalized solutions for equations with "nonsmooth" coefficients, while in practice the presence in the functional of derivatives of lower order raises the stability of numerical methods (see [83], [336], and [59] and reduces the volume of computations (it is possible to choose a shorter Ritz series which is especially essential, for example, in the method of finite elements [88], [149], [343], [355]).

Condition B) makes it possible to use in proofs a well-developed scheme of a minimizing-sequence [311], while in applications it makes it possible to use various methods of minimizing a functional; the latter is important, in particular, if in numerical realization a system of equations with a large number of unknowns is obtained [61], [58], [235]. It has also been noted repeatedly (see [245] and [26]) that direct methods of the calculus of variations are especially effective in those cases where the functional has a unique critical point, and it is a maximum or minimum point of it.

Condition C) is a necessary condition for the application of the variational method, but there are a number of negative results just for it (see [1], [18], [54], and [226]) which explain why the variational method has been developed and applied mainly for equations of elliptic type.

The idea of extending the variational method to equation (1) with a nonsymmetric operator consists in the following. If for some auxiliary operator B, $\overline{D(A,B)} = H$, the property of B-symmetry is satisfied (for more details, see Chapter 1)

$$(Au, Bv) = (Bu, Av) \quad \forall u, v \in D(A, B) \tag{5}$$

as well as the property of B-positivity (4), then the problem of solving equation (1) turns out to be equivalent, under some additional conditions on A and B, to the problem of minimizing the functional (here $u \in D(A, B)$)

$$D(u) = (Au, Bu) - 2(f, Bu) \to \min \tag{6}$$

in a corresponding Hilbert space H_{AB}. If $(A^{-1})^*$ exists, the auxiliary operator B can be represented in the form

$$B = (A^{-1})^* C, \qquad D(C) \supseteq D(A), \tag{7}$$

where C is an arbitrary symmetric and positive (in the usual sense (2), (3)) operator. Of course the representation (7) is nonconstructive and can be used only for the proof of the existence of an operator B with properties (4) and (5) and of the functional (6). Only individual cases of constructing operators B for substantial equations of mathematical physics having theoretical and practical significance are known. For example, A. E. Martynyuk [211] investigated boundary value problems for some weakly elliptic equations of odd order by this method. V. M. Shalov constructed an operator B for a complex boundary value problem for the wave equation [294]. Variational problems (4)–(6) have been constructed by the author for individual local and nonlocal boundary value problems for equations of parabolic [100], [101], elliptic [91], [97], and hyperbolic types [94], [96].

In 1968 J. L. Lions [194] presented an example of an operator B for the biharmonic equation (see also [192]), and in a section of "unsolved problems" he indicated the importance of the problem of a constructive construction of operators B of (4), (5) for various partial differential equations.

The construction of variational principles, i.e., the existence of solutions of inverse problems of variational calculus—functionals with properties A)–C), is closely connected with the choice of the class of functionals. This is occasioned, in particular, by the following result. We consider the equation

$$\mathscr{L}u \equiv a(x,y)\frac{\partial^2 u}{\partial x^2} + 2b(x,y)\frac{\partial^2 u}{\partial x\,\partial y} + c(x,y)\frac{\partial^2 u}{\partial y^2}$$
$$\equiv G(x,y,u,u_x,u_y) \tag{8}$$

and a class of functionals

$$F[u] = \int_\Omega \Phi(x, y, u, u_x, u_y)\, d\Omega, \tag{9}$$

where a, b, c, G, and Φ are sufficiently smooth functions in all their arguments. If equation (8) is of elliptic or hyperbolic type in Ω, then there exists [18] a smooth multiplier $\mu = \mu(x, y, u, u_x, y_y) \neq 0$, and in the class of functionals (9) there is a functional such that the Euler equation for it is equation (8) or the equivalent equation

$$\mu(\mathscr{L}u - G) = 0. \tag{8'}$$

If equation (8) is of parabolic type in Ω, then the indicated multiplier $\mu(x, y, u, u_x, u_y)$ does not exist, and in the class of Euler functionals (9) there does not exist a functional whose Euler equation is of the form (8), of parabolic type, or the equivalent equation (8'). Of course, this negative result can be avoided: it suffices to take, for example, a functional of the method of least squares [9] or a convolution functional [129], [246], [245], [322]. However, they do not satisfy our conditions A)–C).

In a number of sections of the book we demonstrate the necessity of invoking other classes of functionals distinct from the class of Euler functionals. Having established the lack of functionals with properties A)–C) in the class of Euler functionals for nonsymmetric equations of elliptic, hyperbolic, and parabolic types (§§6 and 11)(^{3}) in §§12, 13, and 15 for linear equations of parabolic, hyperbolic, and mixed types we construct functionals with properties A)–C) in a class of functionals of the form

$$\Phi[u] = \int_\Omega F\left(x, y, u, u_x, u_y, \int_a^x u_y\, dx, \int_b^y u_x\, dy,\right.$$
$$\left. \int_a^x u\, dx, \int_b^y u\, dy, \int_a^b \int_b^y u\, dx\, dy\right) d\Omega. \tag{10}$$

In §14 functionals are constructed for a class of linear hypoelliptic PDE with constant coefficients in a class of functionals of the type

$$\tilde{F}[u] = \int_\Omega \dot{P}(x, (u * g)(x), (\nabla u * g)(x))\, dx, \qquad \Omega \subset R_n,$$

where $u(x)$ is an unknown function, $g(x)$ is a derivative of the fundamental solution of the adjoint equation, and $*$ is the operation of convolution of two functions.

_____
(^{3})In the individual sections (§§5–8 and 11) intended also for specialists in applied mathematics have omitted certain details of the proofs and sometimes limited ourselves to heuristic constructions.

Functionals with properties A)–C) in the class of Euler functionals

$$F[u] = \int_\Omega F(x, u(x), u'_{x_1}, \ldots, u'_{x_n}) \, dx \qquad (11)$$

are classical solutions of the inverse problem of the calculus of variations: a familiar example is the Dirichlet functional for the Laplace equation. Functionals possessing properties A)–C) but not belonging to the class of Euler functionals (11)([4]) we shall call *quasiclassical solutions* of the inverse problem of the calculus of variations.

A general approach to the construction of an analogue of a symmetrizing operator B was proposed by V. P. Didenko [75]; a constructive modification of this approach is presented in §16. Unexpectedly, spaces of functions dual to spaces with a dominant mixed derivative turn out to be useful in this case; the investigation of such spaces was first begun by Nikol'skiĭ [239].

It should be noted that spaces of functions which are solutions of variational problems have always played an important role in the development of variational methods. All results new in principle for the variational method of Hilbert [137], Levi [191], Lebesgue [189], Sobolev [311], Nikol'skiĭ [238]–[242], and Kidryavtsev [172]–[174] are based on essentially new results in the theory of functions and function spaces. However, the spaces they considered were generated usually by norms of a class of Euler functionals, while the class of functionals (10), for example, generate norms and, correspondingly, function spaces which require further investigation; results are partially expounded in §§9 and 10.

In the works of Volterra [342], Kerner [159], and Vaĭnberg [330]–[334] (see [324] and [331]) a variational method is developed for the investigation of the nonlinear equation

$$N(u) = q, \quad \overline{D(N)} = H, \quad N(0) = 0 \qquad (12)$$

with a potential operator N. The criterion of the potential property is the condition of symmetry of the linear operator N'_u (the Gâteaux derivative of the operator N)

$$(N'_u\varphi, \psi) = (\varphi, N'_u\psi) \quad \forall u, \varphi, \psi \in D(N). \qquad (13)$$

If this condition is satisfied, the corresponding functional has the form

$$V[u] = \int_0^1 (N(tu), u) \, dt - (q, u).$$

([4])The functional $F[u]$ of (11) is written for a PDE of second order; if the differential operator being investigated is of order m, then integrand F in (11) may depend on derivatives through order $[m/2]$.

Nashed [235] and Lyashko [202] independently proposed generalization of this variational method to equations with a nonpotential operator N: if there exists an auxiliary operator B such that

$$(N_u'\varphi, B\psi) = (B\varphi, N_u'\psi) \quad \forall u, \varphi, \psi \in D(B, N), \tag{14}$$

then under some additional assumptions regarding the operators N and B the corresponding functional can be represented in the form

$$\mathscr{L}[u] = \int_0^1 (N(tu), Bu)\, dt - (q, Bu).$$

In Chapter 4 we further generalize and develop these ideas, using in an essential manner Petryshyn's construction of a solvable extension of a nonlinear operator. It is important that also in this nonlinear case the symmetrizing operator B is constructed for the linear operator N'.

Using a result of Tonti [324], in §17 we prove the existence of a quasi-classical solution of the inverse problem of the calculus of variations for a broad class of nonlinear equations, generally speaking, with a nonpotential operator, so that in the case of elementary domains the construction of a functional in explicit form not connected with the form of the nonlinearity is possible.

In conclusion we remark that in some examples of §§12–17 we have limited ourselves to the solution of the inverse problem of the calculus of variations for the nonsymmetric equations considered—the construction of functionals with properties A)–C)—without investigating further these functionals for a minimum, i.e., without rigorously justifying the variational method by investigating generalized solutions of the equations and solutions of the variational problems. In these sections a variational method is constructed for closely related equations, and for mathematicians analogous constructions occasion no difficulty, while for specialists in applications a constructive procurement of the functional of the variational principle is important.

CHAPTER I

Variational Problems for Linear Equations with B-Symmetric and B-Positive Operators

In this chapter we present the theoretical foundations of the formulation and investigation of variational principles for linear equations worked out mainly by A. E. Martynyuk, W. V. Petryshyn, and V. M. Shalov. The somewhat cumbersome exposition is connected with the fact that we present the construction of a variational principle for the equation $Au = f$, generally speaking, with a nonsymmetric unbounded operator A of variable sign which may also not have a bounded inverse operator on $R(A)$.

Consideration of B-positive operators A required in the introduction of weak closability of the operator A relative to the operator B, which, in turn, complicated the corresponding constructions of the extension of operators. However, in the majority of cases clearer constructions and results are presented, and sometimes the proofs for equations with a B-positive-definite operator A can be repeated.

In §5 we present dual variational principles for the variational problem of minimizing a quadratic functional constructed in §4. Since in the domestic literature considerably less attention than abroad has been devoted to these principles in recent years, we present a generalization of some known approaches to the construction of dual variational principles for the problem of minimizing a quadratic functional constructed for an equation with a B-symmetric and B-positive operator.

§1. Definitions and properties of operators symmetric and positive in a generalized sense

1.1. Suppose in a Hilbert space H with norm $\|\cdot\|_H \equiv \|\cdot\|$ and inner product $(\cdot, \cdot)$ there are given distributive operators A and B. We denote their domains by $D(A)$ and $D(B)$, setting $D(A, B) = D(A) \cap D(B)$ and $R_A(B) = \{Bv : v \in D(A, B)\}$.

Everywhere below in this chapter we assume that $D(A, B)$ is dense in H.([5])

DEFINITION 1.1. A distributive operator A is called *B-symmetric* if

$$(Au, Bv) = (Bu, Av) \tag{1.1}$$

for all $u, v \in D(A, B)$.

DEFINITION 1.2. A distributive operator A is called *B-positive* if

$$(Au, Bu) > 0 \quad \forall u \in D(A, B), \ u \neq 0 \tag{1.2}$$

and for any sequence $\{u_n\} \subset D(A, B)$

$$(Au_n, Bu_n) \to \quad implies \ \|u_n\| \to 0 \ (n \to \infty). \tag{1.3}$$

We present an example of an operator which is *B*-positive but not *B*-symmetric. Let

$$\mathscr{L}u = -\Delta u + \sum_{i=1}^{n} b_i \frac{\partial u}{\partial x_i} + cu + g(x), \qquad c(x) \geq 0, \ x \in \Omega,$$

$$D(\mathscr{L}) = \overset{\circ}{C}^2(\Omega) \subset H \equiv L_2(\Omega), \qquad c(x) \in C(\overline{\Omega}).$$

Then in the case of coefficients b_i, $i = 1, \dots, n$, constant in $\overline{\Omega}$ among which at least one is nonzero

$$(\mathscr{L}u, v) = \int_{\Omega} \left\{ \nabla u \cdot \nabla v + \sum_{i=1}^{n} b_i \frac{\partial u}{\partial x_i} v + cuv \right\} dx \quad \forall u, v \in D(\mathscr{L})$$

and $(\mathscr{L}u, v) \neq (u, \mathscr{L}v)$ for arbitrary $u, v \in \overset{\circ}{C}^2(\Omega)$, but

$$(\mathscr{L}u, u) = \int_{\Omega} \left\{ \sum_{i=1}^{n} (u_{x_i})^2 + cu^2 \right\} dx \geq 0,$$

and $(\mathscr{L}u, u) = 0$ if and only if $u = 0$ almost everywhere in Ω.

On the other hand, the operator P: $Pu \equiv u_{tt} - u_{xx}$, $D(P) = \overset{\circ}{C}^2(\Omega)$, is an example of an operator which is symmetric but not positive.

Thus, neither of the sets of *B*-symmetric and *B*-positive operators contains the other in the case of a real Hilbert space H. However, in a complex Hilbert space, as is easily shown, (1.2) implies *B*-symmetry of the operator A [50].

Another definition of *B*-positivity of an operator A is more customary.

([5])Unless otherwise mentioned, in Chapters 1–3 distributive operators in a real Hilbert space are considered.

DEFINITION 1.2′. A distributive operator A is called B-*positive* if there exists $\alpha = \text{const} > 0$ such that

$$(Au, Bu) \geq \alpha^2 \|u\|^2 \quad \forall u \in D(A, B), \tag{1.4}$$

The last two definitions are equivalent. Indeed, (1.2) and (1.3) follow from (1.4) in an obvious manner. Conversely, suppose (1.2) and (1.3) are satisfied; we shall show that then we have (1.4), or the equivalent relation

$$(Av, Bv) \geq \alpha^2 > 0 \quad \forall v \in D(A, B), \ \|v\| = 1. \tag{1.4′}$$

If there does not exist a constant $\alpha > 0$ such that (1.4′) holds, then it is possible to find a sequence $\{v_n\}: v_n \in D(A, B), \ n = 1, 2, \ldots, (Av_n, Bv_n) \to 0$ $(n \to \infty)$, and then by (1.3) $\|v_n\| \to 0$ $(n \to \infty)$, which contradicts the condition $\|v_n\| = 1$.

DEFINITION 1.3 A distributive operator A is called B-*positive-definite* if there exists an operator $B: H \to H$ closable in H such that (1.4) is satisfied and

$$(Au, Bu) \geq \beta^2 \|Bu\|^2 \quad \forall u \in D(A, B) \tag{1.5}$$

for some constant $\beta > 0$ not depending on u.

REMARK 1.1. We recall that an operator B is called *closable in* H if for any sequence $\{u_n\} \subset D(B)$ the convergence $(Bu_n \to f, u_n \to 0)$ $(n \to \infty)$ in H implies $f = 0$.

Closability in the weak sense is important for B-positive operators: if for any sequence $\{u_n\} \subset D(A, B)$ the relation $\|u_n\| \to 0$ $(n \to \infty)$ implies

$$(Au_n, Bv) \to 0 \ (n \to \infty) \quad \forall v \in D(A, B),$$

then the operator A is called *closable in the weak sense relative to the operator* B.

We shall not present here examples showing that the classes of operators introduced are nonempty; one of the purposes of the book is a constructive formation of operators B with properties (1.1)–(1.5) for various interesting differential operators A. We note only that, first of all, for $B = I$ the familiar definitions of symmetric, positive, and positive-definite operators follow from Definitions 1.1–1.3, and second, Theorem 1.1 below shows that if for a linear operator A there exists an inverse operator A^{-1} on $R(A)$ with $\overline{R(A)} = H$, then there exists an infinite set of auxiliary operators B such that the operator A is B-symmetric and B-positive.

1.2. We now establish some properties of the classes of operators defined which will be useful below.

LEMMA 1.1. *If an operator A is B-symmetric and B-positive, then*
1) *the generalized Schwarz inequality holds*:

$$|(Au, Bv)|^2 \le (Au, Bu)(Av, Bv) \quad \forall u, v \in D(A, B);$$

2) $(A(u + v), B(u + v))^{1/2} \le (Au, Bu)^{1/2} + (Av, Bv)^{1/2} \; \forall u, v \in D(A, B);$
and
3) A^{-1} *exists on* $R(A)$.
If an operator A is B-symmetric and B-positive-definite, then also
4) *the operator* A^{-1} *is bounded on* $R(A)$; *and*
5) *A is closable in H if* $\overline{R_A(B)} = H$.

PROOF. 1) We set $w = u + \lambda(Au, Bv)v$ for arbitrary $u, v \in D(A, B)$.
Then for any real λ

$$0 \le (Aw, Bw) = (Au, Bu) + 2\lambda |(Au, Bv)|^2 + \lambda^2 (Au, Bv)^2 \cdot (Av, Bv).$$

We have obtained a nonnegative polynomial of second order (in λ) with
real coefficients; assertion 1) follows from the nonpositivity of its discrim-
inant.
   2) On the basis of 1)

$$\begin{aligned} |(Au + Av, Bu + Bv)| &\le (Au, Bu) + 2|(Au, Bv)| + (Av, Bv) \\ &\le (Au, Bu) + 2(Au, Bu)^{1/2} \cdot (Av, Bv)^{1/2} + (Av, Bv) \\ &\le [(Au, Bu)^{1/2} + (Av, Bv)^{1/2}]^2, \end{aligned}$$

whence assertion 2) follows.
   3) It is necessary to show that $Au = 0$ implies $u = 0$. But $(Au, Bu) = 0$
for $Au = 0$ and from Definition 1.2' it follows that $u = 0$.
   4) By the Cauchy-Schwarz-Bunyakovskiĭ inequality we have from (1.5)

$$\beta^2 \|Bu\|^2 \le (Au, Bu) \le \|Au\| \cdot \|Bu\|,$$

whence $\|Au\| \ge \beta^2 \|Bu\|$; considering this inequality, from (1.4) we obtain

$$\alpha^2 \cdot \|u\|^2 \le (Au, Bu) \le \|Au\|^2 / \beta^2,$$

i.e., $\|Au\| \ge \alpha\beta\|u\|$, and hence the operator A^{-1} is bounded in H on $R(A)$.
   5) Since by (1.5) $\{Bu_n\}$ converges to some h in H and $u_n \to 0$ in H, it
follows from the closability of B that $h = 0$. Therefore,

$$(f, Bv) = \lim_{n \to \infty} (Au_n, Bv) = \lim_{n \to \infty} (Bu_n, Av) = 0$$

for all $v \in D(A, B)$. Since $\overline{R_A(B)} = H$, it follows that $f = 0$, i.e., A is
closable in H. ∎
   1.3. We consider the question of the existence of operators B with
properties (1.1)–(1.5) for an invertible operator A.

THEOREM 1.1. *If* $\overline{R(A)} = H$ *and there exists the operator* A^{-1} *defined on* $R(A)$, *then there exists a closed operator* B *such that* A *is weakly closable,* B-*symmetric, and* B-*positive;* B *can be represented in the form*

$$B = (A^{-1})^* C, \qquad (1.6)$$

where C *is an arbitrary selfadjoint positive-definite (in the usual sense) operator with* $D(C) \supseteq D(A)$. *If* A^{-1} *is bound, then* A *is* B-*positive-definite.*

PROOF. The existence of A^{-1} on the set $R(A)$ dense in H and the condition $\overline{D(A)} = H$ imply [162] the existence of A^*, $(A^*)^{-1}$, and $(A^{-1})^*$, and

$$(A^*)^{-1} = (A^{-1})^*. \qquad (1.7)$$

We choose an arbitrary operator C satisfying the condition of the theorem. There exists an infinite set of such operators; for example, it suffices to set $C = \kappa \cdot I$, $\kappa = \text{const} > 0$, with $D(C) \supseteq D(A)$. We consider the product $C \cdot A^{-1}$, which is meaningful, since $D(C) \supseteq D(A)$. From this and the fact that the set $D(A^{-1}) = R(A)$ is dense in H it follows that $D(\overline{C \cdot A^{-1}}) = H$, and hence the adjoint operator $B = (C \cdot A^{-1})^*$ exists and $B = (C \cdot A^{-1})^* = (A^{-1})^* C$.

With consideration of (1.7) we have for all $u, v \in D(A, B)$

$$(Au, Bv) = (Au, (A^{-1})^* Cv) = (u, Cv) \qquad (1.8)$$

and because of the properties of symmetry and positive-definiteness of C the operator A is B-symmetric and B-positive.

From (1.8) it is obvious also that if $\|u_n\| \to 0$, then $(Au_n, Bv) = (u_n, Cv) \to 0$, i.e., the operator A is weakly closable (relative to this operator B).

If A^{-1} is bounded on $R(A)$ in H, then, subjecting the arbitrary operator C in (1.6) also to the condition of boundedness in H, we have for all $u, v \in D(A, B)$

$$\|Bu\| = \|(CA^{-1})^* u\| \le \gamma_1 \|u\|,$$

$$(Au, Bu) = (u, Cu) \ge \gamma_2^2 \|u\|^2 \ge \frac{\gamma_2^2}{\gamma_1^2} \|Bu\|^2,$$

where the positive constants γ_1 and γ_2 do not depend on u. ∎

COROLLARY 1.1. *Theorem* 1.1 *and Lemma* 1.1 *directly imply that*

a) *the operator* A $(\overline{R(A)} = H)$ *is* B-*symmetric and* B-*positive for some operator* B *if and only if there exists* A^{-1} *defined on* $R(A)$; *and*

b) *the operator* A $(\overline{R(A)} = H)$ *is* B-*symmetric and* B-*positive-definite for some operator* B *if and only if the operator* A^{-1} *is bounded on* $R(A)$ *in* H.

THEOREM 1.2. *If an operator A is B-symmetric and B-positive-definite and the closed operator B has a bounded inverse on R(B), then the representation*

$$A = A_0 B \qquad (1.9)$$

is possible, where A_0 is the operator admitting extension to a selfadjoint operator.

PROOF.([6]) We observe first that invertibility of B makes it possible to write conditions (1.4) and (1.5) in the form

$$(Au, Bu) \geq \beta^2 \|Bu\|^2 \geq \gamma^2 \|u\|^2, \qquad \gamma = \text{const} > 0. \qquad (1.10)$$

Since $Au = AB^{-1}Bu$ for $u \in D(A, B)$, setting $A_0 = AB^{-1}$, we obtain (1.9). We shall show that A_0 admits extension to a selfadjoint operator.

We choose an arbitrary element $v \in D(A_0)$ and form the inner product (u, v) in H, $u \in D(A, B)$. Since the operator B is invertible and closed, it follows that $(u, v) = (B^*u, dw)$, $w = B^{-1}v \in D(A, B)$.

The bilinear functional $\Phi(u, w) = (B^*u, w)$ is bounded in the space H_{AB} defined([7]) as the completion of the set $D(A, B)$ in the norm

$$\|u\|_{AB} = [u, u]^{1/2}, \qquad [u, v] = (Au, Bv), \quad u, v \in D(A, B). \qquad (1.11)$$

Indeed, applying the Cauchy-Schwarz-Bunyakovskiĭ inequality and considering (1.10) and (1.11), we obtain

$$|\Phi(u, w)| = |(u, v)| \leq \|u\| \cdot \|v\| \leq \|u\|_{AB} \cdot \|v\|/\gamma. \qquad (1.12)$$

Since $v = Bw$, $w \in D(A, B)$, it follows from (1.5) that $\|v\| = \|Bw\| \leq \|w\|_{AB}/\beta$; therefore, $|\Phi(u, w)| \leq \|u\|_{AB}\|v\|_{AB}/\beta\gamma$, which implies the boundedness of the functional $\Phi(u, w)$ in H_{AB}. Then, by the familiar Riesz lemma on the representation of a bounded linear functional in a Hilbert space, there exists a bounded operator G defined on all of H_{AB} such that

$$\Phi(u, w) = (u, v) = [Gu, w] \quad \forall v \in D(A, B), \qquad (1.13)$$

and $\|G\|_{AB} = \|\Phi\|_{AB} \leq 1/\beta\gamma$. Thus, the operator G is an operator imbedding the space H_{AB} in H. If $Gu = 0$, then it follows from (1.13) that $u = 0$, and hence the inverse operator G^{-1} exists. Setting $v = BG\eta$ in (1.13), we obtain

$$(u, BG\eta) = [Gu, G\eta],$$

since $w = B^{-1}v = G\eta$. hence, the operator BG is symmetric. We shall show that it is bounded in H. Indeed, for $v = BGu$ we have $w = Gu$,

---

([6])This theorem is not used below, but we give its somewhat lengthy proof, since part of the constructions used in the process will be needed below.

([7])The space H_{AB} is described in more detail in §3.

and from (1.13) by means of (1.5) and the Cauchy-Schwarz-Bunyakovskiĭ inequality we obtain

$$\|Gu\|_{AB}^2 = [Gu, Gu] = (u, BGu) \le \|u\| \cdot \|BGu\| \le \|u\| \cdot \|Gu\|_{AB}/\beta.$$

From this we obtain

$$\|Gu\|_{AB} \le \|u\|/\beta, \qquad (1.14)$$

i.e., the operator G constructed by Riesz's lemma is bounded not only in H_{AB} but also in H; then the operator BG is also bounded in H, since from (1.5), (1.11), and (1.14)

$$\|BGu\| \le \frac{1}{\beta}\|Gu\|_{AB} \le \frac{1}{\beta^2}\|u\|,$$

i.e., $\|BG\| \le 1/\beta^2$. The symmetric bounded operator BG in H can be extended by continuity to a selfadjoint operator defined on all of H, so that the inverse operator $(BG)^{-1}$ is selfadjoint.

Suppose now that $\tilde{u}$ is an arbitrary element of $D(A_0)$. We set $B^{-1}\tilde{u} = u'$; then $u' \in D(A, B)$. Since by construction the operator G maps all of H_{AB} onto a part of H containing $D(A, B)$, for the element $u' \in D(A, B)$ there exists an element $u \in D(G)$ such that $u' = Gu$. Thus, $Gu = B^{-1}\tilde{u}$, $u = (BG)^{-1}\tilde{u}$, and by (1.13) and (1.11)

$$((BG)^{-1}\tilde{u} - A_0\tilde{u}, v) = 0 \quad \forall v \in D(A, B).$$

Because the set $D(A, B)$ is dense in H, from this we conclude that $(BG)^{-1}\tilde{u} = A_0\tilde{u}$ for all $\tilde{u} \in D(A_0)$, i.e., the operator A_0, just as $(BG)^{-1}$, is selfadjoint. ∎

With consideration of Theorem 1.2 it is possible to give still another definition of a positive operator in the generalized sense which is close to Definition 1.3.

DEFINITION 1.4. Suppose a distributive operator A can be represented in the form (1.9), where A_0 is a selfadjoint operator and B is closed. Such an operator A is called *positive in the generalized sense* if $(Au, Bu) > 0$ for all $u \ne 0$.

From this definition it follows, first of all, that $(Au, Bu) = 0$ if and only if $u = 0$. Second, the property of B-symmetry follows, since for all $u, v \in D(A, B)$

$$(Au, Bv) = (A_0Bu, Bv) = (Bu, A_0Bv) = (Bu, Av).$$

REMARK 1.2. It is possible to give a definition which in a certain sense generalizes the idea of Definitions 1.2–1.4 given above and certain others [186], [313], [92].

DEFINITION 1.5. A distributive operator A is called *B-positive with power* γ, $0 \leq \gamma \leq 2$, if there exists a constant $c > 0$, not depending on u, and a linear closable operator B such that

$$(Au, Bu) \geq c\|Au\|^{\gamma} \cdot \|Bu\|^{2-\gamma} \quad \forall u \in D(A, B). \tag{1.15}$$

For $B = A$ inequality (1.15) is trivial. For $B = I$ we obtain from (1.15) the definitions of [186] regarding strong positivity with power γ and strong positivity (for $\gamma = 2$) of the operator A, and hence the examples of such operators presented in [186] satisfy Definition 1.5. For $\gamma = 0$ we obtain (1.5) from (1.15), and if, moreover, the closed operator B is invertible we completely obtain Definition 1.3 (or 1.4). For $\gamma = 1$ and $D(A) = D(B)$, from (1.15) we obtain the definition of [313] of operators forming an acute angle.

We note that for $B = I$ the operator A of (1.15) satisfies the relations

$$\|Au\| \geq c_{\gamma}\|u\|, \qquad 0 \leq \gamma < 1, \ c_{\gamma} > 0,$$

$$\|Au\| \leq c_{\gamma}\|u\|, \qquad 1 < \gamma \leq 2, \ u \in D(A),$$

i.e., any operator A which is strongly positive with power γ, $1 < \gamma \leq 2$, is bounded. For operators B-positive with power γ defined above boundedness is not necessary. Moreover, many properties established in [186], [208], [313], [292], [250], and [253] extend [92] to operators satisfying Definition 1.5.

To conclude this section we clarify the property of weak closability of an operator. There is the well-known

THEOREM (HELLINGER-TOEPLITZ [70]). *If $D(A) = H$ and A is a symmetric operator, then A is bounded in H.*

We have

THEOREM 1.3. *If $D(A) = H$, the B-symmetric operator A is weakly closable, and $\overline{R_A(B)} = H$, then the operator A is bounded in H.*

PROOF. We first show that A is closable in H. Let $u_n \to 0$ and $Au_n \to f$ in H as $n \to \infty$. By B-symmetry and weak closability of A we then have

$$(f, Bv) = \lim_{n \to \infty} (Au_n, Bv) = 0 \quad \forall v \in D(A, B).$$

From this by the density of $R_A(B)$ in H it follows that $f = 0$, i.e., A is closable in H. Now since $D(A) = H$, it follows that A is closed in H, and a closed operator defined on all of H is a bounded operator by the familiar closed-graph theorem [170]. ∎

It is obvious that the Hellinger-Toeplitz theorem follows from Theorem 1.3 for $B = I$, the identity operator, and Theorem 1.3 encompasses a larger classes of operators.

§2. Extensions of operators in the sense of Sobolev and Friedrichs

Investigation of operator equations by a variational method leads to the necessity of extension of an operator, i.e., to defining it on a broader set of functions, since in the general case a solution of the equivalent variational problem may not belong to the domain of the original operator.

DEFINITION 2.1. On elements $u, v \in H$ *Friedrichs B-extensions* of the operators A and B respectively are defined if there exist sequences $\{u_k\}$ and $\{v_k\}$ of elements in $D(A, B)$ such that $\|u_k - u\| \to 0$ and $\|v_k - v\| \to 0$ as $k \to \infty$ and if there exist elements $\omega_u, \chi_v \in H$ not depending on v and u respectively such that as $k \to \infty$ the following relations are simultaneously satisfied:

$$(\omega_n - Au_k, \chi_v - Bv_k) \to 0, \tag{2.1}$$

$$[(\omega_n, \chi_v) - (Au_k, Bv_k)] \to 0. \tag{2.2}$$

We set

$$\omega_n = \mathscr{A} u, \quad u \in D(\mathscr{A}), \qquad \chi_v = \mathscr{B} v, \quad v \in D(\mathscr{B}).$$

In the general case a Friedrichs B-extension is not well defined in the sense that for fixed u and v the elements ω_u and χ_v may not be unique; but we do have

THEOREM 2.1. *If the operator A is closable in the weak sense relative to the operator B and* $\overline{R_A(B)} = H$, *then the Friedrichs B-extension is unique in H.*

PROOF. If suffices to prove that ω_u does not depend on the choice of the sequence $\{u_k\}$ and for a given sequence $\{u_k\}$ there cannot be two distinct elements ω_u and ω'_u. Indeed, suppose there exist two sequences $\{u_k\}$ and $\{u'_k\}$ converging in H to the same element and leading to elements ω_u and ω'_u respectively, so that by definition for each ω_u and ω'_u (2.1) and (2.2) are satisfied for all v and v_k in $D(A, B)$, $k = 1, 2, \ldots$. Suppose $v_k = v$ for all $k = 1, 2, \ldots$. Then $\chi_v = Bv$ and (2.1) is satisfied while for all $\varepsilon > 0$ we can write (2.2) for $k \geq N(\varepsilon)$ in the form

$$|(\omega_n - Au_k, Bv)| < \varepsilon, \qquad |(\omega'_n - Au'_k, Bv)| < \varepsilon.$$

From the last two inequalities we obtain

$$|(\omega_n - \omega'_n, Bv)| \leq |(Au_k - Au'_k, Bv)| + 2\varepsilon, \tag{2.3}$$

whence by the weak closability of A

$$(\omega_n - \omega'_u, Bv) = 0 \quad \forall v \in D(A, B).$$

Since $\overline{R_A(B)} = H$, it follows that $\omega_u = \omega'_n$, i.e., the element ω_u does not depend on the choice of the sequence $\{u_k\}$.

From (2.3) for $u_k = u'_k$, $k = 1, 2, \ldots$, it thus follows that for a given sequence $\{u_k\}$ there exists only one limit element ω_u. ∎

Uniqueness of a Friedrichs extension of B can be proved similarly if it is weakly closable and $\overline{R_B(A)} = H$. However, in concrete realizations of the variational method cases are possible in which extension of B is not necessary. If extension of A only is necessary, then we use the simpler

DEFINITION 2.1a. A *Friedrichs B-extension* $\mathscr{A}$ of an operator A is defined on an element $u \in H$ if there exist a sequence $\{u_k\}$ of elements in $D(A, B)$ and an element $\tilde{\omega}$ in H such that as $k \to \infty$ the following relations are satisfied simultaneously:

$$\|u_k - u\|_H \to 0,$$

$$(Au_k - \tilde{\omega}, Bv) \to 0 \quad \forall v \in D(A, B).$$

In this case we set $\tilde{\omega} = \mathscr{A} u$, $u \in D(\mathscr{A})$.

We remark that Definitions 2.1 and 2.1a and Theorem 2.1 are valid for operators which are not necessarily B-symmetric or B-positive in the sense of §1. The properties of B-symmetry and B-positivity are invariant relative to the Friedrichs extension (we assume further that the B-positive operator A is weakly closable).

THEOREM 2.2. *The operator* $\mathscr{A}$ *corresponding to a B-symmetric and B-positive operator* A *is* $\mathscr{B}$-*symmetric and* $\mathscr{B}$-*positive.*

PROOF. We consider a sequence $\{u^n\}$ of elements $u^n \in H$, $n = 1, 2, \ldots$, on each of which Friedrichs B-extensions $\mathscr{A}$ and $\mathscr{B}$ corresponding to the original operators A and B are defined. For a fixed n, according to Definition 2.1, for the element u^n there is a sequence $\{u_k^n\}$, $u_k^n \in D(A, B)$ for all k, such that

$$\|u_k^n - u^n\| \to 0 \quad (k \to \infty),$$

$$(\mathscr{A} u^n - Au_k^n, \mathscr{B} u^n - Bu_k^n) \to 0 \quad (k \to \infty), \qquad (2.4)$$

$$(\mathscr{A} u^n, \mathscr{B} u^n) - (Au_k^n, Bu_k^n) \to 0 \quad (k \to \infty),$$

from which by the B-positivity of the operator it follows that $(\mathscr{A} u^n, \mathscr{B} u^n) \geq 0$ for all n.

We shall prove that the condition

$$(\mathscr{A} u^n, \mathscr{B} u^n) \to 0' \quad (n \to \infty)$$

implies that $\|u^n\|_H \to 0$ $(n \to \infty)$.

For fixed n we set

$$(Au^n, \mathscr{B}u^n) = \varepsilon_n/2.$$

By (2.4) for each n it is possible to find a number $k_n = k(n)$ such that

$$\|u_{k_n}^n - u^n\| \le \varepsilon_n/2, \tag{2.5}$$

$$|(\mathscr{A}u^n, \mathscr{B}u^n) - (Au_{k_n}^n, Bu_{k_n}^n)| \le \varepsilon_n/2. \tag{2.6}$$

Hence,

$$\|u^n\| \le \|u_{k_n}^n\| + \varepsilon_n/2, \tag{2.7}$$

$$(Au_{k_n}^n, Bu_{k_n}^n) \le (\mathscr{A}u^n, \mathscr{B}u^n) + \varepsilon_n/2 = \varepsilon_n.$$

Passing now to the limit as $n \to \infty$ and using the fact that $\varepsilon_n \to 0$ and also that $(Au_k^n, Bu_k^n) \to 0$ implies that $\|u_k^n\| \to 0$, we find from (2.7) that $\|u^n\| \to 0$, i.e., the required assertion.

It is also obvious that

$$(\mathscr{A}u, \mathscr{B}u) > 0 \quad \forall u \in D(\mathscr{A}, \mathscr{B}), \ u \ne 0.$$

Indeed, if $(\mathscr{A}u', \mathscr{B}u') = 0$, for some $u' \ne 0$, then, taking in $D(A, B)$ a $\{u_k\}$ converging strongly to u', we would have as $k \to \infty$

$$\|u_k - u'\| \to 0, \qquad (Au_k, Bu_k) \to (\mathscr{A}u' \mathscr{B}u') = 0.$$

From this by the B-positivity of A it would follow that $u' = 0$, which contradicts the choice of it. Thus, $\mathscr{B}$-positive of the operator $\mathscr{A}$ has been established.

$\mathscr{B}$-symmetry of $\mathscr{A}$ follows directly from the limit relation in (2.2), Definition 2.1, and property (1.1). ∎

In a manner similar to the way in which in solving selfadjoint equations by a variational method it is possible to introduce extensions of the operators of differentiation in the sense of Sobolev [311], we define a Sobolev B-extension of an operator A, which is nonsymmetric in the general case, if together with $D(A, B)$ the set $\{Bv: v \in D(A^*B) \cap D(B)\}$ is also dense in H.

DEFINITION 2.2. A *Sobolev B-extension* $\tilde{A}$ of an operator A is defined on an element $u \in H$ if in H there exists an element ω such that

$$(u, A^*Bv) = (\omega, Bv) \quad \forall v \in D(A^*B). \tag{2.8}$$

A Sobolev A-extension of the operator B can be defined in a similar way. It is evident from Definition 2.2 that if A is an operator of differentiation, then, setting $B = I$ on the smooth compactly supported functions, we obtain the familiar definition of a generalized derivative in the sense of Sobolev and of a generalized differential operator.

In complete analogy to the proof of [311] we obtain

THEOREM 2.3. *In order that a Sobolev B-extension $\tilde{A}$ of an operator A be defined on an element $u \in H$ it suffices that there exist a sequence $\{u_k\}$ of elements $u_k \in D(A, B)$, $k = 1, 2, \ldots$, such that*

$$\lim_{k \to \infty} (u_k, A^*Bv) = (u, A^*Bv),$$
$$\|Au_k\|_H \leq C, \qquad k = 1, 2, \ldots, \quad C \equiv \text{const} > 0. \tag{2.9}$$

LEMMA 2.1. *A Sobolev B-extension $\tilde{A}$ of an operator A is unique in H.*

Indeed, suppose that for some element $u \in H$ on which the operator $\tilde{A}$ is defined there exist two elements $\omega, \omega' \in H$ in $\omega = \tilde{A}u$ and $\omega' = \tilde{A}u$ for which the equalities

$$(u, A^*Bv) = (\omega, Bv), \qquad (u, A^*Bv) = (\omega', Bv) \tag{2.10}$$

hold for all $v \in D(A^*B) \cap D(B)$. Subtracting the relations (2.10) from one another, obtain

$$(\omega - \omega', Bv) = 0.$$

But since we assumed earlier that the set $\{Bv: v \in D(A^*B) \cap D(B)\}$ was dense in H, it follows that $\omega = \omega'$. ∎

We now establish some relations between the *B*-extensions of operators A and B under the assumption that these extensions can be constructed simultaneously in a well-defined manner.

THEOREM 2.4. *Between Friedrichs B-extensions $\mathscr{A}$ and $\mathscr{B}$ and the Sobolev B-extensions $\tilde{A}$ and $\tilde{B}$ there are the relations*

$$1) \quad D(A, B) \subseteq D(\mathscr{A}, \mathscr{B}) \subseteq D(\tilde{A}, \tilde{B}); \tag{2.11}$$

$$2) \quad A = \tilde{A} = \mathscr{A}, \quad B = \tilde{B} = \mathscr{B} \quad \text{in the domain } D(A, B); \tag{2.12}$$

$$3) \quad \tilde{A} = \mathscr{A}, \quad \mathscr{B} = \tilde{B} \quad \text{in the domain } D(\mathscr{A}, \mathscr{B}). \tag{2.13}$$

PROOF. The first inclusion in (2.11) follows directly from Definition 2.1. To prove the second we fix an element $u \in D(\mathscr{A}, \mathscr{B})$ and a sequence $\{u_k\} \subset D(A, B)$ converging strongly to u in H. For elements $u_k \in D(A, B)$ and $v \in D(A^*B)$ we consider the identity

$$(u_k, A^*Bv) = (Au_k, Bv). \tag{2.14}$$

Using the fact that $u \in D(\mathscr{A}, \mathscr{B})$, and hence for the sequences $\{u_k\}$ and $\{v\}$ relation (2.2) holds, by passing to the limit we obtain

$$(u, A^*Bv) = (\mathscr{A}u, Bv). \tag{2.15}$$

Thus, there exists an element $\mathscr{A}u \in H$ for which (2.15) holds, i.e., (2.8) holds with $\omega = \mathscr{A}u$, and hence by Definition 2.2 $u \in D(\tilde{A})$, and

$$\mathscr{A}u = \tilde{A}u \quad \forall u \in D(\mathscr{A}, \mathscr{B}). \tag{2.16}$$

By similar arguments it can be established that if $\tilde{u} \in D(\mathscr{A}, \mathscr{B})$, then $\tilde{u} \in D(\tilde{B})$. Therefore, (2.13) follows from (2.16) and the analogous equality for the extension of B. Relations (2.12) are then obtained directly from Definitions 2.1 and 2.2. ∎

§3. The generalized space of Friedrichs

To investigate selfadjoint equations by a variational method, K. O. Friedrichs introduced the Hilbert space with inner product

$$[u, v]_A = (Au, v), \qquad u, v \in D(A) \tag{3.1}$$

and norm

$$\|u\|_A = [u, u]_A^{1/2}. \tag{3.2}$$

Such spaces arise naturally in solving the problem of minimizing the functional

$$I[u] = (Au, u) - 2(f, u), \qquad u \in D(A), \tag{3.3}$$

corresponding to the equation

$$Au = f \tag{3.4}$$

with a semibounded symmetric operator A.

In investigating equation (3.4) with a B-symmetric and B-positive operator A an analogous space H_{AB} is introduced which we call the *generalized Friedrichs space*. We first define the space $\tilde{H}_{AB}$ as the set of elements of $D(A, B)$ with inner product

$$[u, v] = (Au, Bv) \tag{3.5}$$

and norm

$$\|u\|_{AB} = [u, u]^{1/2}. \tag{3.6}$$

The usual properties of the norm can be verified easily with the help of Lemma 1.1:

$$\|\alpha u\|_{AB} = |\alpha| \cdot \|u\|_{AB}, \tag{3.7}$$

$$\|u + v\|_{AB} \leq \|u\|_{AB} + \|v\|_{AB}. \tag{3.8}$$

Moreover, by the B-positivity of A, if $\|u\|_{AB} = 0$, then $u = 0$.

Thus, the space $\tilde{H}_{AB}$ possesses all the properties of a Hilbert space with the possible exception of completeness. We complete it by means of all possible fundamental sequences in $\tilde{H}_{AB}$ in the following usual manner [162].

Suppose there is a fundamental sequence $\{u_n\} \subset D(A, B)$ in $\tilde{H}_{AB}$. By the definition of the norm in the space $\tilde{H}_{AB}$ we have

$$\|u_m - u_n\|_{AB}^2 = (A(u_n - u_m), B(u_n - u_m)) \to 0 \qquad (m, n \to \infty).$$

From this by the *B*-positivity (1.3) of A it follows that

$$\|u_m - u_n\|_H \to 0 \qquad (m, n \to \infty), \tag{3.9}$$

and hence any sequence $\{u_n\}$ fundamental in $\tilde{H}_{AB}$ has a limit $\tilde{u}$ in H, which we adjoin to $\tilde{H}_{AB}$. For an arbitrary element $\tilde{u}$ of the completed space of Friedrichs

$$\|\tilde{u}\|_{AB} = \lim_{n \to \infty} \|u_n\|_{AB}, \tag{3.6'}$$

where the sequence $\{\tilde{u}_n\}$ is some representer of the class of sequences of elements in $\tilde{H}_{AB}$ defining the limit element $\tilde{u}$.

We denote this completion of $\tilde{H}_{AB}$ by H_{AB}.

THEOREM 3.1. *If a weakly closable operator A is B-symmetric and B-positive, then*

a) *$D(A, B)$ is dense in H_{AB}, and*

b) *H_{AB} is a subset of H in the sense of one-to-one identification of elements of H_{AB} with elements of H.*

If a B-symmetric operator A is B-positive-definite, then a) *and* b) *are satisfied without the condition of weak closability of the operator A, and, moreover (for $\overline{R_A(B)} = H$),*

c) *the operator B can be extended to a bound operator B_0 mapping all of H_{AB} into H, so that $B \subseteq B_0 \subseteq \overline{B}$, where $\overline{B}$ is the closure of B in H; and*

d) *for all u of H_{AB} the following inequalities hold:*

$$\|u\|_{AB}| \geq \alpha \|u\|_H, \tag{1.4'}$$

$$\|u\|_{AB} \geq \beta \|B_0 u\|_H. \tag{1.5'}$$

PROOF. a) This follows directly from the construction of H_{AB}.

b) Suppose $h_0 \in H_{AB}$ and $\{u_n\}$ is a sequence of elements in $D(A, B)$ such that $\|u_n - h_0\|_{AB} \to 0$ as $n \to \infty$. Since $\|u_m - u_n\|_{AB} \to 0$, from (1.3) it follows that $\|u_m - u_n\|_H \to 0$ as $n, m \to \infty$. Hence, the sequence $\{u_n\}$ converges in H to some element $h \in H$. We shall show that the correspondence $h_0 \to h$ is one-to-one, i.e., if $h = 0$, then $h_0 = 0$, and conversely. Indeed, if $h = 0$, then from (3.6) and (1.1), considering the weak closability of A and the convergence of the sequence $u_n \to 0$ in H_{AB} and in H, we obtain for all $u \in D(A, B)$

$$[u, h_0] = \lim_{n \to \infty} [u, u_n] = \lim_{n \to \infty} (Au, Bu_n) = \lim_{n \to \infty} (Au_n, Bu) = 0. \qquad (*)$$

Hence, $h_0 = 0$.

The converse is obvious: if $h_0 = 0$, then convergence to zero in H_{AB} of a sequence $\{u_n\} \subset D(A, B)$ implies by (3.6) and (1.3) the convergence

$u_n \to 0$ in H for $n \to \infty$, i.e., $h = 0$, and the identification of H_{AB} with a subset in H is one-to-one.

It is useful to distinguish the last assertion.

COROLLARY 3.1. $H_{AB} \subseteq H$.

If the operator A is B-symmetric and B-positive but not weakly closable, then the relation $(*)$ used in the proof of assertion b) remains valid, since if $h = 0$, then from the convergence $u_n \to 0$ in H_{AB} and H, (1.5), and the closability of B it follows that $Bu_n \to 0$ in H.

We now prove c). By (3.6), (3.5), and (1.5) inequality $(1.5')$ holds for all $u \in D(A,B)$, and hence B is a bound operator in H_{AB} mapping the dense subset $D(A,B)$ in H_{AB} into H. It is then possible to construct by continuity an extension $B_0 \supseteq B$ of B to all of H_{AB}. Suppose $\overline{B}$ is the closure of B in H, which exists because of the condition of closability of B (Definition 1.3). To prove the inclusions $B \subseteq B_0 \subseteq \overline{B}$ it now obviously suffices to show that $H_{AB} \subseteq D(\overline{B})$. Suppose h_0 is an arbitrary element of H_{AB}. Then there exists a sequence of elements $\{u_n\}$ in $D(A,B) \subseteq D(\overline{B})$ such that $\|u_n - h_0\|_{AB} \to 0$ $(n \to \infty)$, and from (3.6), (1.4), and (1.5) we obtain $\|u_n - h_0\|_H \to 0$ and $\|\overline{B}u_n - \overline{B}u_m\|_H \to 0$ as $n, m \to \infty$. But since $\overline{B}$ is a closed operator in H it then follows that $h_0 \in D(\overline{B})$ and $\overline{B}u_n \to \overline{B}h_0$, i.e., $H_{AB} \subseteq D(\overline{B})$.

Inequalities $(1.4')$ and $(1.5')$ extend by continuity to all of H_{AB} because of a), c), and properties (1.4) and (1.5) of B. ∎

COROLLARY 3.2. For a B-symmetric, B-positive-definite operator A the inclusion

$$H_{AB} \to H. \tag{3.10}$$

This is simply a shorter way of writing property $(1.4')$ and Corollary 3.1.

COROLLARY 3.3. If an operator A is B-symmetric, B-positive, and weakly closable, or B-symmetric and B-positive-definite, and $\mathscr{A}$ and $\mathscr{B}$ are the Friedrichs extensions of A and B, then

$$D(\mathscr{A},\mathscr{B}) \subseteq H_{AB}, \tag{3.11}$$

$$\|u\|_{AB} = (\mathscr{A}u, \mathscr{B}u) \quad \forall u \in D(\mathscr{A},\mathscr{B}). \tag{3.12}$$

Indeed, if $\tilde{u}$ is some arbitrary element of $D(\mathscr{A},\mathscr{B})$, then by Definition 2.1 there exists a sequence $\{u_n\} \subset D(A,B)$ such that $\|u_n - \tilde{u}\| \to 0$ $(n \to \infty)$, and the sequence $\{(Au_n, Bu_n)\}$ is fundamental, which means by construction of the space H_{AB} that $\tilde{u} \in H_{AB}$. Moreover, from the definition $(3.6')$

of the norm in H_{AB}

$$\|\tilde{u}\|_{AB} = \lim_{n\to\infty} (Au_n, Bu_n),$$

and from (2.2) we obtain

$$(\mathscr{A}\tilde{u}, \mathscr{B}\tilde{u}) = \lim_{n\to\infty} (Au_n, Bu_n).$$

Combination of these two equalities proves (3.12). ∎

In an earlier similar way, from Definition 2.1a it is not hard to obtain

COROLLARY 3.4. *Under the conditions of Corollary 3.3, $D(\mathscr{A}) \subset H_{AB}$.*

§4. The variational problem for equations with a B-symmetric and B-positive operator

4.1. Suppose we are given the equation

$$Au = f, \tag{4.1}$$

where A is a B-symmetric B-positive operator acting in H (in this section we assume to simplify the exposition that $D(B) \supseteq D(A)$, $\overline{R_A(B)} = H$, and $\overline{D(A)} = H$). Considering the results of §§1–3 for a B-symmetric B-positive operator A, we henceforth everywhere assume that it is weakly closable, while for a B-symmetric B-positive-definite operator this condition is not required.

On $D(A)$ it is possible to define the quadratic functional

$$D_f(u) = (Au, Bu) - 2(f, Bu), \qquad f \in H. \tag{4.2}$$

Suppose Friedrichs extensions $\mathscr{A}$ and $\mathscr{B}$ of A and B have been constructed. We consider the equation

$$\mathscr{A}u = f \tag{4.1'}$$

and the functional

$$\mathscr{D}_f(u) = (\mathscr{A}u, \mathscr{B}u) - 2(f, \mathscr{B}u), \tag{4.2'}$$

defined on $D(\mathscr{A})$, assuming that $D(\mathscr{A}) \subset D(\mathscr{B})$.

REMARK 4.1. We do not consider the case $D(\mathscr{A}) \not\subseteq D(\mathscr{B})$; when a functional containing derivatives of lower order than the original equation (4.1) is constructed the differential operator B practically always has lower order than A, $D(B) \subseteq D(A)$, and (in all cases considered in this book) $D(\mathscr{A}) \subset D(\mathscr{B})$. For B-positive-definite operators A such a question does not arise at all, since in this case B can be extended by continuity to all of H_{AB}, and $D(\mathscr{A}) \subseteq H_{AB}$ (see Theorem 3.1, Corollary 3.4, and Lemma 4.3).

DEFINITION 4.1. An element $u_0 \in D(\mathscr{A})$ for which $\mathscr{A}u_0 - f = 0$ in H is called a *solution of problem $\mathscr{A}_f$*.

DEFINITION 4.2. An element $u_0 \in D(\mathscr{A})$ is called a *solution of problem* $\mathscr{D}_f$ (or of the variational problem of minimizing the functional $\mathscr{D}_f(u)$) if in $D(\mathscr{A})$ there exists a sequence of elements $\{u_n\}$ such that as $n \to \infty$

$$\mathscr{D}_f(u_n) \to \mathscr{D}_f(u_0) \equiv d = \min_{D(\mathscr{A})} \mathscr{D}(u), \qquad (4.3)$$

$$\|u_n - u_0\|_H \to 0. \qquad (4.4)$$

THEOREM 4.1. *If the operator A is B-symmetric and B-positive, then problems $\mathscr{A}_f$ and $\mathscr{D}_f$ are equivalent.*

PROOF. It must be shown that if u_0 is a solution of problem $\mathscr{A}_f$, then this same element u_0 is a solution of problem $\mathscr{D}_f$, and conversely.

Suppose $\mathscr{A} u_0 = f$. Then, using the $\mathscr{B}$-symmetry of the operator $\mathscr{A}$ (Theorem 2.2), the functional (4.2′) can be written in the form

$$\mathscr{D}_f(u) = (\mathscr{A} u = \mathscr{A} u_0, \mathscr{B} u - \mathscr{B} u_0) - (\mathscr{A} u_0, \mathscr{B} u_0). \qquad (4.5)$$

Therefore, on the basis of the $\mathscr{B}$-positivity of $\mathscr{A}$

$$d = \min_{D(\mathscr{A})} \mathscr{D}_f(u) = \mathscr{D}_f(u_0) \leq \mathscr{D}_f(u) \quad \forall u \in D(\mathscr{A}). \qquad (4.6)$$

Choosing as $\{u_n\}$ the sequence $u_n = u_0$ for all n, we ensure that (4.3) and (4.4) are satisfied; hence u_0 is also a solution of problem $\mathscr{D}_f$.

Suppose now that u_0 is a solution of problem $\mathscr{D}_f$, i.e., $\mathscr{D}_f(u) \geq \mathscr{D}_f(u_0)$ for all $u \in D(\mathscr{A})$. Setting $u = u_0 + tv$, where v is an arbitrary element in $D(\mathscr{A})$ and t is a real parameter ($t \in R_1$), we obtain

$$\mathscr{D}_f(u) - \mathscr{D}_f(u_0) = t^2(\mathscr{A} v, \mathscr{B} v) + 2t(\mathscr{A} u_0 - f, \mathscr{B} v) \geq 0, \qquad (4.7)$$

which is obviously possible only if $(\mathscr{A} u_0 - f, \mathscr{B} v) = 0$. Since the element $v \in D(\mathscr{A})$ is arbitrary and the set $R_A(B)$ is dense in H, from this it follows that $\mathscr{A} u_0 - f = 0$. ∎

4.2. Theorem 4.1 was proved under the assumption that there exists either a solution of problem $\mathscr{D}_f$ or a solution of problem $\mathscr{A}_f$. We shall establish for the classes of operators considered a necessary and sufficient condition for the existence and uniqueness of a solution of these problems.

We consider the equation

$$Au = f_0 \qquad (4.8)$$

for some fixed element $f_0 \in H$. Suppose also that an auxiliary operator B has been constructed so that $D(B) \supseteq D(A)$, $\overline{R_A(B)} = H$, and the linear functional (f_0, Bu) is bounded (with respect to u) in H_{AB}, i.e.,

$$|(f_0, Bu)| \leq C_{f_0} \cdot \|u\|_{AB} = C_{f_0} \cdot (Au, Bu)^{1/2} \quad \forall u \in D(A). \qquad (4.9)$$

Then the functional (f_0, Bu) can be extended by continuity to all of H_{AB} with preservation of norm and property (4.9). We denote such an extension by $l(u)$; obviously, $l(u) = (f_0, Bu)$ for all $u \in D(A)$. We now assign to equation (4.8) the functional

$$\tilde{\mathscr{D}}_{f_0}(u) = \|u\|_{AB}^2 - 2l(u), \tag{4.10}$$

defined on all of H_{AB}. By a *solution* of the variational problem $\tilde{\mathscr{D}}_{f_0}$ we mean an element $u_0 \in H_{AB}$ which minimizes the functional (4.10) in H_{AB}.

THEOREM 4.2. *A solution of the variational problem $\tilde{\mathscr{D}}_{f_0}$ in the space H_{AB} exists and is unique if and only if the linear functional (f_0, Bu) is bounded in H_{AB}.*[8]

PROOF. *Sufficiency.* If the functional (f_0, Bu) is bounded in H_{AB}, then, constructing the functional $l(u)$ defined on H_{AB} and bounded in H_{AB} as indicated earlier, by the familiar Riesz lemma we find that there exists a unique element $u_0 \in H_{AB}$ such that

$$l(u) = [u, u_0] \quad \forall u \in H_{AB}. \tag{4.11}$$

Transforming (4.10) with consideration of (4.11) and (3.6), we find that

$$\tilde{\mathscr{D}}_{f_0}(u) = [u, u] - 2[u, u_0] = \|u - u_0\|_{AB}^2 - \|u_0\|_{AB}^2$$

and hence there exists a unique element $u_0 \in H_{AB}$ minimizing the functional $\tilde{\mathscr{D}}_{f_0}$ in H_{AB}.

Necessity. Suppose an element $u_0 \in H_{AB}$ realizes a minimum of the functional $\tilde{\mathscr{D}}_{f_0}(u)$ in H_{AB}. From the necessary condition for a minimum of the functional (4.10) we have

$$0 = \left\{ \frac{d}{dt} \tilde{\mathscr{D}}_{f_0}(u_0 + tu) \right\}_{t=0} = -2l(u) + 2[u_0, u]_{AB}, \tag{4.12}$$

where u is an arbitrary element of H_{AB}. From this we obtain the estimate

$$|l(u)| = |[u_0, u]_{AB}| \le \|u_0\|_{AB} \cdot \|u\|_{AB}.$$

It is obvious that for $u \in D(A)$

$$|(f_0, Bu)| = |l(u)| \le \|u_0\|_{AB} \cdot \|u\|_{AB}$$

and hence the functional (f_0, Bu) is bounded with respect to u in H_{AB}. ∎

---

[8]We shall not always emphasize that the operator A is B-symmetric and B-positive definite; from the foregoing it is clear that the space H_{AB} in the formulations is constructed only for a B-symmetric and B-positive operator A.

COROLLARY 4.1. *For a B-symmetric B-positive operator A the Friedrichs B-extension is defined on any element u_0 which is a solution of the variational problem of minimizing the functional $\tilde{\mathscr{D}}_{f_0}$, i.e., $u_0 \in D(\mathscr{A})$ and $\mathscr{A} u_0 = f_0$.*

Indeed, from the necessary condition for a minimum we have for all $v \in D(A)$

$$0 = \left\{ \frac{d}{dt} \tilde{\mathscr{D}}_{f_0}(u_0 + tv) \right\}_{t=0} = 2[[u_0, v] - (f_0, Bv)] = 0,$$

whence

$$(f_0, Bv) = [u_0, v] \quad \forall v \in D(A). \tag{4.13}$$

Since the set $D(A)$ is dense in H_{AB}, it is possible to select a sequence $\{u_n\} \subset D(A)$ such that

$$\|u_n - u_0\|_{AB} \to 0 \quad (n \to \infty). \tag{4.14}$$

According to (3.5),

$$(Au_n, Bv) = [u_n, v] \quad \forall u_n, v \in D(A). \tag{4.15}$$

From (4.14) and (1.3) we find that

$$\|u_n - u_0\|_H \to 0 \quad (n \to \infty), \tag{4.16}$$

and from (4.13)–(4.15) it follows that

$$(Au_n, Bv) \to (f_0, Bv) \quad (n \to \infty) \tag{4.17}$$

for al $v \in D(A)$, which by Definition 2.1a means that (Friedrichs) B-extension of A is defined on the element u_0 and

$$\mathscr{A} u_0 = f_0. \quad \blacksquare \tag{4.18}$$

From Theorem 4.1 and Corollary 4.1 we obtain

COROLLARY 4.2. *Suppose for equation (4.8) there exists an operator B such that $D(B) \supseteq D(A)$, $\overline{R_A(B)} = H$, the operator A is B-symmetric and B-positive, and*

$$|(f_0, Bu)| \le C_{f_0} \cdot (Au, Bu)^{1/2} \quad \forall u \in D(A), \tag{4.19}$$

where $C_{f_0} > 0$ does not depend on u. Then in $\tilde{H}_{AB}$ there exists a unique element u_0 which is simultaneously the solution of problems $\tilde{\mathscr{D}}_{f_0}$, $\mathscr{A}_{f_0}$, and $\mathscr{D}_{f_0}$.

PROOF. The existence of a unique element $u_0 \in H_{AB}$ which is a solution of the variational problem $\tilde{\mathscr{D}}_{f_0}$ is established in Theorem 4.2. From

Corollary 4.1 we find that $u_0 \in D(\mathscr{A})$ and $\mathscr{A} u_0 = f_0$, i.e., there exists a unique solution of problem $\mathscr{A}_{f_0}$. From Theorem 4.1 we conclude that this same element $u_0 \in D(\mathscr{A})$ is a unique solution of the variational problem $\mathscr{D}_{f_0}$. ∎

With Theorem 2.4 in mind a solution $u_0 \in D(\mathscr{A})$ of (4.1') is naturally called a *generalized solution* of equation (4.1). It will be more convenient below for us to use the following equivalent definition.

LEMMA 4.1. *An element $u_0 \in H_{AB}$ is a generalized solution of equation* (4.1) *if and only if*

$$[u_0, v] = (f, Bv) \quad \forall v \in D(A). \tag{4.20}$$

PROOF. *Necessity.* If $\mathscr{A} u_0 = f$, then $(\mathscr{A} u_0, Bv) = (f, Bv)$ for all $v \in D(A)$. From this with consideration of (3.12) and (2.12) we obtain $[u_0, v] = (f, Bv)$ for all $v \in D(A)$.

Sufficiency. If (4.20) is satisfied, then the linear functional (f, Bv) is bounded with respect to v in H_{AB}:

$$|(f, Bv)| \leq \|u_0\|_{AB} \cdot \|v\|_{AB} \quad \forall v \in D(A).$$

Then the functional (f, Bv) can be extended by continuity to a bounded linear functional $l(v)$ defined on all of H_{AB}; the following functional is such an extension:

$$l(v) = [u_0, v], \quad v \in H_{AB}. \tag{4.21}$$

For a quadratic functional $\tilde{\mathscr{D}}_f(u)$ we have (see (4.10))

$$\tilde{\mathscr{D}}_f(u) = \|u\|_{AB}^2 - 2l(u);$$

setting $v = u$ in (4.21), we obtain

$$\tilde{\mathscr{D}}_f(u) = [u, u] - 2[u_0, u] = \|u - u_0\|_{AB}^2 - \|u_0\|_{AB}^2$$

and hence a minimum of this functional in H_{AB} is achieved precisely on the element u_0. Then, by Corollary 4.1, $u_0 \in D(\mathscr{A})$, and $\mathscr{A} u_0 = f$. ∎

The results of this subsection were established for the case when a solution of (4.8) or (4.1') may not exist for any right side $f_0 \in H$, i.e., it is possible that the operators A^{-1} and $\mathscr{A}^{-1}$ are unbounded in H.

4.3. For the case of a *B*-positive-definite operator A our results can be made concrete in a manner useful for what follows.

If in Definition 1.3 of a *B*-positive-definite operator A we set $B = I$, then conditions (1.4) and (1.5) coincide and express the property of positive definiteness (in the usual sense) of A.

We shall clarify the role of (1.5) for $B \neq I$.

LEMMA 4.2. *If the operator A is B-symmetric and B-positive and f_0 is a fixed (but arbitrary) element of H[9] the condition* (1.5) *is equivalent to the*

---

[9] Or of a set dense in H.

validity of the inequality

$$|(f_0, Bu)| \leq C_2(f_0)(Au, Bu)^{1/2} \quad \forall u \in D(A), \tag{4.21'}$$

where the constant $C_2(f_0) > 0$ does not depend on u.

PROOF. If (1.5) is satisfied, then (4.21') is easily obtained:

$$|(f_0, Bu)| \leq \|f_0\| \cdot \|Bu\| \leq \frac{1}{\beta}\|f_0\|(Au, Bu)^{1/2} \quad \forall u \in D(A).$$

Conversely, if (4.21') is satisfied, then

$$\left|\left(f_0, \frac{Bu}{\|u\|_{AB}}\right)\right| \leq C_2(f_0) \quad \forall u \in D(A),$$

where the constant $C_2(f_0) > 0$ does not depend on u; but then, as is known, the set $\{Bu/\|u\|_{AB}, \ u \in D(A)\}$ is totally bounded in H, i.e., there exists a constant $C_3 > 0$, not depending on u, such that

$$\|Bu/\|u_{AB}\|_H \leq C_3,$$

whence (1.5) follows with $\beta = 1/C_3$. ∎

From the last lemma and Theorem 4.2 we obtain

COROLLARY 4.3. *A solution of the variational problem of minimizing the functional* (4.10) *constructed for equation* (4.8) *with a B-symmetric and B-positive operator A exists and is unique in H_{AB} for an arbitrary $f_0 \in H$ if and only if condition* (1.5) *is satisfied.*

COROLLARY 4.4. *If in equation* (4.8) *the operator A is B-symmetric is B-positive, $f_0 \in H$, and condition* (1.5) *is satisfied, then in the space H_{AB} there exists a unique element u_0 which is simultaneously a solution of problem $\mathscr{A}_{f_0}$ and the variational problem of minimizing the functionals $\tilde{\mathscr{D}}_{f_0}(u)$ of* (4.10) *$(\mathscr{D}_{f_0}(u)$ of* (4.2') *for $f = f_0$).*

The assertion is obtained directly from Corollaries 4.2 and 4.3.

THEOREM 4.3. *If the operator A is B-symmetric and B-positive-definite, then A can be extended to a closed, B_0-symmetric and B_0-positive-definite operator A_0 so that $A_0 \supseteq A$, A_0 is continuously invertible operator,*[10] *and $D(A_0)$ consists of all elements realizing a minimum in H_{AB} of the functional*

$$D_f(u)\|u\|_{AB}^2 - 2(f, B_0 u), \tag{4.22}$$

---

[10] Following [253], we call an operator A *invertible* if there exists A^{-1} on $R(A)$, *densely invertible* if A is invertible and $R(A) = H$, and *continuously invertible* if A is invertible on $R(A) = H$.

when f runs through all of H.

PROOF. We recall that the operator B_0 in the conditions of the theorem is the extension by continuity of the bounded operator B constructed in Theorem 3.1c), so that B_0 maps all of H_{AB} into H, i.e., the functional $(f, B_0 u)$ for all $f \in H$ is a distributive bounded functional with respect to u defined on H_{AB}. Therefore, there exists a bounded linear operator G acting from all of H to H_{AB} such that

$$(f, B_0 u) = [Gf, u] \quad \forall u \in H_{AB}, \ f \in H. \tag{4.23}$$

Considering formulas $(1.4')$ and $(1.5')$ from Theorem 3.1, we find that

$$\|G\|_{AB} = \sup_{\substack{\|f\|=1, \\ \|v\|_{AB}=1}} |[Gf, v]| = \sup_{\substack{\|f\|=1 \\ \|v\|_{AB}=1}} |(f, B_0 v)|$$

$$\leq \sup_{\|v\|_{AB}=1} \|B_0 v\| \leq \frac{1}{\beta} \sup_{\|v\|_{AB}=1} \|v\|_{AB} = \frac{1}{\beta},$$

$$\|G\|_H = \sup_{f \neq 0} \frac{\|Gf\|_H}{\|f\|_H} \leq \frac{1}{\alpha} \sup_{f \neq 0} \frac{\|Gf\|_{AB}}{\|f\|_H} \leq \frac{1}{\alpha\beta}.$$

Thus, the operator is bounded also if it is considered as a mapping from H to H. If $Gf = 0$, then from (4.23) by the density of $R_A(B)$ in H it follows that $f = 0$.

Therefore, the operator $A_0 = G^{-1}$ exists and is closed, since the operator G, defined on all of H, is closed. Moreover, $A_0 \supseteq A$, and A_0 remains B_0-symmetric and B_0-positive-definite.

Indeed, suppose $f = Au$, $u \in D(A)$. By the Riesz lemma there exists a unique element $w \in H_{AB}$ such that

$$(f, B_0 u) = [w, u] \quad \forall u \in H_{AB}. \tag{4.24}$$

Then by (4.23), (4.24), and (3.5) we have $w = Gf = u$, i.e., $u \in D(G^{-1}) = D(A_0)$, and $A_0 u = Au$.

If w and v are arbitrary elements of $D(A_0) \subseteq D(B_0) = H_{AB}$, then, considering (3.5) and the definition of A_0 $(A_0 = G^{-1})$, we obtain

$$(A_0 w, B_0 v) = [GA_0 w, v] = [v, w] = [GA_0 v, w] = (A_0 v, B_0 w),$$

i.e., A_0 is B_0-symmetric.

For any $w \in D(A_0)$ we have from $(1.4')$, $(1.5')$, (3.5), and (3.6)

$$(A_0 w, B_0 w) = [GA_0 w, w] = \|w\|_{AB}^2 \geq \alpha^2 \|w\|_H^2,$$

$$\|B_0 w\|_H^2 \leq \frac{1}{\beta^2}\|w\|_{AB}^2 = \frac{1}{\beta^2}(A_0 w, B_0 w)$$

and hence A_0 is also B_0-positive-definite.

Now if $u_0 \in D(A_0)$, i.e., $u_0 \in D(G^{-1})$, then from the definition of the operator G mapping all of H into H_{AB} we obtain the existence of an element $f \in H$ such that $Gf = u_0$. By this and (4.23)

$$(f, B_0 u) = [Gf, u] = [u_0, u] \quad \forall u \in H_{AB}, \ f \in H$$

and the functional (4.22) is transformed to the form

$$D_f(u) = [u, u] - 2[u_0, u] = \|u - u_0\|_{AB}^2 - \|u_0\|_{AB}^2,$$

i.e., this arbitrary selected element $u_0 \in D(A_0)$ actually realizes a minimum of the functional $D_f(u)$ of (4.22) in the space H_{AB} for some $f \in H$, which proves the last assertion of the theorem. ∎

From this theorem and Corollary 4.3 we obtain

COROLLARY 4.5. *If the operator A is B-symmetric and B-positive, then for equation (4.1) a generalized solution exists and is unique in H_{AB} for any $f \in H$ if and only if condition (1.5) is satisfied.*

THEOREM 4.4. *If A is a B-symmetric and B-positive-definite operator, B is closed, and $D(A) = D(B)$, then $B = B_0$, $A = A_0$, $H_{AB} = D(A) = D(B)$, $R(A) = R(B)$, and the equations $Au = f$ and $Bv = \psi$ have one and only one solution in H_{AB} for all $f, \psi \in H$. Moreover,*

$$\|Au\| \geq \beta\|u\|_{AB} \geq \beta^2\|Bu\| \geq \frac{\beta^2}{\sqrt{\theta}}\|u\|_{AB} \geq \frac{\alpha^2\beta^2}{\sqrt{\theta}}\|u\|, \qquad (4.25)$$

where the constant $\theta > 0$ is such that

$$\|Au\| \leq \theta\|Bu\| \quad \forall u \in H_{AB}. \qquad (4.26)$$

PROOF. Since $\overline{B} = B$, by Theorem 3.1c) we find that $B_0 = B$ and $D(B) = D(A) = H_{AB}$; it then follows from Theorem 4.3 that $A = A_0$, and A has a bounded inverse A^{-1} on all of $R(A) = H$. To prove that in this case B is also continuously invertible we first show that A, as an operator from H_{AB} onto H, is closed. Since $D(A) = H_{AB}$, it suffices to show that A admits closure in H_{AB}, i.e., for any sequence $\{u_n\}$ in H_{AB} such that $\|u_n\|_{AB} \to 0$ and $Au_n \to f$ in H as $n \to \infty$ we have $f = 0$. Since the operator $A = A_0$ is closable in H, while $\|u_n\|_{AB} \to 0$ implies (see (1.4)) that $\|u_n\| \to 0$, it follows that $Au_n \to 0$ in H. Thus, A is a closed operator in H_{AB}. Now a closed operator A defined on the whole space is bounded, i.e., there exists $\theta > 0$ such that

$$\|Au\| \leq \theta\|u\|_{AB} \quad \forall u \in H_{AB}. \qquad (4.27)$$

Considering (3.5) and (3.6), inequality (4.26) is easily obtained from this. Since B is closed and $\overline{R_A(B)} = H$, from (4.22) by the boundedness of A^{-1} on H we conclude that B has a bounded inverse B^{-1} defined on all of H, and hence both equations indicated in the theorem are uniquely solvable in H_{AB} for all $f, \psi \in H$. Inequalities (4.25) are obtained by comparing (1.4), (1.5), and (4.26). ■

COROLLARY 4.6. *If the operator A is B-symmetric and B-positive-definite, B is a closed operator, and $D(B) = D(A)$, then the operators A and B form an acute angle (the definition of Sobolevskiĭ [313]; see also §1).*

Indeed, by (1.4) and (4.26)

$$(Au, Bu) \geq \beta^2 \|Bu\|^2 \geq \frac{\beta^2}{\theta} \|Bu\| \cdot \|Au\|. \tag{4.28}$$

It is obvious that necessarily $\beta^2 < \theta$. ■

COROLLARY 4.7. *Under the assumptions of Theorem 4.4, by (4.25) and (4.17)*

$$\|u\|_{AB} \sim \|Bu\|_H \sim \|Au\|_H. \tag{4.29}$$

These equivalence relations show that the space H_{AB} can actually be constructed only on the basis of the operator A or the operator B if the conditions of Theorem 4.4 are satisfied. Moreover, it follows from the second equivalence relation that if A (or B) is a differential operator and $H = L_2$, then the maximal order of the derivatives in the norm $\|u\|_{AB}$ is equal to the order of the differential operator A (respectively, B).

LEMMA 4.3. *The extension A_0 of the operator A constructed in Theorem 4.3 is a Friedrichs B_0-extension (in the sense of Definition 2.1a) of the B-symmetric and B-positive-definite operator A.*

PROOF. Let u_0 be an arbitrary element of $D(A_0)$. By Theorem 4.3 the element u_0 realizes a minimum of the functional $D_f(u)$ of (4.22) for some $f \in H$. From the necessary condition of a minimum of the functional $D_f(u)$ we have for all $v \in D(A)$

$$\left[\frac{d}{dt} D_f(u_0 + tv)\right]_{t=0} = 2\{[u_0, v] - (f, B_0 v)\} = 0,$$

i.e.,

$$(f, B_0 v) = [u_0, v] \quad \forall v \in D(A).$$

From the proof of sufficiency in Lemma 4.1 we conclude that $u_0 \in D(\mathscr{A})$.

Conversely, suppose $u_0 \in D(\mathscr{A})$; we set $f = \mathscr{A} u_0$. By Theorem 4.1 and Corollary 4.3 this element u_0 then realizes a minimum in H_{AB} of the functional $D_f(u)$ of (4.22) (the functional $D_f(u)$ of (4.22) coincides with $\tilde{D}_{f_0}(u)$ for $f = f_0$). Hence, by Theorem 4.3, $u_0 \in D(A_0)$, as required.

§5. Dual variational principles for nonselfadjoint equations

5.1. For the linear equation

$$Au = f. \tag{5.1}$$

with a B-symmetric and B-positive operator A we present some methods of constructing dual variational principles([11]) which play an important role in establishing as a posteriori approximate solution of the corresponding variational problem of minimization (we call it the direct variational problem).

First, however, we mention some aspects of obtaining a priori estimates of an approximate solution of equation (5.1) with a nonsymmetric (in the general case) operator A in connection with minimization of the corresponding functional

$$D[u] = \|u\|_{AB}^2 - 2(f, B_0 u) \tag{5.2}$$

in the Hilbert space H_{AB}—the completion of $D(A) \subseteq D(B)$ in the norm

$$\|u\|_{AB} = [u, u]^{1/2}, \tag{5.3}$$

$$|u, v| = (Au, Bv), \qquad u, v \in D(A). \tag{5.4}$$

If the operator A is B-symmetric and B-positive-definite, then, according to the results of §4, for each $f \in H$ there exists a unique element $u_0 \in H_{AB}$ minimizing the functional $D(u)$ in H_{AB}, so that

$$d \equiv \min_{H_{AB}} D(u) = \min_{H_{AB}}(\|u - u_0\|_{AB}^2 - \|u_0\|_{AB}^2) = -\|u_0\|_{AB}^2. \tag{5.5}$$

From (5.5) it is not hard to see that obtaining an a priori estimate of $\|u_n - u_0\|_{AB}$ for an approximate solution of equation (5.1) (or of the corresponding variational problem) can be reduced to the following problem of the theory of approximation of functions.

Suppose it is known (from a theorem on smoothness of a solution of (5.1) or of a solution of the variational problem) that u_0 belongs to some class $W^\alpha \subseteq H_{AB}$, while $\{\varphi_n\}$ is a system of coordinate functions on the basis of which an approximate solution is constructed:

$$u_n = \sum_{k=1}^{n} a_k \varphi_k(x). \tag{5.6}$$

([11])These principles are variously called "dual", "supplementary", or "complementary" principles (see [225] and [338]).

Then the problem of obtaining an a priori estimate of $\|u_n - u_0\|_{AB}$ can be formulated as follows: for an element $u_0 \in W^\alpha \subseteq H_{AB}$, estimate its best approximation in the norm $\|\cdot\|_{AB}$ by linear combinations of the form (5.6) (i.e., an approximation of this element by systems of algebraic polynomials, trigonometric series, etc.). For linear differential equations with a selfadjoint and positive operator A the norm (5.3), (5.4) with $B = I$ refers to a known class of Euler functionals, and the indicated problem of the theory of approximation of functions in many cases has a rather complete solution (see [143], [172], and [242]).

However, for nonsymmetric equations the norm (5.3) may not belong to a class of Euler functionals (as, for example, for parabolic equations [1], [18], [53]; see below, §§6, 12, and 16), and obtaining a priori estimates of the solution of the variational problem in the norm $\|\cdot\|_{AB}$ requires new results in the theory of approximation of functions.

Therefore, in the case of nonsymmetric operators A in (5.1) a posteriori estimates of an approximate solution

$$\|u_n - u_0\| \leq \varepsilon_n, \qquad (5.7)$$

where ε_n is a number obtained in the process of constructing u_n, have a still more important significance than for selfadjoint equations.

The estimate (5.7) is usually obtained in the following manner.

From (5.5) it is not hard to conclude that for all $u_n \in H_{AB}$, $n = 1, 2, \ldots$,

$$d_n = D[u_n] \geq d, \qquad (5.8)$$

$$\|u_n - u_0\|_{AB} = \sqrt{D[u_n] - d}. \qquad (5.9)$$

If it is possible to construct a functional $\Psi(v)$ such that

$$\Psi(v) \leq d \quad \forall v \in D(\Psi), \qquad (5.10)$$

$$\sup_{D(\Psi)} \Psi(v) = d, \qquad (5.11)$$

then it is possible to find a sequence $\{v_n\} \subset D(\Psi)$ such that

$$\delta_n \equiv \Psi(v_n) \to d \quad (n \to \infty), \qquad (5.12)$$

$$\delta_n < d, \qquad n = 1, 2, \ldots. \qquad (5.13)$$

From (5.9)–(5.13) we obtain the a posterior estimate

$$\|u_n - u_0\|_{AB} < \sqrt{D[u_n] - \delta_n}. \qquad (5.14)$$

From (1.4) and (5.14) we get the a posterior estimate in the norm in the original Hilbert space

$$\|u_n - u_0\|_H < \frac{1}{\alpha}\sqrt{D[u_n] - \delta_n}. \qquad (5.15)$$

In a number of cases it is possible to construct a variational problem dual to (5.5) of maximizing a functional $\Psi(v)$ such that

$$\min_{H_{AB}} D[u] = D[u_0] = d = \Psi(u_0) = \max_{\tilde{H}} \Psi(v), \qquad (5.16)$$

where $\tilde{H}$ is a linear manifold in $D(\Psi)$ and $\tilde{H} \cap H_{AB} \ni u_0$. For equation (5.1) with a B-symmetric and B-positive-definite operator A the variational problem (5.16) is constructed on $\tilde{H} = D(A)$, for example, in the following manner. From (5.5) for all $u \in H_{AB}$

$$D[u] \geq D[u_0], \qquad (5.17)$$

$$D[u] = D[u_0] + \|u - u_0\|_{AB}^2. \qquad (5.18)$$

We suppose that an element u_0 minimizing the functional $D[u]$ in H_{AB} belongs to $D(A)$, i.e., by Theorem 4.1, $Au_0 = f$. For $u \in D(A)$, by (1.5),

$$\|u\|_{AB}^2 = (Au, Bu) \leq \frac{1}{\beta^2}\|Au\|^2 \qquad (5.19)$$

or with u replaced by $u - u_0$

$$\|u - u_0\|_{AB}^2 \leq \frac{1}{\beta^2}\|Au - f\|^2. \qquad (5.20)$$

From (5.17)–(5.20) it follows that

$$D[u] \geq D[u_0] \geq D[u] - \frac{1}{\beta^2}\|Au - f\|^2, \qquad (5.21)$$

where equality both on the left and on the right is achieved only for $u = u_0$, which makes it possible to pose a problem of the form (5.16):

$$\min_{H_{AB}} D[u] = D[u_0] = d = \Psi[u_0] = \max_{D(A)} \Psi[v], \qquad (5.22)$$

$$\Psi[v] = D[v] - \frac{1}{\beta^2}\|Av - f\|^2, \qquad D(\Psi) = D(A). \qquad (5.23)$$

Thus, an approximate solution u_n of (5.1) can be found as a minimization in H_{AB} of the functional $D[u]$ of (5.2) as well as a maximization in $D(A)$ of the functional $\Psi[v]$ of (5.23).

For $B = I$, i.e., for equation (5.1) with a selfadjoint and positive-definite operator A, the dual variational problem (5.22), (5.23) was obtained by other constructions in [38] and [267] (see also [308] and [310]).

However, the presence in the dual functional (5.23) of the norm $\|Av - f\|_H$ actually eliminates the advantages of the "energy" method as compared with the method of least squares. In the case of a differential operator A and $H = L_2$ the functional $\Psi(v)$ contains derivatives of the same order as (5.1); such functionals do not satisfy condition A) of the Introduction.

5.2. M. G. Slobodyanskii [306] proposed a method of constructing a complementary variational principle for a linear equation with a symmetric and positive operator so that the dual functional $\Psi(v)$ does not contain, as in (5.23), in explicit form the operator A, and in a number of cases (if M is a differential operator of order s) it is not necessary to compute s derivatives of the function being determined.

With the help of Slobodyanskii's approach we construct a dual variational problem for equation (5.1) with a B-symmetric and B-positive-definite operator A.

Suppose first that $u \in D(A)$, and the operator A can be represented as a sum of operators

$$A = \sum_{1}^{l} A_j, \qquad D(A) \subseteq D(A_j) \subseteq H, \qquad (5.24)$$

where for each operator A_j there exists B_j, $D(B_j) \supseteq D(A_j)$, $j = 1, \ldots, l$,

$$B = \sum_{1}^{l} B_j, \qquad D(B) \subseteq D(B_j) \subseteq H, \qquad (5.25)$$

and for all $u, v \in D(A_j)$, $j = 1, \ldots, l$

$$(A_j u, B_j v) = (B_j u, A_j v), \qquad (5.26)$$

$$(A_j u, B_j u) \geq \beta^2 \|B_j u\|^2, \qquad (5.27)$$

$$(A_j u, B_j u) \geq \alpha^2 \|u\|^2. \qquad (5.28)$$

We introduce the linear set W of l-dimensional vector-valued functions $v' = (v_0, \ldots, v_l)$ whose independent components v_j are defined first on $D(A_j)$, respectively, i.e., W is the direct product $D(A_1) \times \cdots \times D(A_l)$. Properties (5.26)–(5.28) allow us to define on W the inner product

$$\langle u', v' \rangle = \sum_{1}^{l} (A_j u_j, B_j v_j) = \sum_{1}^{l} [u_j, v_j]_{A_j B_j}. \qquad (5.29)$$

We denote by H'_{AB} the completion of W in the norm

$$\|u|H'_{AB}\| = \langle u', u' \rangle^{1/2}. \qquad (5.30)$$

By construction, $H'_{AB} \supseteq H_{AB}$ if an element u of H_{AB} is identified with the vector $u = (u, \ldots, u)$.

THEOREM 5.1. *If the operator A is B-symmetric and B-positive-definite and relations (5.24)–(5.28) hold, then for all $f \in H$ there exists a unique $u_0 \in H_{AB}$ such that*

$$\min_{H_{AB}} D[u] = D[u_0] = \mathscr{L}[u_0] = \max_{H'_{AB}} \mathscr{L}(v'), \qquad (5.31)$$

where $\mathscr{L}[v'] = -\|v'\|^2_{H'_{AB}}$, and the vector v' satisfies

$$\langle v', \xi \rangle = (f, B\xi) \quad \forall \xi = (\xi, \ldots, \xi) \in D(A).$$

PROOF. The existence and uniqueness of an element $u_0 \in H_{AB}$ which minimizes the functional $D[u]$ of (5.2) in H_{AB} for a B-symmetric and B-positive-definite operator A were established in §4, i.e., the first equality in (5.31) holds. Here the element u_0 is a generalized solution of problem (5.1):

$$[u_0, v] = (f, B_0 v) \quad \forall v \in H_{AB} \qquad (5.32)$$

(we assume that B has been extended by continuity to an operator $B_0 \supseteq B$, $D(B_0) = H_{AB}$, in virtue of (1.5), as was done in §4).

Obviously, for all $u', v' \in H_{AB}$

$$\|u' = v'\|_{H'_{AB}} = \|u'\|^2_{H'_{AB}} - 2\langle u', v' \rangle + \|v'\|^2_{H'_{AB}} \geq 0. \qquad (5.33)$$

We choose the vector v' so that

$$\langle v', \xi \rangle = (f, B\xi) \quad \xi = (\xi, \ldots, \xi) \in H_{AB}. \qquad (5.34)$$

The existence of at least one such vector v' follows from the existence of a generalized solution u_0: it suffices to choose $v' = (u_0, \ldots, u_0)$, and by (5.32) relation (5.34) will be satisfied. Setting $u' = u = (u, \ldots, u)$, with consideration of (5.34) we obtain from (5.33)

$$\begin{aligned} \|u' = v'\|^2_{H'_{AB}} &= \|u\|^2_{H_{AB}} - 2(f, Bu) + \|v'\|^2_{H'_{AB}} \\ &= D[u] = \mathscr{L}[v'] \geq 0, \end{aligned} \qquad (5.35)$$

i.e.,

$$D[u] \geq \mathscr{L}(v') \quad \forall u \in H_{AB} \qquad (5.36)$$

and for all $v' \in H'_{AB}$ such that (5.34) holds, where $u = u' = v' = u_0, u_0$ being a generalized solution of (5.1); from (5.33)–(5.35) we obtain

$$D[u_0] = \mathscr{L}[u_0]. \qquad (5.37)$$

Relations (5.36) and (5.37) demonstrate the validity of the second and third equalities in (5.31). ∎

COROLLARY 5.1. *If an element u_0 minimizing the functional $D[u]$ belongs to $D(A)$ and the set $R_A(B)$ is dense in H, then condition (5.34) is equivalent to*

$$Av' = f. \tag{5.38}$$

Indeed, in this case, since a minimum of the functional $D[u]$ is achieved on an element $u_0 \in D(A)$, everywhere in (5.32)–(5.34) it is possible to restrict attention to functions u_0, v', and u' in $D(A)$; relation (5.34) is then equivalent to

$$(Av' - f, B\xi) = 0 \quad \forall \xi = (\xi, \dots, \xi) \in D(A),$$

whence the corollary is obtained because $R_A(B)$ is dense in H. ∎

Thus, if the conditions of the theorem and corollary are satisfied, the Slobodyanskiĭ variational principle dual to (5.5) consists in maximizing $L[v'] = -\|v'|H'_{AB}\|^2$ on the set of vectors $v' = (v_1, \dots, v_l)$, $v_j \in D(A_j)$, satisfying the equation([12])

$$A_1 v_1 + \cdots + A_l v_l = f. \tag{5.39}$$

In spite of the formal difference between Slobodyanskiĭ's dual variational principle (5.31) and the dual variational principle (5.22), (5.23), in some cases they coincide.

Suppose the operator A in (5.1) is B-symmetric and B-positive-definite, $D(B) \supseteq D(A)$, $\overline{R_A(B)} = H$, and

$$(Au, Bu) \geq \beta^2 \|Bu\|^2 \geq \alpha^2 \|u\|^2 \quad \forall u \in D(A). \tag{5.40}$$

We set $B_1 = B_2 + B$, $A_1 u = Au - CBu$, and $A_2 u = CBu$, where the number $C > 0$ is such that $\kappa = \beta^2 - C > 0$. It is obvious that the operators A_j are B_j-symmetric ($j = 1, 2$); that they are B_j-positive-definite follows from the validity on $D(A)$ of the relations

$$(A_1 u, B_1 u) = (Au, Bu) - C(Bu, Bu)$$
$$\geq (\beta^2 - C)\|Bu\|^2 \geq \frac{\kappa \alpha^2}{\beta^2} \|u\|^2, \tag{5.41}$$

$$(A_2 u, B_2 u) = C\|Bu\|^2 \geq C \frac{\alpha^2}{\beta^2} \|u\|^2. \tag{5.42}$$

Then for $v_1, v_2 \in D(A)$

$$\mathcal{L}[v'] = -\|v'\|_{H'_{AB}}^2 = (-Av_1 - CBv_1, Bv_1) - C(Bv_2, Bv_2), \tag{5.43}$$

([12])We point out that only one equation (5.39) connects the l independent variables $v_1, \dots, v_l$ (with regard to this question see also [135]). This turns out to be important for applications.

where the components v_1 and v_2 are connected by the relation $A_1v_1 + A_2v_2 = f$, i.e.,

$$Av_1 - CBv_1 + CBv_2 = f. \tag{5.44}$$

With consideration of (5.44) the functional (5.43) can be transformed to the form

$$\mathscr{L}[v'] = (-Av_1 - CBv_1, Bv_1) - \frac{1}{C}(Av_1 - CBv_1 - f, Av_1 - CBv_1 - f)$$

$$= D[v_1] - \frac{1}{C}\|Av_1 - f\|_H^2,$$

$$\tag{5.45}$$

where the functional $D[v_1]$ is defined in (5.2). The last relation shows that in this case the functional $\Psi(v')$ of (5.23) can be obtained from Slobodyanskiĭ's dual functional $\mathscr{L}(v')$.

5.3. We shall apply to the class of B-symmetric, B-positive operators A in (5.1) Velte's construction [337] of a dual variational problem for the direct variational problem of minimizing the functional $D[u]$ of (5.2) and (5.5).

We remark first that in many cases the bilinear form $[u, v]$ (equal to (Au, Bv) for $u, v \in D(A)$) is symmetric and nonnegative: $[u, u] \geq 0$ on some set $\tilde{H}_{AB} \supset H_{AB}$, while positive-definiteness on H_{AB}, $[u, u] = 0 \Rightarrow u = 0$, is achieved by functions in H_{AB} satisfying some homogeneous boundary conditions; for example, for the problem

$$Au \equiv -\Delta u = f(x), \qquad x \in \Omega,$$

$$u(x) = 0, \qquad x \in \partial\Omega,$$

$$D(A) = C^2(\overline{\Omega}) \cap C^0(\Omega)$$

we have

$$[u, v] = \int_\Omega \operatorname{grad} u \cdot \operatorname{grad} v \, d\Omega,$$

$$\|u|\tilde{W}_2^1(\Omega)\|^2 = [u, u]. \tag{5.46}$$

In this case $\tilde{H}_{AB} = \tilde{W}_2^1(\Omega)$, the space with finite seminorm (5.46), and $H_{AB} = W_2^1(\Omega)$.

In a more general case, suppose the operator A of equation (5.1) is B-symmetric and B-positive, i.e., the bilinear form $[u, v]$ (equal to (Au, Bv) for $u, v \in D(A)$) is such that

$$[u, v] = [v, u] \quad \forall u, v \in \tilde{H}_{AB}, \tag{5.47}$$

$$[u, u] \geq 0 \quad \forall u \in \tilde{H}_{AB}, \tag{5.48}$$

$$[u, u] > 0 \quad \forall u \in H_{AB}, \; u \neq 0, \tag{5.49}$$

where H_{AB} is the closure of $D(A)$ in the norm (5.3), while $\tilde{H}_{AB}$ is a normed linear space, $\tilde{H}_{AB} \supset H_{AB}$, on which (5.47) and (5.48) are satisfied. We suppose also that the estimate (4.9) holds for the linear functional (f, Bu); therefore, according to results of §4, the functional (f, Bu) is assumed extended by continuity to all of H_{AB}, and then for all $f \in H$ there exists a unique $u_0 \in H_{AB}$ which in H_{AB} minimizes the quadratic functional

$$D[u] = \|u\|^2_{H_{AB}} - 2(f, Bu). \tag{5.50}$$

By Lemma 4.1 a generalized solution of problem (5.1) is characterized by the identity

$$[\overline{u}, \varphi] = (f, B\varphi) \quad \forall \varphi \in H_{AB}. \tag{5.51}$$

We further introduce the set V orthogonal in $\tilde{H}_{AB}$ to H_{AB}, $H_{AB} \oplus V = \tilde{H}_{AB}$, i.e.,

$$V = \{v \in \tilde{H}_{AB} : [v, \varphi] = 0 \quad \forall \varphi \in H_{AB}\}, \tag{5.52}$$

and we determine an element $w \in \tilde{H}_{AB}$ such that

$$[w, u] = (f, Bu) \quad \forall v \in H_{AB}. \tag{5.53}$$

It is obvious that w in general does not belong to V, but there exists an element $w_0 \in \tilde{H}_{AB}$ such that $w - w_0 \in V$; this element w_0 is determined according to (5.52) and (5.53) by the relation

$$[w_0, \varphi] = (f, B\varphi) \quad \forall \varphi \in H_{AB}. \tag{5.54}$$

THEOREM 5.2. *For a B-symmetric and B-positive operator A under condition* (4.9) *there exists a unique element* $\overline{u} \in H_{AB}$ *such that*

$$\min_{H_{AB}} D[u] D[\overline{u}] = K[\overline{u}] = \max_{w - w_0 \in V} K(w), \tag{5.55}$$

where $K(w) = -\|w\|^2_{AB}$.

PROOF. For any $u, w \in \tilde{H}_{AB}$

$$\|u - w\|^2_{AB} = \|u\|^2_{AB} - 2[u, w] + \|w\|^2_{AB} \geq 0, \tag{5.56}$$

which with consideration of (5.53) we write in the form

$$\|u - w\|^2_{AB} = \|u\|^2_{AB} - 2(f, Bu) - K(w) = D[u] - K(w) \geq 0. \tag{5.57}$$

Thus, $D[u] \geq K(w)$ for all admissible u and w, and as a result of §4 there exists a unique $\overline{u} \in H_{AB}$ such that

$$D[\overline{u}] = \min_{H_{AB}} D[\overline{u}] = -\|\overline{u}\|^2_{AB}. \tag{5.58}$$

As before, it is obvious that $K(w) \leq D[\overline{u}]$. Now the element $\overline{u}$ is such that $\overline{u} - w_0 \in V$, since by (5.51) and (5.54) $[\overline{u} - w_0, \varphi] = 0$ for all $\varphi \in H_{AB}$, and $K(\overline{u}) = -\|\overline{u}\|_{AB}^2 = D[\overline{u}]$, which proves the theorem. ∎

Commentary to Chapter I

To §1. 1. Symmetrization of equations of mathematical physics, i.e., reduction of a nonsymmetric equation to an equivalent symmetric equation by means of multiplication of the original equation by an auxiliary ("symmetrizing") factor, has long been known (see, for example, [208], where symmetrization of an integral equation is considered). Rather complete results were first obtained for equations with symmetrizable, bounded or compact operators A (see [187], [275], [348], and [349]. For a symmetrizable operator $B = J$, where J is an isometric unitary operator, there exists a theory of J-isometric operators [169].

It should be noted that if a symmetrizing operator B is symmetric and positive-definite, then the operator A is symmetric and positive-definite relative to the inner product $[u, v]_B = (u, Bv)$, since

$$[Au, v]_B = (Au, Bv) = (Bu, Av) = [u, Av]_B$$

(regarding applications of such operators, see [250], [254], [256], and [259]).

Friedrichs [108] gave a systematic approach to the investigation of equations with a symmetrizable differential operator (see also [160, [287], and [300]).

Symmetrization of a complex boundary value problem for the neutron-transport equation was carried but by V. S. Vladimirov in his doctoral dissertation [340].

2. As already noted in the Introduction, the idea of considering B-positive operators was first expressed by Academicians M. F. Kravchuk and N. M. Krylov (see [171], and also the monograph [167] and the bibliography presented there), who proposed determining the coefficients a_i^n $(i = 1, \ldots, n)$ of the approximate solution $y_n(x) = \sum_1^n a_i^n \varphi_i(x)$ of a differential equation $\mathscr{L}y(x) = f(x)$ not from the Ritz system

$$(\mathscr{L}y_n - f, \varphi_m) = 0, \qquad m = 1, \ldots, n \tag{1}$$

or from the system of least squares

$$(\mathscr{L}y_n - f, \mathscr{L}\varphi_m) = 0, \qquad m = 1, \ldots, n, \tag{2}$$

but from the condition

$$(\mathscr{L}y_n - f, B\varphi_m) = 0, \qquad m = 1, \ldots, n, \tag{3}$$

for some auxiliary operator B. It was thus established that B-positivity of the operator is sufficient for solvability of the system of linear algebraic equations (3):

$$(\mathscr{L}z, Bz) > 0 \quad \forall z \in D(\mathscr{L}, B), \ z \neq 0. \tag{4}$$

The direct methods developed on the basis of this approach are called Krylov methods or moment methods [167], [210], [213].

It is possible to indicate some closely related ideas used by many authors [130], [313], for example, the "a-b-c" method developed by Friedrichs [108], Morawetz [231], and others, which makes it possible to establish uniqueness of a solution of the equation $A\psi(x, y) = f(x, y)$ by obtaining and analyzing relations $0 = (A\psi, a\psi + b\psi_x + c\psi_y) \geq \|\psi\|_A \geq 0$.

It should be noted that back in 1955 Lions [192] defined for a given elliptic operator A a Hilbert space with inner product $[u, v]_{AB} = (Au, Bv)$ for a certain auxiliary operator B. Lions' was a development of his generalization of results of Leray [190] by means of the concept of a conditionally E-elliptic bilinear form.

The problem of constructing operators B for a linear hyperbolic system of first-order equations $Au = f$ satisfying a generalized energy inequality $(Au, Bu) \geq \|u\|^2$ was also posed by A. A. Dezin ([63], §3.5).

Krylov and Kravchuk's idea was used most purposefully to develop variational methods of solving equations of mathematical physics by Martynyuk [209]–[217], Lyashko [196]–[204], Petryshyn [250]-[263], and Shalov [292]-[295].

3. In this chapter for B-positive operators B-positive operators A we nowhere used inequality (1.4) as an estimate, but applied only properties (1.2) and (1.3), which follow form (1.4).

To §§1–4. 1. The main results of the theory presented in these sections are due to Martynyuk, Petryshyn, and Shalov. In the work of Martynyuk and Petryshyn a class of B-positive-definite (Definition 1.3) closable operators A was investigated: with a closable operator B by Petryshyn [250], [253], [257] and with a closed B by Martynyuk [209], [212]. Shalov [292], [293] investigated a class of B-positive (Definition 1.2) weakly closable operators A.

The results of Lemma 1.1 (for the various classes of operators indicated) are contained in the works of all three authors (see, for example, [209] and [257]); Theorem 1.1 is a generalization of a result of Shalov [293]. Theorem 1.2 was established by Martynyuk [212], and in connection with this we mention that in a recent paper [356] Dincă and Mateescu proved a stronger result for a B-symmetric B-positive-definite (in the sense of

Petryshyn) operator A there exists a unique symmetric positive-definite operator K such that $A = K \cdot B$.

Shalov's results are expounded in §2.

The construction of the space H_{AB} in §3 is a direct generalization of results of Friedrich's [106], [107]; this construction and Theorem 3.1 were present in the works of all three authors indicated; in the proof of properties b) and c) we followed Petryshyn [250], [275]. Corollary 3.3, which will be useful later on, was established by us, as were a number of other results.

The exposition of §4.1 is carried out mainly according to the scheme of Shalov [293], to whom Theorem 4.1 for B-positive operators is due. For B-positive-definite operators an analogous scheme was used by Martynyuk [209] and Petryshyn [250], [257]. Theorem 4.2 refines the corresponding result of Shalov [293]. We established Lemmas 4.1, 4.2, and 4.3.

2. Comparison of Theorem 1.1 with the results of §4 shows that if the equation

$$Au = f \tag{5}$$

is well-posed in the sense of Hadamard, then there exists a set of symmetrizing operators B of (1.6) with properties (1.1)–(1.5) and hence a set of corresponding quadratic functionals

$$\Phi[u] = (Au, Bu) - 2(f, Bu). \tag{6}$$

Substituting (1.6) into (6), we find that

$$\Phi[u] = (u, Cu) - 2(f, (A^{-1} \cdot Cu). \tag{7}$$

If A is a well-posed differential operator in the sense of Hadamard, then $(A^{-1})^*$ is frequently an integral operator; the operator C in the conditions of Theorem 1.1 can also be chosen to be an integral operator. Then the functional (7) does not contain derivatives of the unknown function. This fact is especially important in applications: to minimize such a functional it is possible to take a shorter Ritz series, which is especially essential in the finite element method [343], [355].

In §13 an example is given of the construction of a functional not containing derivatives of the unknown function for a complex boundary value problem generalizing the Cauchy, Goursat, Darboux, and mixed boundary value problem for the wave equation.

However, the construction of the operator B in the form (1.6) is non-constructive for the majority of boundary value problems for partial differential equations of interest in practice.

3. Petryshyn's Theorem 4.4 and its corollary pertain to known methods (see [224], [313], [250], [251], and [257] of investigating complex operator equations by comparing them with much simpler operator equations whose properties have been well studied. The sharing of various important properties such as the existence and uniqueness of solutions, applicability of various approximate methods and their convergence, etc. follows from the "closeness" in a particular sense of two operators. These methods are not variational methods and are not considered here although many concrete results on construction of operators B with properties (1.1)–(1.5) presented in Chapter 3 can be used in applications of these methods.

4. There is another possible treatment of the approach presented in this chapter to solution of equation (5) with a nonsymmetric operator A by a variational method; namely a treatment that generalizes the following familiar argument [225], [319]. Acting on both sides of equation (5) with the operator B^* (assuming $f \in D(B^*)$), we obtain

$$\tilde{A}u = \tilde{f}, \tag{1'}$$

where $\tilde{f} = B^* f$, and the operator B should be such that the operator $\tilde{A} = B^* A$ is symmetric and positive-definite in the usual sense. This leads by the equality $(\tilde{A}u, v) = (Au, Bv)$ to definitions analogous to Definitions 1.1–1.3 of §1. However, such a treatment affords little either at a theoretical or, especially, a practical level: theoretically it is more convenient to investigate a variational problem with a Friedrichs B-extension of the operator A than an extension of the operator B^*A, while in practice the construction of operators B is frequently based precisely on the analysis of the identities obtained by integrating by parts when conditions (1.1)–(1.5) are satisfied. It is known, by the way, that the usual variational method (the "energy" method [225]) for equation (5) with a selfadjoint and positive-definite operator can also be considered as a method of least squares in application to the equation $Bu = B^{-1}f$, where $B^2 = A$. In connection with this we mention [350], where a variational method for equation (5) with a B-symmetric and B-positive operator is presented, but a solution is determined in the space H_A of the method of least squares, $\|u|H_A\| = (Au, Au)^{1/2}$.

To §5. Complementary variational principles have a long history (for more details on this see [225], [14], and [278], and since in the foreign mathematical literature, especially in recent years, considerably more attention has been devoted to them than in the domestic literature, we indicate the main trends of their development and applications.

The ideas of transforming the original variational problem of minimization of a functional into a corresponding problem of maximization and of obtaining a posteriori estimates of approximate solutions of variational problems go back to Zaremba [353], [354], Trefftz [325], [326], and Friedrichs [105]. The work of these authors forms the basis of the three principal directions of obtaining complementary variational principles; these directions can tentatively be called geometric, operator, and functional directions. The first two of them are very close in the modern treatment, since they use the concept of an orthogonal decomposition of Hilbert spaces or a corresponding representation of the operator. Complementary variational principles and a posteriori estimates of an approximation by variational methods of solutions of various boundary value problems were obtained within the framework of these two approaches already in the forties and fifties [303], [70], [66], [68], [125], [315], [344], [345]. The method of Trefftz was generalized to a broad class of boundary value problems by Birman [34]–[36]. We note that in applications the results from these directions were used mainly for equations with symmetric and positive operators, since only for them was it possible to constructively attack the initial variational problem of minimizing a quadratic functional. In recent years with the help of operator theory interesting results have been obtained in the application of complementary variational principles to the investigation of variational inequalities [122], [55].

The third direction, the functional direction, is based on the duality transformation introduced by Friedrichs [105] and generalizing the Legendre transformation for the class of Euler functionals; it is discussed in detail in [57], Vol. I, and [14]. Generalizing this transformation, Slobodyanskii [302]–[305] proposed a rather simple method of reducing the problem of minimizing an "energy" functional to a problem of maximizing a corresponding functional while obtaining an a posteriori estimate of the approximation (see also [135]). Beginning from the works of Noble([13]) and Synge [316], in which the canonical transformation of Courant were generalized, this direction received intensive development by many foreign mathematicians (see [12]–[14], [21], [22], [38], [51], [148]–[150], [266], [267], [271], [272], [278], [291], [55], and [338]), since it was possible to construct simultaneously direct and complementary variational principles for broad classes of linear and nonlinear equations. We remark, however, that the majority of results in this direction were obtained (see the surveys

---

([13])See [243] and [244].

in [14] and [278] for equations

$$Au = F(u), \qquad (8)$$

where F is a potential operator in the case where the linear operator A can be represented in the form

$$A = T \cdot T^* + Q, \qquad (9)$$

where Q is positive, and A is either bounded in the original space H or

$$\mu_1 \|u\|^2 \leq (Au, u) \leq \mu_2 \|u\|^2, \qquad \mu_1, \mu_2 \equiv \text{const} > 0.$$

It is possible to extend some results of this direction (see [267] and §7) to a class of functionals

$$\Phi[u] = \int_\Omega \Phi(x, u(x), \mathscr{L}_1 u(x), \dots, \mathscr{L}_N u) \, dx,$$

where the $\mathscr{L}_i$, $i = 1, \dots, N$, are linear operators. Just with the help of these generalizations it is possible to obtain complementary variational principles for variational problems constructed for equations with a nonsymmetric, nonpositive, unbounded, and, generally speaking, nonlinear, nonpotential operator (see Chapter 4).

The important significance of dual variational principles and of obtaining them by means of a posteriori estimates was emphasized by Courant in the report [56], where among the three basic drawbacks of the variational method he called the first the fact that approximate solutions may fail to converge to the exact solution even if their derivatives converge. In a number of cases dual variational principles give not only an a posteriori integral estimate of the form (5.14), (5.15) but also a pointwise two-sided estimate of the approximate solution (see [53], [66], [67], [69], [70], [111], [337], and [344]).

CHAPTER II

Classes of Functionals and Function Spaces

§6. The role of classes of functionals for the variational method

For a long period of time the variational method was applied for solving mainly equations with symmetric and positive operators, among partial differential equations primarily elliptic equations. It was not only their important theoretical and practical value which played a decisive role here: results of a negative character were known which demonstrated the impossibility of solving by a variational method equations of certain other types of equations while remaining within the framework of the classical variational method of the class of Euler functionals.

6.1. Suppose there is given a general multidimensional, linear, homogeneous differential equation of second order with variable coefficients

$$\mathscr{L}_u u = \sum_{\beta=1}^{n}\sum_{\gamma=1}^{n} p^{\beta\gamma} D_\beta D_\gamma u + \sum_{\beta=1}^{n} q^\beta D_\beta u + r u(x) = 0. \tag{6.1}$$

Here $x = (x_1, \ldots, x_n) \in \Omega \subset R_n$; $p^{\beta\gamma}(x), q^\beta(x), r(x)$ are smooth functions in $\overline{\Omega}$ depending only on the independent variable x; $p^{\beta\gamma} = p^{\gamma\beta}$ for all $\gamma, \beta = 1, \ldots, n$; $D_\beta = \partial/\partial x_\beta$; and $u(x) \in \overset{\circ}{C}{}^2(\Omega)$. Only real-valued functions of a real variable are used in this chapter.

We consider the question of whether some sufficiently well-studied class of Euler functionals

$$F[u] = \int_\Omega F(x, u(x), u'_{x_1}, \ldots, u'_{x_n})\, dx \tag{6.2}$$

contains a functional $F_0[u]$ such that (6.1) or an equivalent equality follows from the condition

$$\delta F_0[u] = 0. \tag{6.3}$$

Since the differential equation of second order (6.1) is linear, it suffices to restrict attention to a special form of functionals (6.2)—a quadratic functional

$$V[u] = \int_\Omega \exp \Phi \cdot \left\{ \sum_{\beta=1}^n \sum_{\gamma=1}^n a^{\beta\gamma} \cdot D_\beta u \cdot D_\gamma u + cu^2 \right.$$

$$\left. +2\sum_{\beta=1}^n b^\beta u D_\beta u \right\} dx, \qquad (6.4)$$

where the smooth functions $a^{\beta\gamma}$, b^β, c, and Φ in $\overline{\Omega}$ depend only on x, and $a^{\beta\gamma} = a^{\gamma\beta}$ $(\beta, \gamma = 1, \ldots, n)$.

The first variation of the functional (6.4) is

$$\delta V = -2 \int_\Omega \delta u \cdot \exp \Phi \left\{ \sum_{\beta=1}^n \sum_{\gamma=1}^n a^{\beta\gamma} D_\beta D_\gamma u \right.$$

$$+ \sum_{\beta=1}^n \sum_{\gamma=1}^n (a^{\beta\gamma} \cdot D_\gamma \Phi + D_\gamma a^{\beta\gamma}) D_\beta u$$

$$\left. + \left[\sum_{\beta=1}^n (D_\beta b^\beta + b^\beta D_\beta \Phi) - c \right] u \right\} dx. \qquad (6.5)$$

This is shown by integrating by parts and setting the resulting boundary integrals equal to zero; this can be achieved for an appropriate choice of an admissible set (with regard to the boundary conditions) of functions $u(x)$.[14]

From the condition (6.3) that the functional (6.4) be stationary, since the variation δu in (6.5) is arbitrary, it follows that

$$\exp \Phi \mathscr{L}_\Phi u \equiv \exp \Phi \left\{ \sum_{\beta=1}^n \sum_{\gamma=1}^n a^{\beta\gamma} D_\beta D_\gamma u \right.$$

$$+ \sum_{\beta=1}^n \sum_{\gamma=1}^n (a^{\beta\gamma} D_\gamma \Phi + D_\gamma a^{\beta\gamma}) D_\beta u$$

$$\left. + \left[\sum_{\beta=1}^n (D_\beta b^\beta + b^\beta D_\beta \Phi) - c \right] u \right\} = 0. \qquad (6.6)$$

Since $\exp \Phi \neq 0$,

$$\mathscr{L}_\Phi u = 0. \qquad (6.7)$$

---

[14]Since we are considering the problem of finding a functional of the form (6.4) such that $\delta V = 0$ entails (6.1), we are so far not concerned with boundary conditions, although this sometimes turns out to be essential: a boundary value problem may be "nonvariational" (see [156] and [280]) due to the boundary conditions. We shall discuss this question later.

It is not hard to verify by direct integration by parts that the operator $\exp\Phi\mathscr{L}_\Phi$ in (6.6) is selfadjoint, while, generally speaking $\mathscr{L}_\Phi \neq (\mathscr{L}_\Phi)^*$ on $\overset{\circ}{C}{}^2(\Omega)$.

Generalizing this result, the English mathematician E. T. Copson showed back in 1925 [54] that for broad classes of equations (6.1) of elliptic and hyperbolic type in Ω there exists a smooth function $\Phi(x)$, and in the class of functionals (6.4) there exists a functional $V_0[u]$ such that equation (6.1) of elliptic hyperbolic type or an equation of the form $\exp\Phi\mathscr{L}_n u = 0$ equivalent to it follows from the condition $\delta V_0[u] = 0$. If (6.1) is of parabolic type in Ω, then a function $\Phi(x)$ does not exist, and in the class of functionals of the form (6.4) there is none for which the original equation (6.1) of parabolic type or one equivalent to it in the above indicated sense follows from the condition $\delta V[u] = 0$.([15])

For the quasilinear equation

$$\mathscr{L}u \equiv au_{xx} + 2bu_{xy} + cu_{yy} = G(x,y,u,u_x,u_y), \qquad (x,y) \in \Omega, \qquad (6.8)$$

where $u = u(x,y)$ and $a(x,y)$, $b(x,y)$, $c(x,y)$, and $G(x,y,u,u_x,u_y)$ are sufficient smooth functions in $\overline{\Omega}$ in all their variables, in 1960 Balatoni obtained the following result.

By introducing an auxiliary smooth function $\mu = \mu(x,y,u,u_x,u_y) \neq 0$, he showed that if equation (6.8) is of elliptic or hyperbolic type in Ω, then there exists a "variational multiplier" $\mu = \mu(x,y,u,u_x,u_y)$, $u \in \overset{\circ}{C}{}^2_0(\Omega)$, and in the class of Euler functionals (6.2) ($n = 2$) there exists a functional $F_0[u]$ such that condition (6.3) entails (6.8) or the equivalent equation

$$\mu(\mathscr{L}u - G) = 0. \qquad (6.8')$$

If the quasilinear equation (6.8) is of parabolic type in Ω, then a multiplier $\mu = \mu(x,y,u,u_x,u_y) \neq 0$ does not exist, and in the class of functionals (6.2) there is no functional for which (6.8) or (6.8') would follow from the vanishing of its first variation.

Thus, the results of [1], [18], and [54] demonstrate the impossibility in principle of constructing a functional with properties A–C of the Introduction in the class of Euler functionals (6.2) for parabolic equations. We also note the negative result of Millikan [226] on the nonexistence in the class of Euler functionals of a solution of the inverse problem of the calculus of variations for the Navier-Stokes equation (see also [102], §8.6).

([15])This negative result for parabolic equations was rediscovered in 1957 by Adler [1] for the one-dimensional equation $u_t - u_{xx} = g(x,t)$ (see also [245]). It should be noted that Copson [54] established his results in a more general formulation than that presented here, in particular, for systems of second-order differential equations.

6.2. However, classes of functionals are known in which the negative result [1], [18], and [54] for parabolic equations is inconsequential. For example, for an equation (6.1) of parabolic type it suffices to take a functional of the method of least squares [9]. However, this functional, generally speaking, contains derivatives of the unknown function of the same order as the original equation, and so it does not belong to the class we are considering with properties A–C from the Introduction. In §16, we present D. P. Didenko's generalization of the method of least squares to negative spaces, which is such that the functionals constructed for a broad class of systems of PDE possess properties A–C. The functionals there are defined by the norm $\| \cdot |W_2^{-l}| \|$ of the negative space, and, in general, they do not belong to the class of Euler functionals.

Functionals are also known which contain derivatives of the unknown function of lower order than in the parabolic equation being investigated, and the parabolic equation or one equivalent to it follows from the vanishing of its first variation. First of all, this is a functional of Fourier's "integral variational principle" in thermodynamics, from which the heat equation is derived [131]. However, this functional is an integral functional only in the space variables; with respect to the time axis this principle is a differential principle, since there is no integral with respect to time and the derivative of the temperature function is not varied ([131], Chapter VI, §1a).

Secondly, this is the variational principle introduced by Gurtin [129] with convolution on the time axis. Indeed, if, for example, we define the convolution

$$(u * v)(x, t) = \int_0^t u(x, \tau)v(x, t - \tau) \, d\tau, \tag{6.9}$$

then the system of equations

$$u_t - a^2 \Delta u = g(x, t), \qquad (x, t) \in \Omega \times (0, T), \tag{6.10}$$

$$u(x, 0) = u_0(x), \qquad x \in \Omega, \ \Omega \subset R_n \tag{6.11}$$

can be reduced to the equivalent integro-differential equation

$$u - u_0 = a^2 * \Delta u. \tag{6.12}$$

With the help of integration by parts it is not hard to show that for the functional

$$G[u] = \int_\Omega \{u * u + a^2 * \operatorname{grad} u * \operatorname{grad} u - 2u_0 * u\} \, dx \tag{6.13}$$

equation (6.12) follows from condition (6.3) (of course, assuming that the boundary integrals vanish). However, as Gurtin noted [129], investigation of a solution constructed by this variational principle is impeded by the fact that in the general case it is only a stationary point of a corresponding functional of the form (6.13); in general, such convolution functionals are not bounded either above or below.

Other approaches are possible to the construction for a given linear equation

$$Au = f \qquad (6.14)$$

of a functional $F_0[u]$ such that the original equation (6.14) or an equation equivalent to it follows from condition (6.3) (see, for example, [2], [110], [134], [139], [163], [280], [284], [322], and [323]).

In connection with this we note two facts. First of all, for a general nonsymmetric equation (6.14) the construction of such functionals (if it was not based on the scheme of Chapter 1) led to nonclassical functionals distinct in their variational properties from the Dirichlet functional for the Laplace equation; in general, these functionals did not satisfy conditions A–C of the Introduction, i.e., they contained derivatives of the same order as the original differential equation or were unbounded and had only a stationary point.

Secondly, known constructions of functionals with property C are based on the following general result.

We introduce two conditions.

CONDITION α. We suppose that on the domain $D(A)$ and range $R(A)$ of the operator A ($D(A)$ and $R(A)$ are linear sets) there is defined an inner product $\langle \cdot, \cdot \rangle$, and that some Hilbert space H is the completion of $D(A)$ in the norm $\|u\| = \langle u, u \rangle^{1/2}$ so that $A: D(A) \subset H \to R(A) \subset H$; $f \in H$.

CONDITION β. Suppose there exists an auxiliary linear operator $B: H \to H$, $D(B) \supseteq D(A)$, such that the operator A is a B-symmetric (1.1) and $\overline{R_A(B)} = H$.

We then have

LEMMA 6.1. *Under Conditions α and β, the functional*

$$D[u] = \langle Au, Bu \rangle - 2\langle f, Bu \rangle, \quad u \in D(A) \qquad (6.15)$$

is such that

$$\delta D[\bar{u}] = 0 \Leftrightarrow A\bar{u} = f, \quad \bar{u} \in D(A). \qquad (6.16)$$

Indeed, for any real parameter $t \in R^1$ and for all $\eta \in D(A)$ we find, using the B-symmetry (1.1) of the operator A, that

$$\delta D[u] = \left\{ \frac{d}{dt} D[\bar{u} + t\eta] \right\}_{t=0} = 2\langle A\bar{u} - f, B\eta \rangle = 0. \qquad (6.17)$$

(6.16) follows from this, since the set $R_A(B)$ is dense in H. ∎

With the help of this lemma the equivalence of passing from a solution of the equation to the variational problem can actually be established. We remark that if the operator A is selfadjoint in H, then for $B = I$ Condition β becomes superfluous. If the operator A is nonsymmetric, for example, relative to the inner product in L_2, then B is necessarily distinct from the identity operator: it is possible to achieve that Conditions α and β are satisfied by choosing a suitable inner product, as was actually done in [129], [134], [246], [284], [285], [321], and [322].([16]) Directly from Lemma 6.1 and Theorem 1.1 we obtain

COROLLARY 6.1. *If* $\overline{D(A)} = H$, $\overline{R(A)} = H$, *and the operator* A^{-1}, *defined on* $R(A)$, *exists, then there exist an operator* B *satisfying Condition* β *and a functional* (6.15) *with property* (6.16).

At a constructive level it is important to bear in mind that the functionals (6.15) in the general case of equation (6.14) (even a differential equation) need not belong to the class of Euler functionals.

In [91], [94], [100], [101], and [294] functionals (6.15) with properties A–C of the Introduction were constructed for concrete equations of elliptic, hyperbolic, and parabolic types in a class of functionals of the form

$$\Phi[u] = \int_\Omega \tilde{\Phi}\left(x, y, u(x,y), u_x, u_y, \int_a^x u(\xi, y)\,d\xi, \int_b^y u(x, \eta)\,d\eta,\right.$$
$$\left. \int_a^x d\xi \int_b^y u(\xi, \eta)\,d\eta, \int_a^x u_y(\xi, y)\,d\xi, \int_b^y u_x(x, \eta)\,d\eta\right) dx\,dy$$
$$(6.18)$$

(for simplicity the functional is written for a function $u = u(x,y), (x,y) \in \Omega \subset R_2$).

If in (6.18) $\tilde{\Phi} = \tilde{\Phi}(x, y, u, u_x, u_y)$, then such functionals generate the Sobolev norms and function spaces. If

$$\tilde{\Phi} = \tilde{\Phi}\left(x, y, \int_a^x u\,d\xi, \int_b^y u\,d\eta\right),$$

then the functionals (6.18) generate norms and function spaces of finite order; some properties of these spaces which will be useful below are considered in §9. In the case of a function $\tilde{\Phi}(\cdots)$ of general form the functionals generate some new function spaces; their definitions and properties are expounded in §10.

([16])The role of the inner product in solving the inverse problem of the calculus of variations was especially emphasized by Tonti [322] and Magri [205].

§7. On a class of functionals depending on linear operators

7.1. Suppose that the functional

$$D[u] = \int_\Omega F(x, u(x), \mathscr{L}_1 u, \ldots, \mathscr{L}_N u)\, dx \tag{7.1}$$

is defined on a linear, open, dense set in $L_2(\Omega)$ of admissible functions M; $x = (x_1, \ldots, x_n) \in \Omega \subset R_n$; the boundary $\Gamma = \partial\Omega$ is piecewise smooth; $\mathscr{L}_i: M_i \to L_2(\Omega)$ are linear operators acting in $L_2(\Omega)$, so that $M_i = D(\mathscr{L}_i) \supseteq M$, $i = 1, \ldots, N$; and $\bigcap_1^N M_i = M$. We shall assume that the function F is continuous in all its arguments, is continuously differentiable with respect to x and u, and is twice continuous differentiable with respect to $\mathscr{L}_i u$ ($i = 1, \ldots, N$), and, moreover, the operators $\mathscr{L}_i$ are such that for all $u(x) \in M$ the function $F(x, u(x), \mathscr{L}_1 u, \ldots, \mathscr{L}_N u)$ is continuous with respect to x in Ω. Below for brevity we shall sometimes write integrand in (7.1) in the form $F(x, u, \mathscr{L} u)$.

In the case where $\mathscr{L}_i u = \partial u / \partial x_i$ ($D(\mathscr{L}_i) = C^2(\Omega)$), $i = 1, \ldots, n$, $N = n$, and the functions $u \in M$ satisfy the condition

$$u(x) = h(x), \qquad x \in \Gamma_1 \subset \partial\Omega \tag{7.2}$$

($h(x)$ is a prescribed admissible function in $C(\partial\Omega)$) [119] it follows from the condition

$$\delta D(u) = 0 \tag{7.3}$$

that

$$\frac{\partial F}{\partial u} - \sum_{i=1}^n \frac{\partial}{\partial x_i}\left(\frac{\partial F}{\partial u_{x_i}}\right) = 0, \qquad x \in \Omega, \tag{7.4}$$

$$\sum_{i=1}^n \frac{\partial F}{\partial u_{x_i}} n_i = 0, \qquad x \in \Gamma_2, \tag{7.5}$$

where $n = (n_1, \ldots, n_n)$ is the vector of the outer normal to $\Gamma_2 = \partial\Omega \backslash \Gamma_1$. Here equations (7.4) and (7.5) and the variation of the functional $D(u)$ are considered on the set of functions $u(x)$ satisfying (7.2).

Equation (7.4) is the classical Euler equation for the class of functionals (6.2). In the general case the Euler equation for some functional $D(u)$, considered on a set of functions M, is the equation which follows from (7.3), written more constructively in the form

$$\left[\frac{d}{d\alpha} D(u + \alpha\eta)\right]_{\alpha=0} = 0, \tag{7.6}$$

where η is an arbitrary admissible function of M and α is an arbitrary real parameter.

Since the domains of the operators $\mathscr{L}_i$ are assumed to be dense in $L_2(\Omega)$, we can define the adjoint operators

$$(\mathscr{L}_i u, \varphi) = (u, \mathscr{L}_i^* \varphi) \quad \forall \varphi \in D(\mathscr{L}_i^*), \ u \in M_i. \tag{7.7}$$

In the case of differential operators $\mathscr{L}_i$ boundary conditions make an appearance in obtaining $\mathscr{L}_i^*$; we assume that they are included in the relation $(u, \mathscr{L}_i^* \varphi)$. Suppose also that $(\mathscr{L}_i^*)^* = \mathscr{L}_i$ for all $i = 1, \ldots, N$.

We then have

LEMMA 7.1. *The Euler equation for the functional* (7.1) *is*

$$\frac{\partial F}{\partial u} + \sum_{i=1}^{N} \mathscr{L}_i^* \frac{\partial F}{\partial(\mathscr{L}_i u)} = 0. \tag{7.8}$$

Indeed,

$$\delta D(u) = \int_{\Omega} \left\{ \frac{\partial F}{\partial u} \delta u + \sum_{1}^{N} \frac{\partial F}{\partial(\mathscr{L}_i u)} \mathscr{L}_i \delta u \right\} dx$$

$$= \int_{\Omega} \left\{ \frac{\partial F}{\partial u} + \sum_{1}^{N} \mathscr{L}_i^* \frac{\partial F}{\partial(\mathscr{L}_i u)} \right\} \delta u \, dx = 0. \tag{7.9}$$

where δu is an arbitrary admissible variation of the function $u \in M$. Since $\delta u \in M$ is arbitrary and the set M is dense in $L_2(\Omega)$, (7.8) then follows from (7.9). ∎

Thus, if some function $U(x) \in M$ realizes an extremal value (a largest or least value) of the functional (7.1) in M, then the following identity must be satisfied:

$$\frac{\partial F(x, U, \mathscr{L}U)}{\partial U} + \sum_{1}^{N} \mathscr{L}_i^* \frac{\partial F(x, U, \mathscr{L}U)}{\partial(\mathscr{L}_i U)} \equiv 0. \tag{7.10}$$

If a function $U(x) \in M$ realizes a smallest value of the functional (7.1), then the following condition must also be satisfied [119] (for $u = U(x)$):

$$\delta^2 D(u) = \int_{\Omega} \left\{ F_{uu}(\delta u)^2 + 2 \sum_{i=1}^{N} \frac{\partial^2 F}{\partial u \partial(\mathscr{L}_i u)} \delta u \delta \mathscr{L}_i u \right.$$

$$\left. + \sum_{i=1}^{N} \sum_{j=1}^{N} \frac{\partial^2 F}{\partial \mathscr{L}_j u \cdot \partial \mathscr{L}_i u} \delta \mathscr{L}_j u \delta \mathscr{L}_i u \right\} dx > 0. \tag{7.11}$$

Sufficient conditions for (7.11) to be satisfied are

$$\frac{\partial^2 F}{(\partial \mathscr{L}_1 u)^2} > 0, \quad \begin{vmatrix} \dfrac{\partial^2 F}{(\partial \mathscr{L}_1 u)^2} & \dfrac{\partial^2 F}{\partial \mathscr{L}_1 u \partial \mathscr{L}_2 u} \\ \dfrac{\partial^2 F}{\partial \mathscr{L}_2 \partial \mathscr{L}_1 u} & \dfrac{\partial^2 F}{(\partial \mathscr{L}_2 u)^2} \end{vmatrix} > 0, \dots, \qquad (7.12)$$

$$\begin{vmatrix} \dfrac{\partial^2 F}{(\partial \mathscr{L}_1 u)^2} & \dfrac{\partial^2 F}{\partial \mathscr{L}_1 u \partial \mathscr{L}_2 u} & \cdots & \dfrac{\partial^2 F}{\partial \mathscr{L}_1 u \partial \mathscr{L}_N u} & \dfrac{\partial^2 F}{\partial \mathscr{L}_1 u \partial u} \\ \hdotsfor{5} \\ \dfrac{\partial^2 F}{\partial \mathscr{L}_N u \partial \mathscr{L}_1 u} & \dfrac{\partial^2 F}{\partial \mathscr{L}_N u \partial \mathscr{L}_2 u} & \cdots & \dfrac{\partial^2 F}{(\partial \mathscr{L}_N u)^2} & \dfrac{\partial^2 F}{\partial \mathscr{L}_N u \partial u} \\ \dfrac{\partial^2 F}{\partial u \partial \mathscr{L}_1 u} & \dfrac{\partial^2 F}{\partial u \partial \mathscr{L}_2 u} & \cdots & \dfrac{\partial^2 F}{\partial u \partial \mathscr{L}_N u} & \dfrac{\partial^2 F}{(\partial u)^2} \end{vmatrix} > 0. \qquad (7.13)$$

These conditions may be called generalized Legendre conditions, since, for example, for $n = N = 1$ they take the familiar form [118]

$$F_{u_x u_x} > 0, \, F_{u,u} \cdot F_{u_x,u_x} - F_{u,u_x}^2 > 0.$$

7.2. As noted in the preceding section, the construction of variational problems for equations with nonsymmetric and nonpositive operators may lead to the consideration of functionals not of the class of Euler functionals, i.e., the operators $\mathscr{L}_i$ in (7.1) must sometimes be different from operators of differentiation even in the investigation of differential equations.

Assuming that for a given equation a variational problem of minimizing some functional (7.1) has already been constructed, by transforming this functional, we construct a corresponding dual variational problem. This approach to the construction of a dual variational problem (proceeding from the form of the functional (7.1)) has an advantage over the methods expounded in §5, in which properties of the operator of the original equation were used in an essential manner: linearity, B-symmetry, B-positivity, etc. The transformations carried out below include also the case where the variational problem of minimizing the functional (7.1) is equivalent to solving some nonlinear equation: the function F need not be a polynomial of second degree.

In the present section we follow Pomraning [267] while refining and making concrete his results for the classes of operators we are considering.

We suppose that there exists a function $U(x) \in M$ realizing in M a least value of the functional (7.1), i.e., the variational problem of minimization

$$d \equiv D(U) = \min_M D(u) \qquad (7.14)$$

is meaningful, and for all $u \in M$ the generalized Legendre conditions (7.12) and (7.13) are satisfied.

We consider the functional

$$D_1[u, v_1, \dots, v_n] \equiv D_1(u, v) = \int_\Omega F(x, u, v_1, \dots, v_n) \, dx, \qquad (7.15)$$

where
$$v_i = \mathscr{L}_i u, \qquad i = 1, 2, \ldots, N. \tag{7.16}$$

The functional (7.15) is now considered on the set $C = \{u, v_1, \ldots, v_N : v_i = \mathscr{L}_i u, u \in M, i = 1, \ldots, N\}$, and the variational problem (7.14) is written in the form
$$d = \min_{(u,v)\in C} D_1(u, v). \tag{7.17}$$

Conditions (7.16), under which minimization of the functionals $D_1(u, v)$ is carried out, can be taken into account in the usual manner by Lagrange multipliers $\lambda_1(x), \ldots, \lambda_N(x)$, with formation of the functional
$$D_2(u, v, \lambda) \equiv D_2(u, v_1, \ldots, v_N, \lambda_1, \ldots, \lambda_N)$$
$$= \int_\Omega \left\{ F(x, u, v) + \sum_1^N \lambda_i(v_i - \mathscr{L}_i u) \right\} dx. \tag{7.18}$$

In the functional (7.18) $u, v_1, \ldots, v_N$ are now independent functions, and it is necessary to determine whether the functional $D_2(u, v, \lambda)$ has a minimum value([17]) (with respect to u and v) in M.

We consider the behavior (depending on λ) of the functional $D_2(u, v, \lambda)$ at a critical point. For this we express all functions $u, v_1, \ldots, v_N$ in terms of λ, which can be done by considering the system of $N+1$ Euler equations for the functional $D_2(u, v, \lambda)$:
$$\frac{\partial F(x, u, v)}{\partial u} - \sum_1^N \mathscr{L}_i^* \lambda_i = 0, \tag{7.19}$$
$$\frac{\partial F(x, u, v)}{\partial v_i} + \lambda_i = 0, \qquad i = 1, \ldots, N. \tag{7.20}$$

Since $D_2(u, v, \lambda)$ is considered at a critical point, the system (7.19), (7.20) is solvable, i.e., from this it is possible to determine the functions
$$\bar{u} = \bar{u}(x, \lambda, \mathscr{L}^* \lambda) \equiv \bar{u}(x, \lambda_1, \ldots, \lambda_N, \mathscr{L}_1^* \lambda_1, \ldots, \mathscr{L}_N^* \lambda_N), \tag{7.21$_1$}$$
$$\bar{v}_i = \bar{v}_i(x, \lambda, \mathscr{L}^* \lambda)$$
$$\equiv \bar{v}_i(x, \lambda_1, \ldots, \lambda_N, \mathscr{L}_1^* \lambda_1, \ldots, \mathscr{L}_N^* \lambda_N), \qquad i = 1, \ldots, N. \tag{7.21$_2$}$$

([17])In quite general cases there will be no minimum. For example, for the equation $Au = f$ with a symmetric and positive-definite operator A we consider the functional $D(u) = (Au, u) - 2(f, u)$, $u \in D(A)$. It is obvious that $\delta^2 D = (A\delta u, \delta u) > 0$, and this functional is bounded below on the set $D(A)$. However, if by means of a Lagrange multiplier λ we transform $D(u)$ setting $v = Au$ to the form $D_2(u, v, \lambda) = (u, v) - 2(f, u) + (\lambda, v - Au)$, then $\delta^2 D_2(u, v, \lambda) = 2(\delta u, \delta v)$. The second variation does not have definite sign, and $D_2(u, v, \lambda)$ is not bounded either above or below.

The second variation (with respect to u and v) of the functional $D_2(u, v, \lambda)$ is

$$\delta^2 D_2(u, v, \lambda) = \int_\Omega \left\{ \frac{\partial^2 F}{\partial u \, \partial u}(\delta u)^2 + 2 \sum_{i=1}^N \frac{\partial^2 F}{\partial u \, \partial v_i} \delta u \delta v_i \right.$$
$$\left. + \sum_{i=1}^N \sum_{j=1}^N \frac{\partial^2 F}{\partial v_j \, \partial v_i} \delta v_j \delta v_i \right\} \, dx. \qquad (7.22)$$

Since it was assumed earlier that there exists a function $v(x) \in M$ minimizing $D(u)$ and the sufficient conditions (7.12) and (7.13) are satisfied for all admissible functions, it follows that these conditions obviously suffice in order that $\delta^2 D_2(u, v, \lambda) > 0$ (cf. (7.22) and (7.11)). Thus, $D_2(u, v, \lambda)$ is bounded below, and

$$d = \min_{(u,v) \in C} D_1(u, v) = \min_{(u,v) \in C} D_2(u, v, \lambda)$$
$$\geq \min_{\substack{u \in M \\ v \in D \, (D_2)}} D_2(u, v, \lambda) \qquad (7.23)$$

for all admissible λ. Further on we shall show that there exist functions $\lambda_1, \ldots, \lambda_N$ (Lagrange multipliers) for which the quantity d in (7.23) is achieved, i.e., it is possible to pose the problem

$$d = \max_{\lambda \in D \, (D_2)} \min_{\substack{u \in M \\ v \in D \, (D_2)}} D_2(u, v, \lambda). \qquad (7.24)$$

7.3. We consider the first form of the dual variational problem for the direct variational problem (7.14).

With the help of the functions (7.21) minimizing the functional $D_2(u, v, \lambda)$ we define the function

$$\psi(x, \lambda, \mathscr{L}^*\lambda) = F(x, \bar{u}(x, \lambda, \mathscr{L}^*\lambda), \bar{v}(x, \lambda, \mathscr{L}^*\lambda))$$
$$+ \sum_1^N [\lambda_i \bar{v}_i(x, \lambda, \mathscr{L}^*\lambda) - \bar{u}(x, \lambda, \mathscr{L}^*\lambda) \cdot \mathscr{L}_i^* \lambda_i] \qquad (7.25)$$

and the functional

$$D_3(\lambda) = \int_\Omega \psi(x, \lambda_1, \ldots, \lambda_N, \mathscr{L}_1^* \lambda_1, \ldots, \mathscr{L}_N^* \lambda_N) \, dx. \qquad (7.26)$$

From (7.18) and (7.24)–(7.26) we then obtain the dual variational problem

$$d = \max_{\lambda \in D \, (D_2)} D_3(\lambda). \qquad (7.27)$$

Thus, the original variational problem of minimizing the functional $D(u)$ of (7.1) has been transformed to a maximization problem (7.27), so that the upper and lower bounds of the functionals $D_3(\lambda)$ and $D(u)$ coincide.

Obtaining the functional $D_3(\lambda)$ in explicit form depends, of course, on the possibility of solving the system (7.19), (7.20). If, for example, the functional (7.1) is constructed for some linear equation, then $F(x, u, v_1, \ldots, v_N)$ is a polynomial of second order in $u, v_1, \ldots, v_N$, and the functions (7.21) can be found in explicit form from the system of linear equations (7.19), (7.20).

We now denote by $\bar{\lambda}_1, \ldots, \bar{\lambda}_N$ the functions maximizing the functional $D_3(\lambda)$ of (7.26). These functions satisfy the system of Euler equations for the functional $D_3(\lambda)$ (cf. (7.8)):

$$\frac{\partial \psi(x, \bar{\lambda}_1, \ldots, \bar{\lambda}_N, \mathscr{L}_1^* \bar{\lambda}_1, \ldots, \mathscr{L}_N^* \bar{\lambda}_N)}{\partial \bar{\lambda}_j}$$
$$+ \mathscr{L}_j \frac{\partial \psi(x, \bar{\lambda}_1, \ldots, \bar{\lambda}_N, \mathscr{L}_1^* \bar{\lambda}_1, \ldots, \mathscr{L}_N^* \bar{\lambda}_N)}{\partial (\mathscr{L}_j^* \bar{\lambda}_j)} \equiv 0 \qquad (j = 1, \ldots, N). \tag{7.28}$$

Replacing ψ by its representation (7.25) and using (7.19) and (7.20), we find that

$$\bar{v}_j(x, \bar{\lambda}, \mathscr{L}^* \bar{\lambda}) = \mathscr{L}_j \bar{u}(x, \bar{\lambda}, \mathscr{L}^* \bar{\lambda}), \qquad j = 1, \ldots, N. \tag{7.29}$$

The functions $\bar{u}$ and $\bar{v}_j$ $(j = 1, \ldots, N)$ of (7.21) satisfy (7.19) and (7.20); using the representation (7.29), we reduce this system to a single equation for the function $\bar{u}(x, \lambda, \mathscr{L}^* \lambda)$ in the following manner.

Acting on each ith equation in (7.20) with the operator $\mathscr{L}_i^*$, adding the result with (7.19), and using (7.29) to eliminate the variables $v_1, \ldots, v_N$, we obtain

$$\frac{\partial F(x, \bar{u}, \overline{\mathscr{L}u})}{\partial \bar{u}} + \sum_1^N \mathscr{L}_i^* \frac{\partial F(x, \bar{u}, \overline{\mathscr{L}u})}{\partial (\mathscr{L}_j \bar{u})} \equiv 0. \tag{7.30}$$

Since the function U minimizing the functional $D(u)$ of (7.14) satisfies (7.10), from comparison of (7.30) and (7.10) we get

$$\bar{u}(x, \bar{\lambda}, \mathscr{L}^* \bar{\lambda}) = U, \tag{7.31}$$

and, considering (7.29),

$$\bar{v}_j(x, \bar{\lambda}, \mathscr{L}^* \bar{\lambda}) = \mathscr{L}_j U, \qquad j = 1, \ldots, N. \tag{7.32}$$

Substituting the functions $\bar{u}$ of (7.31) and $\bar{v}_j$ $(j = 1, \ldots, N)$ of (7.32) into (7.25), from (7.26) and (7.14) we find that

$$D_3(\bar{\lambda}_1, \ldots, \bar{\lambda}_N) = \int_\Omega F(x, U, \mathscr{L}U) \, dx = D(U) = d,$$

i.e., there actually exists a collection of Lagrange multipliers $\lambda_1, \ldots, \lambda_N$ delivering a solution of the dual variational problem (7.27).

7.4. To obtain the second form of the dual variational problem for (7.14) we transform the problem (7.27) of maximizing the functional (7.26) in the variables $\lambda_1, \ldots, \lambda_N$ to a maximization problem on the original set of functions $u \in M$ so that the maximum of this functional should also be equal to d and should be achieved on functions U realizing a minimum value of the functional (7.1). This alternative form is more convenient in applications, since the unknown function $u(x)$ in (7.1) usually has a physical meaning and can be estimated qualitatively [150], [267], while the variables $\lambda_1, \ldots, \lambda_N$ (the Lagrange multipliers) were introduced arbitrarily in order to carry out the corresponding mathematical transformations.

We shall first find a relation between the function $U(x)$—a solution of the variational problem (7.14)—and the functions $\lambda_1, \ldots, \lambda_N$—a solution of the variational problem (7.27). Substituting (7.31) and (7.32) into (7.20), we find that

$$\bar{\lambda}_j = -\frac{\partial F(x, U, \mathscr{L}_1 U, \ldots, \mathscr{L}_N U)}{\partial(\mathscr{L}_j U)}, \qquad j = 1, \ldots, N. \tag{7.33}$$

Using this connection between the extremal values $U(x)$ and $\lambda_1, \ldots, \lambda_N$, we can maximize the functional $D_3(\lambda)$ in (7.27) on the set of functions $\lambda_1, \ldots, \lambda_N$ satisfying

$$\lambda_j = -\frac{\partial F(x, u, \mathscr{L}_1 u, \ldots, \mathscr{L}_N u)}{\partial(\mathscr{L}_j u)}, \qquad u \in M, \ j = 1, \ldots, N. \tag{7.34}$$

Substituting (7.34) into (7.25), we see that

$$\psi(x, \lambda_1, \ldots, \lambda_N, \mathscr{L}_1^* \lambda_1, \ldots, \mathscr{L}_N^* \lambda_N)$$

actually depends only on x, $u(x)$, $\mathscr{L}_1, \ldots, \mathscr{L}_N$, and $\mathscr{L}_1^*, \ldots, \mathscr{L}_N^*$ which can be expressed by introducing the function

$$\chi(x, u, \mathscr{L}, \mathscr{L}^*) = \psi(x, u, \lambda_1, \ldots, \lambda_N, \mathscr{L}_1^* \lambda_1, \ldots, \mathscr{L}_N^* \lambda_N), \tag{7.35}$$

where the λ_j $(j = 1, \ldots, N)$ are defined by (7.34).

The maximization problem (7.27) can then be written

$$d = \max_{u \in M} D_4(u), \tag{7.36}$$

where

$$\begin{aligned} D_4(u) &= D_3\left(-\frac{\partial F(x, u, \mathscr{L}_1 u, \ldots, \mathscr{L}_N u)}{\partial(\mathscr{L}_1 u)}, \ldots, \frac{\partial F(x, u, \mathscr{L}_1 u, \ldots, \mathscr{L}_N u)}{\partial(\mathscr{L}_N u)}\right) \\ &= \int_\Omega \chi(x, u, \mathscr{L}, \mathscr{L}^*) \, dx. \end{aligned} \tag{7.37}$$

Relations (7.36) and (7.37) give the second form of the dual variational problem for the minimization problem (7.14). As in the first case, the functional $D_4(u)$ of (7.37) can be written in explicit form very easily if $F(x, u, v_1, \ldots, v_N)$ is a polynomial of second order in u and $v_1, \ldots, v_N$.

7.5. We shall apply these results to the functional of a variational problem with a B-symmetric and B-positive-definite operator.

We consider the equation

$$Au = f, \tag{7.38}$$

where A is a linear operator acting in the Hilbert space H, $\overline{D(A)} = H$, $\|u\|_H \equiv \|u\| = (u, u)^{1/2}$, and $f \in H$.

We assume that there exists a closable linear operator $B: H \to H$, $D(B) \supseteq D(A)$, such that $\overline{R_A(B)} = H$ and for all $u, v \in D(A)$

$$(Au, Bv) = (Bu, Av), \tag{7.39}$$

$$(Au, Bu) \geq \beta^2 \|Bu\|^2 \geq \alpha^2 \|u\|^2. \tag{7.40}$$

From Lemma 6.1 we conclude that an element $u \in D(A)$ is a solution of (7.38) if and only if it is a critical point of the functional

$$\Phi_{AB}(u) = (Au, Bu) - 2(f, Bu). \tag{7.41}$$

We assume that there exists an element $U \in D(A)$ which satisfies (7.38). According to the results of §4, the following variational problem is then meaningful:

$$\Phi_{AB}(u) \to \min, \qquad u \in D(A),$$
$$d = \min_{D(A)} \Phi_{AB}(u) = \Phi_{AB}(U). \tag{7.42}$$

Here we encounter precisely the situation in which we cannot directly apply the method of Lagrange multipliers to immediately obtain a dual variational problem. Indeed, if we set $v = Au$ and take account of this condition by means of a Lagrange multiplier λ, then we obtain the functional

$$\Phi_0(u, v, \lambda) = (v, Bu) - 2(f, Bu) + (\lambda, v - Au), \tag{7.43}$$

which in general will not be bounded; here the second variation $\delta^2 \Phi_0(u, v, \lambda) = 2(\delta v, B \delta u)$ does not have definite sign.

We transform the functional (7.41) to a more convenient form. From (7.39) and (7.40) it follows that the operator $C = B^*A$ is symmetric and positive-definite; it can therefore be represented in the form $C = L' + L$, where L' and L are also symmetric positive-definite operators. Indeed, since $(Cu, u) \geq \alpha^2 \|u\|^2$ it is possible to set, for example, $L \equiv q$ and $L' = C - q$, where $q \equiv \text{const}$, $0 < q < \alpha^2$.

Further, we represent the symmetric positive-definite operator L' in the form $L' = TT^*$, which is always possible.([18]) Considering all these transformations, we can write the functional $\Phi_{AB}(u)$ of (7.41) in the form

$$\Phi_1(u) = (Tu, Tu) + (u, Lu) - 2(f, Bu). \qquad (7.44)$$

We find a lower bound for (7.44) on $D(A)$:

$$\Phi_1(u) = (T^*Tu, u) + (Lu, u) - 2(f, Bu)$$
$$= ((L' + L)u, u) - 2(f, Bu) = (Cu, u) - 2(f, Bu)$$
$$\geq \alpha^2\|u\|^2 - 2\|f\| \cdot \|Bu\| \geq \alpha^2\|u\|^2 - 2\|f\| \cdot \|u\|\frac{\alpha}{\beta}$$
$$= \left(\alpha\|u\| - \frac{\|f\|}{\beta}\right)^2 - \frac{\|f\|^2}{\beta^2} \geq -\frac{\|f\|^2}{\beta^2}. \qquad (7.45)$$

A similar estimate can be obtained by obtaining a lower bound for the functional (7.41); therefore,

$$d = \min_{D(A)} \Phi_1(u) \geq -\frac{\|f\|^2}{\beta^2}. \qquad (7.46)$$

It is not hard to verify (by considering the transformations of the operator A performed) that the condition $\delta\Phi_1(u) = 0$ implies (7.38), while if $U \in D(A)$ is such that

$$AU = f, \qquad (7.47)$$

then

$$\Phi_1(U) = -(f, BU) = d. \qquad (7.48)$$

Now setting

$$v = Tu, \qquad (7.49)$$

with the help of a Lagrange multiplier we obtain

$$\Phi_2(u, v, \lambda) = (v, v) + (u, Lu) - 2(f, Bu) + (\lambda, v - Tu). \qquad (7.50)$$

The Euler equations for this functional have the form

$$Lu = \tfrac{1}{2}(T^*\lambda + 2B^*f), \qquad (7.51)$$
$$v = -\tfrac{1}{2}\lambda. \qquad (7.52)$$

From this we find that

$$\bar{u} = \tfrac{1}{2}L^{-1}(T^*\lambda + B^*f), \qquad (7.53)$$
$$\bar{v} = -\tfrac{1}{2}\lambda, \qquad (7.54)$$

([18])The transformation carried out here is well known; see, for example, [14], [225], [267], and [337]. The operators C, T, and T^* will not participate in the final formulation; this is important, since it is difficult to express them explicitly.

which because of the character of the critical point minimize the functional (7.50). With the help of the functions $\bar{u}$ and $\bar{v}$, according to the general scheme (7.25)–(7.27), we obtain the first form of the dual variational problem:

$$d = \max_{D(\Phi_3)} \Phi_3(\lambda), \qquad (7.55)$$

where

$$\Phi_3(\lambda) = -\tfrac{1}{4}(\lambda, \lambda) - \tfrac{1}{4}(T^*\lambda + 2B^* f, L^{-1}(T^*\lambda + 2B^* f)). \qquad (7.56)$$

In order to obtain the second form of the dual variational problem, from the equation analogous to (7.33) in the general scheme we express the element λ in terms of u:

$$\lambda = -\frac{\partial F(\lambda, u, Tu)}{\partial(Tu)} = -2Tu. \qquad (7.57)$$

Substituting this λ into (7.55) and (7.56), we obtain

$$d = \max_{D(A)} \Phi_4(u),$$

where $\Phi_4(u)$ can be written (after a number of inverse transformations) in the form

$$\Phi_4(u) = \Phi_{AB}(u) - [Au - f, L^{-1}(Au - f)]_B, \qquad (7.58)$$

if we set

$$[u, v]_B = (Bu, Bv), \qquad \|u\|_B = [u, u]_B^{1/2}. \qquad (7.59)$$

Since the operator L is symmetric and positive-definite, so is L^{-1}, and we can define the inner product and norm

$$\langle u, v \rangle_{L^{-1}B} = [L^{-1}u, v]_B, \qquad \|u\|_{L^{-1}B} = (\langle u, u \rangle_{L^{-1}B})^{1/2}. \qquad (7.60)$$

Then (7.58) can be written in the form

$$\Phi_4(u) = \Phi_{AB}(u) - \|Au - f\|_{L^{-1}B}^2. \qquad (7.61)$$

For $B = I$ (i.e., for (7.38) with a symmetric positive-definite operator A) functionals analogous to (7.61) were obtained in [38] and [267].

For a differential operator A the operators B (see Theorem 1.1) and L^{-1} by construction may be integral operators, and derivatives of the unknown function of lower order than in (7.38) may actually be contained in (7.61). As remarked above, it would be simplest to take $L = q = \mathrm{const}$, $\alpha^2 > q > 0$; then the second term on the right side of (7.61) coincides with the functional of the method of least squares in H_B of (7.59), and the functional of the dual variational problem has the form

$$\Phi_4(u) = \Phi_{AB}(u) - \frac{1}{q}\|Au - f\|_B^2$$

(compare this with the functional (5.23) and the functional (5.45) obtained from the functional of M. G. Slobodyanskii).

§8. On a quasiclassical solution of the inverse problem of the calculus of variations for linear differential equations

8.1. We consider the linear homogeneous partial differential equation of second order

$$A_1 u \equiv \sum_{i,j=1}^{n} a_{ij} \frac{\partial^2 u}{\partial x_i \partial x_j} + \sum_{i=1}^{n} b_i \frac{\partial u}{\partial x_i} + cu = 0, \qquad x \in \Omega \subset R_n \qquad (8.1)$$

with smooth coefficients in $\overline{\Omega}$ satisfying the conditions of subsection 6.1.

As already noted, Copson [54] proved that if equation (8.1) is of elliptic or hyperbolic type in Ω, then there exists a multiplier $\mu(x)$, $\mu(x) \neq 0$ in Ω, and in the class of Euler functionals

$$F[u] = \int_{\Omega} F(x, u(x), u_{x_1}, \dots, u_{x_n}) \, dx \qquad (8.2)$$

there is a functional $F[u]$ such that the Euler equation for it is (8.1) or the equivalent equation

$$\mu(x) A_1 u = 0. \qquad (8.3)$$

If equation (8.1) is of parabolic type in Ω, then there does not exist a variational multiplier $\mu(x)$, and in the class of Euler functionals there is no functional $F[u]$ with the property

$$\delta F[u] = \int_{\Omega} \mu(x) A_1 u \delta u \, dx. \qquad (8.4)$$

The results of Balatoni, who proved analogous assertions for a quasilinear equation (6.8) with a variational multiplier $\mu = \mu(x, y, u(x, y), u_x, u_y)$, are also important.

It should be noted that, first of all, for many other differential equations the question of the existence of functionals for which the original equation (8.1) or an equivalent equation would follow from condition (8.4) remains open.

Secondly, the inverse problem of the calculus of variations is usually considered in the following formulation: establish for a given equation the existence of an equivalent functional with property (8.4). We shall consider the inverse problem of the calculus of variationals in a more classical formulation.

Suppose the operator A_1 of (8.1) is defined on a linear manifold of functions M dense in $H = L_2(\Omega)$; the condition $u \in M$ implies not only particular smoothness of function $u(x)$ but also that it satisfies certain boundary conditions (sometimes $u - u_0 \in M$ is written). On the set M we introduce the functional

$$\Phi[u] = \int_\Omega \Phi(x, u(x), \mathscr{L}_1 u, \ldots, \mathscr{L}_N u) \, dx, \qquad (8.5)$$

where the operators $\mathscr{L}_i$ ($i = 1, \ldots, N$), the function $\Phi(\cdots)$, the set M, and Ω satisfy the conditions of §7.1.

The inverse problem of calculus of variations for equation (8.1): Find a functional (8.5) defined on M such that A') the operators $\mathscr{L}_i$ ($i = 1, \ldots, N$) can contain operators of differentiation of the unknown function of no more than first order; B') the functional is bounded below on the set M; and

$$\delta\Phi[u] = \int_\Omega BA_1 u \delta u \, dx, \qquad u \in M, \qquad (8.6)$$

where B is some linear invertible operator, $D(B) \supseteq R(A_1)$.

A solution of this problem is called a *quasiclassical solution of the inverse problem of the calculus of variations*, since the prototype of such a solution is the Dirichlet functional for the Laplace equation.

THEOREM 8.1. *If on the set $R(A_1) = \{A_1 u : u \in M\}$, $\overline{R(A_1)} = H$ the inverse operator A_1^{-1} exists, then there exists a quasiclassical solution of the inverse problem of the calculus of variations (i.e., a functional* (8.5) *with properties* A'–C'). *The operator B in* (8.6) *and the operators $\mathscr{L}_1, \ldots, \mathscr{L}_N$ in* (8.5) *are here determined from the relation*

$$2 \sum_{i,j=1}^N \mathscr{L}_i^*(p_{ij}\mathscr{L}_j u) + 2 \sum_{i=1}^N [q_i \mathscr{L}_i u + \mathscr{L}_i^*(q_i u)] + 2au = BA_1 u, \qquad (8.7)$$

valid for all $u \in M$, where the coefficients $p_{ij}(x), q_i(x) \in D(\mathscr{L}_i^*)$, $i, j = 1, \ldots, N$; and $a(x)$ are arbitrary functions which are also determined only by the condition that (8.7) *hold.*[19]

PROOF. Equation (8.1) is linear and homogeneous; therefore it suffices to define the function $\Phi(\cdots)$ in (8.5) as a homogeneous polynomial of second order:

$$\Phi(x, u, \mathscr{L}u) \equiv \sum_{i,j=1}^N p_{ij}(x)\mathscr{L}_i u \mathscr{L}_j u + 2 \sum_{i=1}^N q_i(x) u \mathscr{L}_i u + a(x) u^2 \qquad (8.8)$$

with coefficients $p_{ij} = p_{ji}$ ($i, j = 1, \ldots, N$).

---

[19] Since (8.7) must be satisfied for the entire collection of functions in M, in each concrete situation this, of course, requires particular smoothness of the functions $p_{ij}(x)$, $q_i(x)$, and $a(x)$; we shall not give special consideration to this question.

The Euler equation for the functional (8.5) by Lemma 7.1 has the form

$$\frac{\partial \Phi}{\partial u} + \sum_{i=1}^{N} \mathscr{L}_i^* \frac{\partial \Phi}{\partial (\mathscr{L}_i u)} = 0. \qquad (8.9)$$

Determining the derivatives

$$\left.\begin{aligned}
\frac{\partial \Phi}{\partial u} &= 2au + 2\sum_{i=1}^{N} q_i \mathscr{L}_i u, \\
\frac{\partial \Phi}{\partial (\mathscr{L}_i u)} &= 2\sum_{i=1}^{N} p_{ij} \mathscr{L}_j u + 2q_i u,
\end{aligned}\right\} \qquad (8.10)$$

from (8.8), we find the Euler equation for the functional (8.5) with quadratic form $\Phi(x, u, \mathscr{L}u)$ of (8.8):

$$E_1 u \equiv 2\sum_{i,j=1}^{N} \mathscr{L}_i^*(p_{ij}\mathscr{L}_j u) + 2\sum_{1}^{N}[\mathscr{L}_i^*(q_i u) + a_i \mathscr{L}_i u] + 2au = 0. \qquad (8.11)$$

From the necessity of satisfying condition C' we obtain (8.7): in our notation

$$E_1(u) = BA_1 u \quad \forall u \in M. \qquad (8.12)$$

The proof that there exist an operator B and a set of operators $\mathscr{L}_1, \ldots, \mathscr{L}_N$ such that (8.7) is satisfied can be carried out in various ways. One way is theoretical and suitable only for establishing the existence of a quasiclassical solution of the inverse problem of the calculus of variations for equation (8.1).

The operator E_1 of (8.11) is symmetric on M; this is a consequence of the familiar fact ([57], vol. II) that the operator of the Euler equation is symmetric on the set on which the variation of the functional is considered. Since the operators $\mathscr{L}_1, \ldots, \mathscr{L}_N$ and the coefficients p_{ij}, q_i, and $a(x)$ in (8.8) and (8.11) are so far arbitrary, we can choose them so that $\mathscr{L}_1, \ldots, \mathscr{L}_N$ contain differential operators of more than first order and the operator E_1 is not only symmetric on M but is also positive-definite:

$$(E_1 u, u) \geq \gamma \|u\|^2 \quad \forall u \in M, \ \gamma \equiv \text{const} > 0; \qquad (8.13)$$

it suffices to set $E_1 u \equiv u$, $D(E_1) = M$. Having chosen the operators $\mathscr{L}_1, \ldots, \mathscr{L}_N$ and the coefficients $p_{ij}(x)$, $q_i(x)$ $(i, j = 1, \ldots, N)$, and $a(x)$, i.e., having actually determined the form (8.8), the functional (8.5), and the operator E_1 in (8.11), it is possible to find the linear operator B of (8.12),

$$B = E_1 A_1^{-1}, \qquad (8.14)$$

so that according to the hypothesis of the theorem the operator A_1^{-1} exists on $R(A_1)$.

We shall show that with this choice of B equation (8.1) is equivalent to $BA_1 u = 0$, i.e., we establish invertibility of B. If $B\varphi = 0$ for $\varphi \in R(A_1)$, i.e., $\varphi = A_1 u$ for some $u \in M$, then

$$0 = (B\varphi, u) = (E_1 A_1^{-1} \varphi, u) = (E_1 u, u) \geq \gamma \|u\|^2,$$

whence $u = 0$ and $A_1 u = \varphi = 0$.

It remains to show that with this choice of the operators $\mathcal{L}_1, \ldots, \mathcal{L}_N$ and the coefficients p_{ij}, q_i $(i, j = 1, \ldots, N)$, and a the functional (8.5) is bounded below on M. Since the symmetric equation (8.11) is the Euler equation for the functional (8.5), it is not hard to see that the functional $\Phi[u]$ of (8.5) and (8.8) can be represented in the form

$$\Phi[u] = \frac{1}{2}(E_1 u, u), \qquad u \in M, \tag{8.15}$$

for which we have from (8.13)

$$\Phi[u] = \frac{1}{2}(E_1 u, u) \geq \frac{\gamma}{2}\|u\|^2 \geq 0. \quad \blacksquare \tag{8.16}$$

It is also not hard to generalize Theorem 8.1 to inhomogeneous equations. If we consider the equation

$$A_1 u = f, \qquad f \in R(A_1), \ u \in M, \tag{8.1'}$$

then the quadratic form (8.8) should be chosen in the form

$$\Phi(x, u, \mathcal{L}u) = \sum_{i,j=1}^{N} p_{ij}(x)\mathcal{L}_i u \mathcal{L}_j u + a(x)u^2$$
$$+ \sum_{1}^{N}(q_i u \mathcal{L}_i u + \beta_i f \mathcal{L}_i u) + 2\alpha f u,$$

and the basic identity (8.7) will have the form

$$2\sum_{i,j=1}^{N} \mathcal{L}_i^*(p_{ij}\mathcal{L}_j u) + 2\sum_{1}^{N}[q_i \mathcal{L}_i u + \mathcal{L}_i^*(q_i u)] + 2\alpha f$$
$$+ 2au + 2\sum_{1}^{N} \mathcal{L}_i^*(\beta_i f)$$
$$= B(A_1 u - f) \quad \forall u \in M. \tag{8.7'}$$

In this generalization it is also necessary to impose some additional conditions on the operators A_1 and B and the function f in order to prove that the functional $\Phi[u]$ is bounded below (see §4).

8.2. We present auxiliary arguments which may sometimes assist (at a heuristic level) in the choice of the class of functionals (8.5) and the symmetrizing operator B for a given differential equation.

We shall analyze relation (8.7), which serves to determine the operators $\mathscr{L}_1, \ldots, \mathscr{L}_N$ and B and the coefficients $p_{ij}(x), q_i(x)$ $(i, j = 1, \ldots, N)$, and $a(x)$. In the proof of the theorem we chose a simple attack: having chosen almost arbitrarily (in any case without regard for (8.1)) the operators $\mathscr{L}_1, \ldots, \mathscr{L}_N$ and the coefficients p_{ij}, q_i $(i = 1, \ldots, N)$, and a, we defined the operator B uniquely by relation (8.14) which, of course, is nonconstructive.

Equality (8.7) affords another approach: simultaneous choice of the operators B and $\mathscr{L}_1, \ldots, \mathscr{L}_N$ and also the coefficients $p_{ij}(x), q_i(x)$ $(i, j = 1, \ldots, N)$, and $a(x)$.

To illustrate the above we limit ourselves to consideration of the heat equation

$$A_2 u \equiv u_t - u_{xx} = f(x, t), \qquad (x, t) \in \Omega \subset R_2, \qquad (8.17)$$

for which a negative result is known [1], [18], [54]: if in (8.5) only differential operators $\mathscr{L}_1, \ldots, \mathscr{L}_N$ of order no higher than first are considered, then in this class of Euler functionals (8.5), (8.2) the inverse problem of the variational calculus in the Copson-Balatoni formulation [18], [54] has no solution.

It is known that if the operator A_2 in (8.17) is considered on the set

$$M_2 = \left\{ u(x, t) \in C^2(\Omega) \cap C^1(\Omega \cup \Gamma_l) \cap C(\overline{\Omega}) : u \Big|_{\Gamma_r} = \frac{\partial u}{\partial x} \Big|_{\Gamma_r} = 0; \right.$$

$$\left. u(x, 0) = u(x, T), \ 0 < x < a \right\},$$

where

$$\Omega = \{x, t : 0 < x < a, \ 0 < t < T\}, \qquad f(x, t) \in C(\overline{\Omega}),$$
$$\Gamma_r = \{x, t : x = a, \ 0 < t < T\}, \qquad \Gamma_l = \{x, t : x = 0, \ 0 < t < T\},$$

then there exists the operator A_2^{-1} defined on the set $C_T = \{\varphi(x, t) \in C^1(\overline{\Omega}), \ \varphi(x, 0) = \varphi(x, T)\}$.

In this case (8.7') has the form

$$E_2 u \equiv 2 \sum_{i,j=1}^{N} \mathscr{L}_i^*(p_{ij}\mathscr{L}_j u) + 2 \sum_{1}^{N} [q_i \mathscr{L}_i u + \mathscr{L}_i^*(q_i u)] + 2au$$

$$+ 2\alpha f + 2 \sum_{1}^{N} \mathscr{L}_i^*(\beta_i f)$$

$$= B_2(u_t - u_{xx} - f) \quad \forall u \in M_2. \tag{8.18}$$

Setting $p_{11} = 1/2$, $p_{ij} = 0$ for all $i, j > 1$, $q_i = 0$ for all $i = 1, \ldots, N$, $a \equiv 0$, $\beta_2 = -1$, $\beta_i = 0$ for all $i \neq 2$, and $\mathscr{L}_2^* = B_2$, we obtain

$$\mathscr{L}_1^* \mathscr{L}_1 = B_2 \left(\frac{\partial}{\partial t} - \frac{\partial^2}{\partial x^2} \right). \tag{8.19}$$

The operator $\mathscr{L}_1$ can be defined by

$$\mathscr{L}_1 u = \int_0^x \frac{\partial u(\xi, t)}{\partial t} \, d\xi - \frac{\partial u}{\partial x}, \qquad D(\mathscr{L}_1) = M_2; \tag{8.20}$$

this choice is explained as follows. First of all, because of condition A′ the operator $\mathscr{L}_1$ must contain operators of differentiation of the unknown function of order no higher than first; second, the term $\partial u / \partial x$ is naturally included, since it is necessary to bear in mind the case $\partial u / \partial t \equiv 0$ (i.e., the equation $u_{xx} = f$); third, if we take the term $\partial u / \partial t$ "in pure form", then we obtain a functional of the class of Euler functionals (8.2); therefore, the second term should be chosen in the form $I_0(\partial u / \partial t)$, where I_0 is some nondifferential operator, and we have taken I_0 in (8.20) to be the simple integral operator $I_0 = \int_0^x (\cdots)(\xi, t) \, d\xi$.

For such an operator $\mathscr{L}_1$ of (8.20) we find, using (8.19), that

$$\mathscr{L}_1^* \mathscr{L}_1 u = \int_a^x \frac{\partial}{\partial t} \left(\int_0^\theta \frac{\partial u(\xi, t)}{\partial t} \, d\xi \right) d\theta - \frac{\partial^2 u}{\partial x^2} + \frac{\partial u}{\partial t} - \int_a^x \frac{\partial^2 u(\xi, t)}{\partial t \partial \xi} \, d\xi$$

$$= \left(\int_a^x d\theta \int_0^\theta \frac{\partial}{\partial t} \, d\xi + I \right) \left(\frac{\partial u}{\partial t} - \frac{\partial^2 u}{\partial x^2} \right) \equiv B_2 \left(\frac{\partial u}{\partial t} - \frac{\partial^2 u}{\partial x^2} \right)$$

and hence B_2 can be defined by

$$B_2 \varphi = \int_a^x d\theta \int_0^\theta \frac{\partial \varphi(\xi, t)}{\partial t} \, d\xi + \varphi(x, t)$$

on the set of functions $C_T^{2,1}(\Omega) = \{\varphi \in C^{2,1}\Omega \cap C(\Omega) \cap C^1(\Omega \cup \Gamma_l); \varphi(x, 0) = \varphi(x, T), a \leq x \leq b\}$.

On this set $C_T^{2,1}(\Omega)$ the equation $B_2\varphi = 0$ is equivalent to the system

$$\varphi_t + \varphi_{xx} = 0, \qquad (x,t) \in \Omega,$$
$$\varphi(x,0) = \varphi(x,T), \qquad 0 \le x \le b,$$
$$\partial\varphi/\partial x|_{\Gamma_l} = \varphi|_{\Gamma_r} = 0$$

and hence [64] is $B_2\varphi = 0$, then $\varphi = 0$.

In this case the functional $\Phi_2(u)$ has the form

$$\Phi_2(u) = \int_\Omega \left\{ \frac{1}{2}\left(\int_0^x u_t(\xi,t)\,d\xi - u_x\right)^2 \right.$$
$$\left. - f\left(u - \int_a^x d\theta \int_0^\theta u_t(\xi,t)\,d\xi\right)\right\} dx\,dt.$$

In conclusion we emphasize that in this section we have not considered the variational problem of minimization of functionals—this is the direct problem of the calculus of variations—but we have rather discussed an approach to the construction of quasiclassical solution of the inverse problem. Of course, to construct functionals bounded below on the set M it is possible to formulate the corresponding minimization problem, but, generally speaking, a solution of it cannot be obtained while restricting attention to the representation of these functionals and the space $H = L_2(\Omega)$; it is necessary to invoke the theory expounded in Chapter 1.

§9. The function spaces $S_{\hat{a}}^{-l}\dot{W}_p(\Omega)$

In this section we briefly present some facts regarding function spaces which will be used below mainly as spaces of functions for the right sides of equations. It is known (see [30] and [194]) that differential operators can realize an isomorphic mapping of a certain "positive" space W^+ onto some "negative" space W^-. Here the norm of W^- frequently turns out to be nonconstructive (difficult to compute explicitly), but, by choosing special classes of subspaces SW^+ with one dominant mixed derivative it is possible to obtain a constructive norm of the corresponding spaces $(SW^+)^*$.

9.1. We present some notation and facts used below.

Throughout §§9 and 10, $l = (l_1, \dots, l_n)$ is a vector with nonnegative integral components, $|l| = l_1 + \cdots + l_n$, and

$$D^l \equiv \frac{\partial^{|l|}}{\partial x^l} \equiv \frac{\partial^{|l|}}{\partial x_1^{l_1} \cdots \partial x_n^{l_n}}. \tag{9.1}$$

We shall write $\bar{p} \leq \bar{r}$ if $\rho_i < r_1$ for some (one or more) i, $1 \leq i \leq n$, while $\rho_k \leq r_k$ for the remaining k. Suppose $\Gamma = \bigcup_1^M \Gamma_k = \partial \Omega$ for some integer $M > 0$, and $\Gamma_k \subset \Gamma$ for all k. We consider all possible subdomains $\Omega' \subset \Omega$ having a boundary in common with Γ only on Γ_k. We denote by $L_{p,\text{loc}}(\Omega, \Gamma_k)$ the set of locally summable functions $u(x)$ in Ω such that $u \in L_p(\Omega')$ for some subdomain Ω'.

The anisotropic Sobolev function space[20] $W_p^l(\Omega)$ is defined as the set of functions $u(x) \in L_p(\Omega)$, $1 \leq p < \infty$, for which one of the following norms exists and is finite:

$$\|u(x)|W_p^l(\Omega)\| = \|u|L_p(\Omega)\| + \sum_1^n \left\| \frac{\partial^{l_j} u}{\partial x_j^{l_j}} \right| L_p(\Omega) \right\|, \qquad (9.2)$$

$$\|u(x)|W_p^l(\Omega)\| = \left[\int_\Omega \left\{ |u(x)|^p + \sum_1^n \left| \frac{\partial^{l_j} u}{\partial x_j^{l_j}} \right|^p \right\} dx \right]^{1/p}, \qquad (9.3)$$

where the derivatives are understood as generalized derivatives in the Sobolev sense ([311] and [242]; see also Definition 2.2). Henceforth in this chapter, unless specially mentioned, the measure and integral are understood in the Lebesgue sense [175].

The norms (9.2) and (9.3) are equivalent, and the space $W_p^l(\Omega)$ can also be defined as the closure of the set of functions $C^l(\overline{\Omega})$ in the norm (9.2) or (9.3).

We shall need the spaces dual to $W_p^l(\Omega)$; these are the spaces $W_q^{-l}(\Omega)$ of generalized functions in the sense of Schwartz (see [30] and [289]) for which the following norm exists and is finite:

$$\|u|W_q^{-l}(\Omega)\| = \sup_{\substack{v \neq 0, \\ v \in W_p^l}} \frac{|(u,v)|}{\|v|W_p^l(\Omega)\|},$$

$$\frac{1}{p} + \frac{1}{q} = 1, \quad 1 < p, q < \infty. \quad (9.4)$$

---

[20] Since we shall mainly consider nonsymmetric differential operators, many results will be obtained in terms of various anisotropic spaces; the definitions and properties of the anisotropic spaces $H_p^l(\Omega)$, $B_p^l(\Omega)$, and $W_p^l(\Omega)$ are described in more detail in [241] and [33]. In the latter book the anisotropic property is extensively investigated also for different powers of summability $p = (p_1, \ldots, p_n)$, $1 \leq p_j \leq \infty$, $j = 1, \ldots, n$; we take $p = (p, \ldots, p)$, $1 \leq p < \infty$, and $p = 2$ in application to a variational problem.

We present the known imbeddings of Sobolev spaces $W_p^{\bar{l}}(\Omega)$, Nikol'skiĭ spaces $H_p^{\bar{l}}(\Omega)$, and Besov spaces $B_{p,\theta}^{\bar{l}}(\Omega)$ $(1 \le p < \infty, 0 \le \bar{l} < \infty)$:

$$B_p^{\bar{l}}(\Omega) \to W_p^{\bar{l}}(\Omega) \qquad (1 \le p \le 2), \tag{9.5}$$

$$B_2^{\bar{l}}(\Omega) = W_2^{\bar{l}}(\Omega), \tag{9.6}$$

$$H_p^{\bar{l}+\bar{\varepsilon}}(\Omega) \to B_{p,\theta}^{\bar{l}}(\Omega) \to H_p^{\bar{l}}(\Omega) \quad \forall \bar{\varepsilon} > 0, \tag{9.7}$$

$$H_p^{\bar{l}+\bar{\varepsilon}}(\Omega) \to W_p^{\bar{l}}(\Omega) \to H_p^{\bar{l}}(\Omega) \quad \forall \varepsilon > 0. \tag{9.8}$$

9.2. Investigation of differential equations by functional methods makes it possible to consider in the general case equations with right sides in the spaces of generalized Sobolev-Schwartz functions W_q^{-l}. Here to eliminate pathological phenomena we often consider not the entire space W_q^{-l} but rather some subset of it which may not be a complete space. However, in practice it is often not convenient to invoke the spaces W_q^{-l}: a norm of the form (9.4) is hard to compute even for elementary functions.([21])

We shall define function spaces $\tilde{S}^{-l} W_p(\Omega)$ which are "close" to certain spaces $[S^l W_q(\Omega)]^*$ dual to spaces of functions $S^l W_q(\Omega)$ with a dominant mixed derivative. Closeness is understood in the sense that for all $u(x) \in \tilde{S}^{-l} W_p(\Omega)$ we have the equality of norms

$$\|u|\tilde{S}^{-l} W_p(\Omega)\| = \|u|(S^l W_q(\Omega))^*\|;$$

the space $\tilde{S}^{-l} W_p(\Omega)$ may not be complete, but its completion gives the space $[S^l W_q(\Omega)]^*$. Here the spaces $\tilde{S}^{-l} W_p(\Omega)$ are sufficiently ample, while the norm of functions in these spaces can be computed constructively.

Let $\Omega = \{x \in R_n : \psi_i(\bar{x}_i) < x_i < \varphi_i(\bar{x}_i), i = 1, \dots, n\}$ be a bounded domain in R_n with piecewise smooth boundary Γ; the functions

$$\varphi_i(\bar{x}_i) \equiv \varphi_i(x_1, \dots, x_{i-1}, x_{i+1}, \dots, x_n),$$

$$\psi_i(\bar{x}_i) \equiv \psi_i(x_1, \dots, x_{i-1}, x_{i+1}, \dots, x_n)$$

define the parts Γ_{φ_i} and Γ_{ψ_i} of the boundary Γ, $\Gamma_\varphi = \bigcup_i \Gamma_{\varphi_i}$, $\Gamma_\psi = \bigcup_1^n \Gamma_{\psi_i}$, $\Gamma = \Gamma_\varphi \cup \Gamma_\psi$. Below to simplify the exposition we shall set $\psi_i(\bar{x}_i) \equiv a_i$, where the a_i $(i = 1, \dots, n)$ are constants, and $\Gamma_{\bar{a}} = \Gamma_\psi$; many of the results remain valid also in the case when $\psi_i(\bar{x}_i) \not\equiv$ const. We suppose also that if an integral $\int_{a_i}^{x_i} (\cdots) d\xi_i$ is included in the norm of a function defined on Ω, then the corresponding interval (a_i, x_i) belongs to Ω; here x_i is the ith component of a point $\bar{x} \in \Omega$.

([21])In some cases [33] it is possible to compute an equivalent norm, but the exact value of the equivalence constant is hard to determine.

We introduce the notation

$$I_{\bar{a}}^{\bar{l}} u(x) \equiv I_{a_1,\ldots,a_n}^{l_1,\ldots,l_n} u(x)$$

$$= \int_{a_1}^{x_1} d\xi_{1,1} \int_{a_1}^{\xi_{1,1}} d\xi_{1,2} \cdots \int_{a_1}^{\xi_{1,l_1-1}} d\xi_{1,l_1} \int_{a_2}^{x_2} d\xi_{2,1} \cdots \int_{a_2}^{\xi_{2,l_2-1}} d\xi_{2,l_2}$$

$$\cdots \int_{a_n}^{x_n} d\xi_{n,1} \cdots \int_{a_n}^{\xi_{n,l_n-1}} u(\xi_{1,l_1},\ldots,\xi_{n,l_n}) \, d\xi_{n,l_n}. \qquad (9.9)$$

The space $\tilde{S}_{\bar{a}}^{-\bar{l}} W_p(\Omega)$ $(\infty > p \geq 1)$ is defined as the set of functions in $L_{p,\mathrm{loc}}(\Omega, \Gamma_{\bar{a}})$ for which the following norm exists and is finite:

$$\|u(x)|\tilde{S}_{\bar{a}}^{-\bar{l}} W_p(\Omega)\| = \left\{ \int_{\Omega} |I_{\bar{a}}^{\bar{l}} u(x)|^p \, dx \right\}^{1/p}. \qquad (9.10)$$

It is straightforward to verify that the properties of a norm are satisfied for (9.10). For $\bar{l} = (0,\ldots,0)$ it is natural to set $\| \cdot |\tilde{S}_{\bar{a}}^{0} W_p(\Omega)\| = \| \cdot |L_p(\Omega)\|$. In the case $l_j = 0$ at the site of the jth component of the vectorial index $\bar{a}$ in $\tilde{S}_{\bar{a}}^{-\bar{l}}$ we insert a dash (–).

We present some examples of spaces $\tilde{S}_{\bar{a}}^{-\bar{l}} W_p(\Omega)$.

EXAMPLE 1. Suppose that $\Omega_2 = \{x,t: a < x < \gamma(t), 0 < t < T\} \subset R_2$, and $\gamma(t)$ is a piecewise smooth function such that each line $t = \tau$, $0 < \tau < T$, parallel to the $0x$ axis intersects $\Gamma = \partial\Omega_2$ in no more than two points. Here

$$\Gamma_{\psi} = \Gamma_{\bar{a}} = \{x,t: x = a, \ 0 \leq t \leq T\} \cup \{x,t: t = 0, \ a \leq x \leq \gamma(0)\},$$

$$\Gamma_{\varphi} = \{x,t: x = \gamma(t), \ 0 \leq t \leq T\} \cup \{x,t: t = T, \ a \leq x \leq \gamma(T)\},$$

and

$$\|u|\tilde{S}_{a,-}^{-1,0} W_p(\Omega_2)\| = \left[\int_{\Omega} \left| \int_a^x u(\xi,t) \, d\xi \right|^p dx \, dt \right]^{1/p}. \qquad (9.11)$$

EXAMPLE 2. Let $\Omega_3 = \{x,y,t: a < x < A, b < y < B, 0 < t < T\}$; then

$$\|u|\tilde{S}_{a,b,-}^{-2,-1,0} W_p(\Omega_3)\| = \left(\int_{\Omega_3} \left| \int_a^x d0 \int_a^0 d\xi \int_b^y u(\xi,\eta,t) \, d\eta \right|^p d\Omega_3 \right)^{1/p}. \qquad (9.12)$$

EXAMPLE 3. $\Omega_4 = \{x,t: \psi(t) < x < \varphi(t), 0 < t < T\}$ where the piecewise smooth functions $\psi(t)$ and $\varphi(t)$ are such that each line $t = \tau$, $0 < \tau < T$, parallel to the $0x$ axis intersects $\Gamma = \partial\Omega_4$ in no more than two points. Then

$$\|u(x,t)|\tilde{S}_{\psi,-}^{-k,0} W_2(\Omega_4)\|$$

$$= \left[\int_{\Omega_4} \left(\int_{\psi(t)}^x d\theta_1 \int_{\psi(t)}^{\theta_1} d\theta_2 \cdots \int_{\psi(t)}^{\theta_{k-1}} u(\theta_k,t) \, d\theta_k \right) d\Omega_4 \right]^{1/2} \qquad (9.13)$$

REMARK 9.1. The function spaces $\tilde{S}_a^{-l}W_p(\Omega)$ are generated by norms in the class of functionals of the form (6.18), where the integrand Φ in (6.19) does not depend on derivatives of the function $u(x)$:

$$\Phi = \Phi\left(x, y, \int_a^x u\,d\xi, \int_b^y u\,d\eta\right).$$

Moreover, from the examples presented above it is clear that (in analogy with the anisotropic space $L_{\bar{p}}(\Omega)$) it is expedient to consider not norms, for example, of the form

$$\|u\| = \left[\int_\Omega \left(\left|\int_a^x u(\xi, y)\,d\xi\right|^p + \left|\int_b^y u(x,\eta)\,d\eta\right|^p\right) dx\,dy\right]^{1/p},$$

but norms of the form

$$\|u|\tilde{S}_{a,b}^{-1,-1}W_p(\Omega)\| = \left[\int_\Omega \left|\int_a^x d\xi \int_b^y u(\xi,\eta)\,d\eta\right|^p dx\,dy\right]^{1/p}.$$

In a manner similar to that in which the Sobolev spaces $W_p^{\bar{l}}(\Omega)$ are defined over the space $L_p(\Omega)$, below (§10) we consider over $\tilde{S}_a^{-l}W_p(\Omega)$ spaces of functions differentiable in several variables which are needed in investigating differential operators with range in $\tilde{S}_a^{-l}W_p(\Omega)$.

9.3. We introduce the space $S^l\dot{W}_q(\Omega, \Gamma_\varphi)$ $(1 < q < \infty)$ with a dominant mixed generalized derivative as the set of functions $v(x) \in L_q(\Omega)$ such that the norm

$$\|v|S^l\dot{W}_q(\Omega,\Gamma_\varphi)\| = \left[\int_\Omega \left|\frac{\partial^{|\bar{l}|}v(x)}{\partial x_1^{l_1}\cdots\partial x_n^{l_n}}\right|^q dx\right]^{1/q} \tag{9.14}$$

exists and is finite, where the derivatives of order less than $\bar{l}$ exist and vanish in the sense of $L_s(\Gamma)$ (for some $s = s(q) \geq 1$) on the corresponding portions of Γ $(x \in \Gamma\backslash\Gamma_a)$:

$$\begin{cases}\dfrac{\partial^{|\bar{l}|-1}v(x)}{\partial x_1^{l_1-1}\partial x_2^{l_2}\cdots\partial x_n^{l_n}}\cos(n,x_1) = \dfrac{\partial^{|\bar{l}|-2}v(x)}{\partial x_1^{l_1-2}\partial x_2^{l_2}\cdots\partial x_n^{l_n}}\cos(n,x_1) \\ \qquad\qquad = \cdots = \dfrac{\partial^{|\bar{l}|-l_1}v(x)}{\partial x_2^{l_2}\cdots\partial x_n^{l_n}}\cos(n,x_1) = 0, \\ \cdots\cdots\cdots\cdots\cdots\cdots\cdots\cdots\cdots\cdots\cdots\cdots\cdots\cdots \\ \dfrac{\partial^{l_n-1}v(x)}{\partial x_n^{l_n-1}}\cos(n,x) = \cdots = \dfrac{\partial v}{\partial x_n}\cos(n,x_n) = v\cos(n,x_n) = 0.\end{cases} \tag{9.15}$$

The space $S^l\dot{W}_q(\Omega,\Gamma_\varphi)$ need not be complete due to conditions (9.15); since it is closely related to the space $\tilde{S}^{-l}W_p(\Omega,\Gamma_\varphi)$ defined earlier, we present some properties of spaces of the type $S^l W_q(\Omega,\Gamma_\varphi)$.

A purposeful study of spaces of functions with a dominant mixed derivative was begun after the work of S. M. Nikol'skiĭ (see [238] and [239]—see [6], [32], [80], [81], and [195]). Authier [17] considered a space of functions associated with only one mixed derivative of the form (9.14).

Following Nikol'skiĭ [239], we denote by e_n the set of natural numbers $1, 2, \ldots, n$ and by e any subset of it; in particular, e_m will denote some subset $e \subset e_n$ consisting of m numbers; e_0 is the empty set; and $\bar{l}^e = (l_1^e, \ldots, l_n^e)$ is a vector such that $l_s^e = l_s$ if $s \in e$, and $l_s^e = 0$ if $s \in e_n \backslash e$. In particular, $\bar{l}^{e_0} = (0, \ldots, 0)$ and $\bar{l}^{e_n} = (l_1, \ldots, l_n)$. A function $u(x)$ belongs to $S^l W_p(\Omega)$, according to Nikol'skiĭ's definition [239], if for it there exists the finite norm

$$\|u(x) | S^l W_p(\Omega)\| = \sum_{e \subset e_n} \|D^{\bar{l}^e} u(x) | L_p(\Omega)\|, \tag{9.16}$$

where the sum goes over all possible $e \subset e_n$. Here, as in (9.14), the derivative $D^{\bar{l}^e} u(x)$ is understood as the generalized derivative

$$D^{\bar{l}^{e_n}} u(x) = \frac{\partial^{l_1}}{\partial x_1^{l_1}} \frac{\partial^{l_2}}{\partial x_2^{l_2}} \cdots \frac{\partial^{l_n} u(x)}{\partial x_n^{l_n}}. \tag{9.17}$$

In defining $S^l W_p(\Omega)$, Nikol'skiĭ emphasized [239], [195] that to avoid pathological phenomena it is necessary to require a priori that, together with p-summability of the dominant mixed derivative $D^l u(x)$, all "support" derivatives of $D^l u(x)$ p-summable, i.e., the derivatives corresponding to the projection of $\bar{l}$ onto the coordinate hyperplanes. This means that if some vector $\overline{K}^0 = (K_1^0, \ldots, K_n^0) \in e$, where $K_j^0 > 0$ $(j = 1, \ldots, n)$, then the K^e-projections of the vector K^0 onto the coordinate hyperplanes of any number of dimensions $0, 1, \ldots, n$ belong to e. For example, for the dominant mixed derivative

$$\frac{\partial^{r_1 + r_2} u(x)}{\partial x_1^{r_1} \partial x_2^{r_2}} \in L_p(\Omega), \qquad x = (x_1, x_2) \tag{9.18}$$

it is necessary to require a priori that

$$u \in L_p(\Omega), \quad \frac{\partial^{r_1} u}{\partial x_1^{r_1}} \in L_p(\Omega), \frac{\partial^{r_2} u}{\partial x_2^{r_2}} \in L_p(\Omega), \tag{9.19}$$

since [240] in the general case (9.19) does not follow from (9.18). It is clear, therefore, that for $S^{(r_1, r_2)} W_p(\Omega)$,

$$\|u | S^{(r_1, r_2)} W_p(\Omega)\|$$

$$\left[\int_\Omega \{ |u|^p + |D_{x_1}^{r_1} u|^p + |D_{x_2}^{r_2} u|^p + D_{x_1 x_2}^{r_1 + r_2} u|^p \} \, dx \right]^{1/p}, \tag{9.20}$$

we have $S^{(r_1, r_2)} W_p(\Omega) \to W_{p x_1, x_2}^{(r_1, r_2)}(\Omega)$.

It is shown in [240] that if $u \in S^l W_p(\Omega)$, then for any $\bar{p}$, $0 \le \bar{p} \le \bar{l}$, there exists $u^{\bar{p}} \in L_p(\Omega)$ (while the order of differentiation in the generalized derivative may vary in any manner), and, moreover,

$$\|u^{\bar{p}}|L_p(\Omega)\| \le \text{const} \cdot \|u|S^l W_p(\Omega)\|. \tag{9.21}$$

Generalizing the approach of Kudryavtsev [174], Nikol'skiĭ investigated the important and complicated question of stable boundary values of functions in $S^l W_p(\Omega)$. Before proceeding to a formulation of the result in general form, we first consider his idea for the example of the space

$$S^{1,1} W_2(\Pi), \quad \Pi \equiv \Omega = \{x, y : 0 < x < a, \ 0 < y < b\} \subset R_2, \quad u = u(x, y),$$

$$\|u|S^{(1,1)} W_2(\Pi)\| = \left(\iint_{\Pi} \{u^2 + u_x^2 + u_y^2 + u_{xy}^2\} \, dx \, dy \right)^{1/2}. \tag{9.22}$$

This example, by the way, demonstrates the possibility of passing from the norm (9.22) (of the form (9.16)) to a norm of the form (9.14) where only one dominant mixed derivative is present.

From an analysis of the Newton-Leibniz equalities

$$\left.\begin{aligned}
\frac{\partial u}{\partial y}(0, y) &= -\int_0^x \frac{\partial^2 u(\xi, y)}{\partial \xi \, \partial y} \, d\xi + \frac{\partial u(x, y)}{\partial y}, & 0 < y < b, \\
u(x, 0) &= -\int_0^y \frac{\partial u(x, y)}{\partial y} \, dy + u(x, y), & 0 < x < a, \\
u(0, y) &= -\int_0^x \frac{\partial u(\xi, y)}{\partial \xi} \, d\xi + u(x, y), & 0 < y < b
\end{aligned}\right\} \tag{9.23}$$

it follows that if $u \in S^{(1,1)} W_2(\Pi)$, then there exist $u_y(0, y) \in L_2(0, b)$, $u(x, 0) \in L_2(0, a)$, $u(0, y) \in L_2(0, b)$, and

$$\begin{aligned}
\|u_y(0, y)|L_2(0, b)\| &\le C_1 \|u|S^{(1,1)} W_2(\Pi)\|, \\
\|u(x, 0)|L_2(0, a)\| &\le C_2 \|u|S^{(1,1)} W_2(\Pi)\|, \\
\|u(0, y)|L_2(0, b)\| &\le C_3 \|u|S^{(1,1)} W_2(\Pi)\|.
\end{aligned} \tag{9.24}$$

The last relations show that for $u \in S^{(1,1)} W_2(\Pi)$ the boundary values

$$\begin{aligned}
\partial u/\partial y &= 0, & (x, y) &\in \Gamma_1 = \{x, y : x = 0, \ 0 < y < b\}, \\
u(x, 0) &= 0, & (x, y) &\in \Gamma_2 = \{x, y : y = 0, \ 0 < x < a\}, \\
u(0, y) &= 0, & (x, y) &\in \Gamma_1
\end{aligned} \tag{9.25}$$

are stable in mean on Γ_1 and Γ_2 respectively.

From (9.23) for functions $u(x, y)$ satisfying (9.25) we obtain

$$\int_\Omega u_y^2 \, d\Omega \leq C_4 \int_\Omega u_{xy}^2 \, d\Omega, \tag{9.26}$$

$$\int_\Omega u_x^2 \, d\Omega \leq C_5 \int_\Omega u_{xy}^2 \, d\Omega, \tag{9.27}$$

$$\int_\Omega u^2 \, d\Omega \leq C_6 \int_\Omega u_{xy}^2 \, d\Omega, \tag{9.28}$$

Having now defined the space $S^{(1,1)}\mathring{W}_2(\Pi, \Gamma_1, \Gamma_2)$ as the set of functions in $S^{(1,1)}W_2(\Pi)$ having boundary values (9.25) which are stable (in the sense of $L_2(\Gamma)$), it is not hard to obtain from (9.26)–(9.28) that in $S^{(1,1)}\mathring{W}_2(\Pi, \Gamma_1, \Gamma_2)$ the norm (9.22) is equivalent to

$$\|u|S^{(1,1)}\mathring{W}_2(\Pi; \Gamma_1, \Gamma_2)\| = \left(\int_\Omega u_{xy}^2 \, dx \, dy \right)^{1/2}. \tag{9.29}$$

This norm is obviously a special case of (9.14), and the passage from the norm (9.22) of the form (9.16) to the norm (9.29) of the form (9.14) is thus justified.

In the general case of a domain Ω with curvilinear boundaries, boundary conditions of the form (9.25) will no longer be stable in the sense of $L_2(\partial\Omega)$ for functions in $S^l W_p(\Omega)$, but they will be stable in the following weaker sense.[22]

Suppose the boundary Γ of Ω is regular in the sense of Nikol'skiĭ [241]; we denote by

$$x_i = \psi_i(\bar{y}_i), \qquad \bar{y}_i = (x_1, \ldots, x_{i-1}, x_{i+1}, \ldots, x_n),$$
$$\bar{y}_i \in \Delta_i = np_i(\gamma), \qquad i \in 1, \ldots, n \tag{9.30}$$

the equation of some part γ of the boundary Γ, $x \in \gamma \in \Gamma$.

Suppose $\bar{p} \leq \bar{l}$, i.e., $\rho_j \subset l_j$ for some $j \in [1, n]$, and $\rho_i \leq l_i$ for the remaining $i \in [1, n]$. If $u \in S^l W_p(\Omega)$, then the derivative $u^{(\bar{p})}$ has no γ an "x_j-boundary function", i.e., there exists a function

$$\lambda_{\gamma, \bar{p}, j}(\bar{y}) = u^{\bar{p}}|_\gamma^j = \lim_{x_j \to \psi_j(\bar{y})} D^{\bar{p}}(x_1, \ldots, x_n), \tag{9.31}$$

where the limit is understood in the sense of convergence almost everywhere on Δ_i or in the mean (in the sense of L_p) in the direction x_j; here

$$\|\lambda_{\gamma, \bar{p}, j}|L_p(\Delta_i)\| \leq \text{const} \, \|u|S^l W_p(\Omega)\|, \tag{9.32}$$

where const depends on p, $\bar{l}$, and Ω, but not on $u(x)$.

---

[22] This result is part of a general theorem established by Nikol'skiĭ [240] and generalizing a corresponding result of Kudryavtsev [174].

Thus, for functions in the space $S^l \dot{W}_q(\Omega, \Gamma_\varphi)$ of (9.14) in the case of a general domain Ω it is not possible to guarantee stability of the boundary values (9.25) in the sense of $L_s(\Gamma_\varphi)$, $s = s(q, \bar{l})$, but these boundary conditions will be stable in the sense of $L_s(\Gamma_\varphi)$ if all parts Γ_φ of the boundary $\Gamma = \partial\Omega$ are $(n-1)$-dimensional hyperplanes parallel to the coordinate planes, $\varphi_i(\bar{x}_i) \equiv$ const, $i = 1, \ldots, n$, since in this case $\gamma_i \equiv np_i\gamma = \Delta_i$ $(i = 1, \ldots, n)$.

We introduce the space $S^{-l} \dot{W}_p(\Omega, \Gamma_\varphi)$ dual to $\dot{S}^l W_q(\Omega, \Gamma_\varphi)$ $(1/p + 1/q = 1, 1 < p, q < \infty)$ as the set of generalized functions in the Schwartz sense for which the following norm exists and is finite:

$$\|u|S^{-l} \dot{W}_p(\Omega, \Gamma_\varphi)\| = \sup_{v \neq 0} \frac{|(u, v)|}{\|v|S^l \dot{W}_q(\Omega, \Gamma_\varphi)\|}. \tag{9.33}$$

As is known [30], the set $C_0^\infty(\Omega)$ of compactly supported functions in Ω is dense in $S^{-l} \dot{W}_p(\Omega, \Gamma_\varphi)$, and the dual space is complete even if the original space (in the present case $S^l \dot{W}_q(\Omega, \Gamma_\varphi)$) is not complete.

LEMMA 9.1. *There is the embedding*

$$\tilde{S}_{\bar{a}}^{-l} W_p(\Omega \to S^{-l} \dot{W}_p(\Omega, \Gamma_\varphi), \tag{9.34}$$

while if $\varphi_i(x) \equiv b_i$, $i = 1, \ldots, n$, *(i.e.,* $\Omega \equiv \Pi = \{x \in R_n : a_i < x_i < b_i,$ $i = 1, \ldots, n\}$ *is a parallelepiped), then for all* $u(x) \in \tilde{S}_{\bar{a}}^{-l} W_p(\Omega)$

$$\|u|\tilde{S}_{\bar{a}}^{-l} W_p(\Omega)\| = \|u|S^{-l} \dot{W}_p(\Omega, \Gamma_\varphi)\|. \tag{9.35}$$

PROOF. Suppose $u(x) \in \tilde{S}_{\bar{a}}^{-l} W_p(\Omega)$ and $v(x) \in S^l W_q(\Omega, \Gamma_\varphi)$. Integrating by parts using (9.15), and then applying the Cauchy-Schwarz-Bunyakovskiĭ inequality, we obtain (using the notation (9.9))

$$\left| \int_\Omega uv \, dx \right| = \left| \int_\Omega \int_{a_n}^{x_n} u(x_1, \ldots, x_{n-1}, \xi_{n,1}) \, d\xi_{n,1} \frac{\partial v}{\partial x_n} \, dx \right|$$

$$= \cdots = \left| \int_\Omega I_{\bar{a}}^l u(x) \frac{\partial^{|\bar{l}|} v(x)}{\partial x_1^{l_1} \cdots \partial x_n^{l_n}} \, dx \right|$$

$$\leq \|u|\tilde{S}_{\bar{a}}^{-l} W_p(\Omega)\| \, \|v|S^{-l} \dot{W}_q(\Omega, \Gamma_\varphi)\|, \tag{9.36}$$

i.e.,

$$\|u|S^{-l} \dot{W}_p(\Omega, \Gamma_\varphi)\| = \sup_{v \neq 0} \frac{|(u, v)|}{\|v|S^l \dot{W}_q(\Omega, \Gamma_\varphi)\|}$$

$$\leq \|u|\tilde{S}_{\bar{a}}^{-l} W_p(\Omega)\| \quad \forall u \in \tilde{S}_{\bar{a}}^{-l} W_p(\Omega). \tag{9.37}$$

The imbedding (9.34) follows from (9.37). If $\varphi_i = b_i$ $(i = 1, \ldots, n)$, then for the function

$$
v(x) = \int_{b_n}^{x_n} d\xi_{n,1} \int_{b_n}^{\xi_{n,1}} d\xi_{n,2} \cdots \int_{b_n}^{\xi_{n,l_n-1}} d\xi_{n,l_n} \cdots \int_{b_1}^{x_1} d\xi_{1,1}
$$
$$
\cdots \int_{b_1}^{\xi_{1,l_1-1}} |l_a^l u(\xi_{1,l_1}, \xi_{2,l_2}, \ldots, \xi_{n,l_n})|^{p/q} \, d\xi_{1,l_1}, \tag{9.38}
$$

belonging to $S^l \dot{W}_q(\Pi, \Gamma_\varphi)$ for all $u \in \tilde{S}_a^{-l} W_p(\Omega)$, equality is achieved in (9.36) and hence also in (9.37), which proves the second assertion of the lemma. ∎

COROLLARY 9.1. *Since the set of functions $C_0^\infty(\Omega)$ is dense in $S^{-l}\dot{W}_p(\Omega, \Gamma_\varphi)$, for all $u(x) \in \tilde{S}_a^{-l}\dot{W}_p(\Omega)$ there exists a sequence $\{u_p\} \subset C_0^\infty(\Omega)$ such that*

$$
\|u - u_n |S^{-l}\dot{W}_p(\Omega, \Gamma_\varphi)\| \to 0 \qquad (n \to \infty).
$$

For $\Omega = \Pi$, also

$$
\|u - u_n |\tilde{S}_a^{-l}W_p(\Omega)\| \to 0 \qquad (n \to \infty).
$$

COROLLARY 9.2. *Equality (9.35) makes it possible to constructively determine by means of (9.10) the norm (9.33), which is difficult to compute.*

We shall establish the relation of the space $\tilde{S}_a^{-l}W_p(\Omega)$ to the weighted Lebesgue space $L_p(\Omega, \rho_a^l(x))$[23] which is defined [172] as the set of locally summable functions $u(x)$ in Ω for which the norm

$$
\|u|L_p(\Omega, \rho_a^l(x))\| = \left(\int_\Omega \rho_a^l(x) |u(x)|^p \, dx \right)^{1/p} \tag{9.39}
$$

exists and is finite, where

$$
\rho_a^l(x) = \prod_{i=1}^n (x_i - a_i)^{l_i} \equiv (x - \bar{a})^l. \tag{9.40}
$$

LEMMA 9.2. *If $1 < p < \infty$, then*

$$
L_p(\Omega, \rho_a^l(x)) \to \tilde{S}_a^{-l}W_p(\Omega) \tag{9.41}
$$

and the imbedding noninvertible for $\bar{l} \geq \bar{0}$.

---

[23] The author is grateful to Professor O. V. Besov, who pointed out the expediency of establishing by means of Hardy's inequality the relation of the spaces $\tilde{S}^{-l}W_p(\Omega)$ and more general spaces of differentiable functions over $\tilde{S}^{-l}(\Omega)$ to known weighted function spaces.

PROOF. We use the following representation of Hardy's inequality on the segment $[a, b]$:

$$\left\|(x_k - a_k)^{-\beta} \int_{a_k}^{x_k} f(y_k)\,dy_k\right\|_{L_p(\bar{a},\bar{b})} \leq C\|(x_k - a_k)^{-\beta+1} f(x)\|_{L_p(\bar{a},\bar{b})}, \quad (9.42)$$

where the constant $C > 0$ does not depend on $f(x)$. We prove (9.42) in a manner similar to the way this is done in [33] (§2.15) for $L_p(0, \infty)$. By means of the change of variable $y_k = t(x_k - a_k) + a_k$ and the Minkowski inequality we obtain

$$\left\|(x_k - a_k)^{-\beta} \int_{a_k}^{x_k} f(y_k)\,dy_k\right\|_{L_p(\bar{a},\bar{b})}$$

$$= \left\|(x_k - a_k)^{-\beta} \int_0^1 f(t(x_k - a_k) + a_k)(x_k - a_k)\,dt\right\|_{L_p(\bar{a},\bar{b})}$$

$$= \left\|\int_0^1 f(t(x_k - a_k) + a_k)(x_k - a_k)^{-\beta+1}\,dt\right\|_{l_p(\bar{a},\bar{b})}$$

$$\leq \int_0^1 \|f(t(x_k - a_k) + a_k)(x_k - a_k)^{-\beta+1}\|_{L_p(\bar{a},\bar{b})}\,dt$$

$$= \int_0^1 t^{\beta-1-1/p}\|(y_k - a_k)^{-\beta+1} f(y_k)\|_{L_p(\bar{a},\bar{b})}$$

$$= \frac{1}{|\beta - 1/p|}\|(x_k - a_k)^{-\beta+1} f(x)\|_{L_p(\bar{a},\bar{b})}, \qquad \beta \neq \frac{1}{p}.$$

Successively applying (9.42) to estimate the integrals with variable upper limit in (9.10) and (9.9), we obtain

$$\|u|\tilde{S}_{\bar{a}}^{-l}W_p(\Omega)\| \leq C\|(x_1 - a_1)^{l_1} \cdots (x_n - a_n)^{l_n} \cdot f(x)\|_{L_p(\Omega)}$$

$$\equiv C \cdot \|(x - \bar{a})^l f(x)\|_{L_p(\Omega)}.$$

Now if for some $j \in [1, n]$ the integer l_j is positive, then obviously

$$L_p(\Omega) \to L_p(\Omega, \rho_{\bar{a}}^l) \to \tilde{S}_{\bar{a}}^{-l}W_p(\Omega),$$

and the imbeddings are noninvertible for $p > 1$, since $(x_j - a_j)^{-\alpha} \in \tilde{S}_{\bar{a}}^{-l}W_p(\Omega)$ for $\alpha < l_j + 1/p$, while $(x_j - a_j)^{-\alpha} \in L_p(\Omega, \rho_{\bar{a}}^l(x))$ only for $\alpha < 1/p + l_j/p$. ∎

We note that the space $\tilde{S}_{\bar{a}}^{-l}W_p(\Omega)$ itself can be interpreted for $\bar{l} \geq 1$ as the "weight" space $\tilde{S}_{\bar{a}}^{-l}W_p(\Omega; \rho^{l-1}(x, \xi))$, where the "weight"

$$\rho^{l-1}(x, \xi) = \frac{(x - \xi)^{l-1}}{(\bar{l} - 1)!} \equiv \prod_{j=1}^n \frac{(x_j - \xi_j)^{l_j-1}}{(l_j - 1)!}$$

Indeed, using the familiar Cauchy formula [296] for multiple integrals with a variable upper limit, we find that

$$\|u|\tilde{S}_{\bar{a}}^{-l}W_p(\Omega)\| = \|I_{\bar{a}}^l u(x)|L_p(\Omega)\|$$

$$= \left\{ \int_{\Omega} \left| \int_{a_1}^{x_1} d\xi_1 \cdots \int_{a_n}^{x_n} \prod_1^n \frac{(x_j - \xi_j)^{l_j - 1}}{(l_j - 1)!} u(\xi_1, \ldots, \xi_n) \, d\xi_n \right|^p dx \right\}^{1/p}$$

$$= \|u \cdot \rho^{l-1}(x, \xi)|\tilde{S}_{\bar{a}}^{-l}W_p(\Omega)\|. \tag{9.43}$$

By means of the Schwarz inequality, from (9.10) and (9.9) it is easy to obtain the following properties:

a) If $l_1 \geq l_2$, then $\tilde{S}_{\bar{a}}^{-l_2}W_p(\Omega) \to \tilde{S}_{\bar{a}}^{-l_1}W_p(\Omega)$.

b) If $p_1 \geq p_2$, then $\tilde{S}_{\bar{a}}^{-l}W_{p_1}(\Omega) \to \tilde{S}_{\bar{a}}^{-l}W_{p_2}(\Omega)$.

It is also not hard to show that functions $u(x) \in \tilde{S}_{\bar{a}}^{-l}W_p(\Omega)$ are globally continuous in the norm (9.10); for this it suffices to use a result of Burenkov [49] that for the set $C^{\infty}(\Omega)$ to be dense in some Banach space $Z(\Omega)$ it is necessary and sufficient (under certain other weak conditions) that compactly supported functions in $Z(\Omega)$ be globally continuous.

We note that if $S^l \dot{W}_q(\Omega, \Gamma_\varphi) \to C(\Omega)$ for some l, q, and n, then the dual space $S^{-l} \dot{W}_p(\Omega, \Gamma_\varphi)$ contains [30] the Dirac delta function with finite norm (9.33) for which the norm (9.10) is not defined. Thus, in this case even if $\Omega = \Pi$ (a parallelepiped) it follows from Lemma 9.1 that the space $\tilde{S}_{\bar{a}}^{-l}W_p(\Omega)$ is not complete.

More substantial imbeddings for the space $\tilde{S}_{\bar{a}}^{-l}W_p(\Omega)$ can be obtained by means of Lemma 9.1 as imbeddings dual to the imbeddings of the spaces of functions $S^l \dot{W}_q(\Omega)$ with a dominant mixed derivative. We shall not present these results here; it is expedient to consider such imbeddings for spaces of differentiable functions formed over $\tilde{S}_{\bar{a}}^{-l}W_p(\Omega)$ in which generalized solutions of variational problems will be determined.

It is sometimes convenient to consider the question of completeness of the spaces $\tilde{S}_{\bar{a}}^l W_p(\Omega)$ and more general spaces of differentiable functions over $\tilde{S}_{\bar{a}}^{-l}W_p(\Omega)$ by means of the following simple property.

LEMMA 9.3. *Suppose that W_1 and W_2 are normed linear spaces, and A and A^{-1} are bounded linear operators such that*

$$A: W_1 \to W_2, \qquad D(A) = W_1, \tag{9.44}$$

$$A^{-1}: W_2 \to W_1, \qquad R(A) = W_2. \tag{9.45}$$

The space W_2 is complete if and only if W_1 is complete.

PROOF. Suppose W_1 is a Banach space and $\{v_n\}$ is a fundamental sequence in W_2: $\|v_n - v_m|W_2\| \to 0$ $(m, n \to \infty)$. Then the sequence $\{u_n\}$,

$u_n = A^{-1}v_n$, $n = 1, 2, \ldots$, because of the relation

$$\|u_n - u_m|W_1\| = \|A^{-1}(v_n - v_m)|W_1\| \le C\|v_n - v_m\|W_2\|$$

will be a fundamental sequence in W_1. Because of the completeness of W_1 there exists $u \in W_1$ such that $\|u - u_n|W_1\| \to 0$ $(n \to \infty)$, and, since $u \in D(A) = W_1$, for $v = Au \in W_2$ we obtain

$$\|v - v_n|W_2\| = \|Au - Au_n|W_2\| \le C_2\|u - u_n|W_1\| \to 0 \qquad (n \to \infty),$$

i.e., W_2 is a complete space.

The converse assertion is proved similarly; in place of the condition $D(A) = W_1$ it is only necessary to use the condition $R(A) = D(A^{-1}) = W_2$.

We emphasize that a homeomorphism of the spaces W_1 and W_2 alone is insufficient to obtain the lemma; boundedness of the operators A and A^{-1} is important. For example [162], the unbounded mapping $x \to \tan x$ realizes a one-to-one correspondence of the segment $W_1 = (0, 1)$ and the line $W_2 = R_1 = (-\infty, +\infty)$. Here W_2 is a complete space, but W_1, an open interval, is not complete.

By means of the mapping $I_{\hat{a}}^{l}$,

$$v(x) = I_{\hat{a}}^{l}u(x), \qquad (9.46)$$

defined by (9.9) for any $u \in \tilde{S}_{\hat{a}}^{-l}W_p(\Omega)$, it is possible to define by (9.10) the space $\tilde{L}_p(\Omega, \Gamma_a)$ as the set of functions $v(x) \in L_p(\Omega)$ such that $v(x)$ satisfies the boundary conditions

$$v = \frac{\partial v}{\partial x_1} = \cdots = \frac{\partial^{l_1-1}v}{\partial x_1^{l_1-1}} = 0, \qquad x \in \partial\Omega: x_1 = a_1;$$

$$\frac{\partial^{l_1}v}{\partial x_1^{l_1}} = \frac{\partial^{l_1+1}v}{\partial x_2 \partial x_1^{l_1}} = \cdots = \frac{\partial^{l_1+l_2-1}v(x)}{\partial x_1^{l_1}\partial x_2^{l_2-1}} = 0, \qquad x \in \partial\Omega: x_2 = a_2; \qquad (9.47)$$

$$\ldots\ldots\ldots\ldots\ldots\ldots\ldots\ldots\ldots\ldots\ldots\ldots\ldots\ldots\ldots\ldots\ldots\ldots$$

$$\frac{\partial^{\sum_1^{n-1}l_k}v(x)}{\partial x_1^{l_1}\cdots\partial x_{n-1}^{l_{n-1}}} = \cdots = \frac{\partial^{|\vec{l}|-1}v(x)}{\partial x_1^{l_1}\cdots\partial x_n^{l_{n-1}}} = 0, \qquad x \in \partial\Omega: x_n = a_n.$$

It is not hard to see that for all $u(x) \in \tilde{S}_{\hat{a}}^{-l}W_p(\Omega)$ the function $v(x)$ belongs to $\tilde{L}_p(\Omega, \Gamma_a)$, and

$$\|v(x)|\tilde{L}_p(\Omega, \Gamma_a)\| = \|I_{\hat{a}}^{l}u(x)|\tilde{L}_p(\Omega)\| = \|u(x)|\tilde{S}_{\hat{a}}^{-l}W_p(\Omega)\|, \quad (9.48)$$

$$\|u(x)|\tilde{S}_{\hat{a}}^{-l}W_p(\Omega)\| = \|(I_{\hat{a}}^{l})^{-1}v|\tilde{S}_{\hat{a}}^{-l}W_p(\Omega)\| = \|v|\tilde{L}_p(\Omega, \Gamma_a)\|. \quad (9.49)$$

It follows from (9.46)–(9.49) that the operators $I_{\hat{a}}^{l}$ and $(I_{\hat{a}}^{l})^{-1}$ are bounded and realize an isometric correspondence of the spaces $L_p(\Omega, \Gamma_a)$

and $\tilde{S}_{\bar{a}}^{-l} W_p(\Omega)$, so that

$$D(I_{\bar{a}}^{\bar{l}}) = \tilde{S}_{\bar{a}}^{-l} W_p(\Omega), \qquad R(I_{\bar{a}}^{\bar{l}}) = D[(I_{\bar{a}}^{\bar{l}})^{-1}] = \tilde{L}_p(\Omega, \Gamma_a).$$

It is obvious that if $l_j > 0$ for some $j \in [1, n]$, then the space $\tilde{L}_p(\Omega, \Gamma_a)$ of functions $v(x) \in L_p(\Omega)$ satisfying conditions of the form (9.47) will be an incomplete space. By Lemma 9.3 we then conclude that the isometric space $\tilde{S}_{\bar{a}}^{-l} W_p(\Omega)$ is also incomplete.

Relations (9.48) and (9.49) show also that the operator $I_{\bar{a}}^{\bar{l}}$ can be extended by continuity to an isometric operator mapping the closure of $\tilde{S}_{\bar{a}}^{-l} W_p(\Omega)$ onto all of $L_p(\Omega)$.

§10. Spaces of differentiable functions over $\tilde{S}_{\bar{a}}^{-l} W_p(\Omega)$

In this section we consider some properties of the function spaces in which solutions of variational problems for differential equations will subsequently be determined. The norms of the function spaces considered in the preceding section pertained to the class of functionals (6.18) not containing derivatives of the unknown function. Here we consider the general case where some derivatives as well as some integrals (of the unknown function) with a variable upper limit are present in the norm of the class of functionals of the form (6.18).

10.1. We introduce an $m \times n$ matrix A defined by m vectors $\bar{\alpha}^1, \ldots \bar{\alpha}^m$. The components of each n-dimensional vector $\bar{\alpha}^j = (\alpha_1^j, \ldots, \alpha_n^j)$ are integers, so that any vector $\bar{\alpha}^j$, $j \in [1, m]$, determines two sets of numbers $\{\bar{\alpha}_+^j\}$ and $\{\bar{\alpha}_-^j\}$: the set $\{\bar{\alpha}_+^j\}$ is the collection $(^+\alpha_{i_1}^j, \ldots, ^+\alpha_{i_n}^j)$ of all positive components of the vector $\bar{\alpha}^j$, while the set $\{\bar{\alpha}_-^j\}$ is the collection $(^-\alpha_{k_1}^j, \ldots, ^-\alpha_{k_n}^j)$ of all nonpositive components of the vector $\bar{\alpha}^j$. It is obvious that the indices $i_1, \ldots, i_n$ and $k_1, \ldots, k_n$—integers in the segment $[1, n]$—form nonintersecting sets $\{i_1, \ldots, i_n\}$ and $\{k_1, \ldots, k_n\}$ whose union is the set $e_n = \{1, 2, \ldots, n\}$.

We assume that the domain $\Omega \subset R_n$ satisfies the conditions of §9, i.e., to be specific, $\Omega = \{x \in R_n : a_i < x_i < \psi_i(\bar{x}_i), i = 1, \ldots, n\}$, where the a_i $(i = 1, \ldots, n)$ are constants and the $\psi_i(\bar{x}_i) \equiv \psi_i(x_1, \ldots, x_{i-1}, x_{i+1}, \ldots, x_n)$ are piecewise smooth functions such that the interval (a_i, x_i) is contained in Ω if for at least one $j \in [1, m]$ there is a negative number α_i^j, $i \in 1, \ldots, n$, in $\bar{\alpha}^j$. If to some $k \in [1, n]$ in all vectors $\bar{\alpha}^j$, $j = 1, \ldots, m$ there correspond nonnegative components $\alpha_k^j \geq 0$ for all $j = 1, \ldots, m$, then this condition (that the interval (a_k, x_k) belong to the domain Ω) is unnecessary.

With the help of the collection of numbers $\{\bar{\alpha}_+^j\}$ it is possible to introduce the generalized derivative

$$D^{\bar{\alpha}_+^j} u(x) = \frac{\partial^{|\bar{\alpha}_+^j|} u(x)}{\partial x_{i_1}^{+\alpha_{i_1}^j} \cdots \partial x_{i_n}^{+\alpha_{i_n}^j}}, \tag{10.1}$$

which is understood in the sense of (9.17) (we assume that all preceding generalized derivatives exist); hence $|\bar{\alpha}_+^j| = {}^+\alpha_{i_1}^j + \cdots + {}^+\alpha_{i_n}^j$.

The collection of numbers $\{\bar{\alpha}_-^j\}$ determines a vector of indefinite integration

$$I^{\bar{\alpha}^j} - v(x) = \int_{a_{k_1}}^{x_{k_1}} d\xi_{k_1} \cdots \int_{a_{k_n}}^{x_{k_n}} \prod_{p=1}^{n} \frac{(x_{k_p} - \xi_{k_p})^{-({}^-\alpha_{k_p}^j + 1)}}{(-{}^-\alpha_{k_p}^j - 1)!}$$

$$\times u(x_{i_1}, \ldots, \xi_{k_1}, \ldots, x_{i_n}, \ldots, \xi_{k_n}) \, d\xi_{k_n}, \tag{10.2}$$

where the independent variables of the function $u(\cdots)$ must be arranged in order of increase of the indices, $1 \leq i_s \leq k_s \leq i_p \leq n$.

We say that a function $u(x) \in W_p^{\bar{\alpha}^j}(\Omega)$ if the function $I^{\bar{\alpha}^j}_- D^{\bar{\alpha}^j}_+ u(x) \in L_p(\Omega)$, and we set

$$\|u|W_p^{\bar{\alpha}^j}(\Omega)\| = \left\{ \int_\Omega |I^{\bar{\alpha}^j}_- D^{\bar{\alpha}^j}_+ u(x)|^p \, dx \right\}^{1/p}. \tag{10.3}$$

Considering the definition of the space $\tilde{S}_{\bar{a}}^{-l} W_p(\Omega)$ in §9, we can also say that $u \in W_p^{\bar{\alpha}^j}(\Omega)$ if $D^{\bar{\alpha}^j}_+ u \in \tilde{S}_{\bar{a}_-^j}^{\bar{\alpha}^j_-} W_p(\Omega)$,

$$\|u|W_p^{\bar{\alpha}^j}(\Omega)\| = \|D^{\bar{\alpha}^j}_+ u|\tilde{S}_{\bar{a}_-^j}^{\bar{\alpha}^j_-} W_p(\Omega)\|, \tag{10.3'}$$

where the lower limits of integration $\bar{a}_-^j = \{a_{k_1}, \ldots, a_{k_n}\}_j$ in (9.10) and (9.9) are determined by the collection $\{k_1, \ldots, k_n\}$ corresponding to the collection $\{\bar{\alpha}_-^j\}$.

We now define the space $W_p^A(\Omega) = W^{\bar{\alpha}^1 \cdots \bar{\alpha}^m}(\Omega)$ as the set of functions for which the following norm exists and is finite:

$$\|u|W_p^A(\Omega)\| = \left(\sum_{j=1}^{m} \|u|W_p^{\bar{\alpha}^j}(\Omega)\|^p \right)^{1/p}. \tag{10.4}$$

The properties of the norm in (10.4) can easily be verified; in order that it be positive-definite we require only the existence among the vectors $\bar{\alpha}^j$ of a vector with nonpositive components $\alpha_i^j \leq 0$ for all $i = 1, \ldots, n$ for some $j \in [1, m]$.

We present some special cases of the spaces introduced.

1. If $\alpha_i^j = 0$ for all $i = 1, \ldots, n$ and $j = 1, \ldots, m$, then $W_p^A(\Omega) \equiv L_p(\Omega)$, where Ω is an arbitrary domain in R_n.

2. If $\bar{\alpha}^j = (0, \ldots, 0, {}^+\alpha_j^j, 0, \ldots, 0)$, then (10.4) defines a seminorm (a norm if among the vectors $\bar{\alpha}^j$ there is the zero vector $\alpha_i^j = 0$ for all $i = 1, \ldots, n$, and some $j \in 1, \ldots, m$) of the space $W_p^{l_1, \ldots, l_n}(\Omega)$, where $l_j = {}^+\alpha_j^j$ and Ω is an arbitrary domain in R_n.

In the general case when all components α_i^j of the vectors $\bar{\alpha}^j$ ($i = 1, \ldots, n$, $j = 1, \ldots, m$) are nonnegative numbers of the norm (10.4) defines spaces of the type $S^l W_p(\Omega)$, which were investigated by Nikol'skiĭ under the condition that the "support" derivatives are included in the norm (see 9.3).

3. Let $m = 3$, $n = 2$, and

$$A = \begin{pmatrix} 0, & 0 \\ -1, & 0 \\ -1, & +1 \end{pmatrix}, \qquad \Omega_2 = \{x, t : a < x < \gamma(t), \ 0 < t < T\},$$

where $\gamma(t)$ is a piecewise smooth function such that each line parallel to the $0x$ axis intersects the curve $x = \gamma(t)$, $0 < t < T$, in only one point. In this case

$$\|u | W_p^A(\Omega_2)\| = \left[\int_{\Omega_2} \left\{ |u|^p + |u_x|^p + \left| \int_a^x u_t(\xi, t)\, d\xi \right|^p \right\} dx \right]^{1/p}.$$

4. $n = 3$, $m = 4$; $\Omega_3 = \{x, y, t : a < x < A_0, \ b < y < B_0, \ 0 < t < T\}$,

$$\|u | W_2^A(\Omega_3)\| = \left\{ \int_{\Omega_3} \left[u^2 + \left(\int_a^x d\xi \int_b^y u_t(\xi, \eta, t)\, d\eta \right)^2 \right. \right.$$

$$\left. \left. + \left(\int_a^x u_y(\xi, y, t)\, d\xi \right)^2 + \left(\int_b^j u_x(x, \eta, t)\, d\eta \right)^2 \right] d\Omega_3 \right\}^{1/2};$$

here

$$A = \begin{pmatrix} 0, & 0, & 0 \\ -1, & -1, & +1 \\ -1, & +1, & 0 \\ +1, & -1, & 0 \end{pmatrix}.$$

It is obvious that a permutation of the rows in the matrix A does not affect the value of $\| \cdot \, |W_p^A(\Omega)\|$; this can mean only a simple permutation of the terms in the sum (10.4).

The kth element of the jth row of the matrix A denotes the operation which is performed in the jth term in (10.4) on the kth variable of the function $u(x_1,\ldots,x_n)$; and since elements in a row are arranged in order of increase of indices of the independent variables $x_1,\ldots,x_n$, it is not possible to permute the columns of the matrix A.

We note that for the case where the sets $\bigcup_{j=1}^{m}\{i_1,\ldots,i_n\}_j$ and $\bigcup_{j=1}^{m}\{k_1,\ldots,k_n\}_j$ determined by the respective components $\bar{\alpha}_+^j$ and $\bar{\alpha}_-^j$ of the vectors $\bar{\alpha}^j$, $j = 1,\ldots,m$, do not intersect closely related spaces $W_p^{\bar{m}}$ were considered by Paneyakh [249]; "the elements of such a space, roughly speaking, are functionals of finite order in one group of variables and ordinary (differentiable) functions in the other group" [341].

10.2. We now establish the relation of the spaces $W_p^A(\Omega)$ to some spaces of the type of Sobolev spaces.

We introduce the vector $\tilde{\alpha} = (\tilde{\alpha}_1,\ldots,\tilde{\alpha}_n)$ with components $\tilde{\alpha}_k = \min 1 \le j \le m\{\alpha_k^j\}$ if there exists $\alpha_k^j \le 0$, $j \in [1,m]$, (k fixed), and $\tilde{\alpha}_k = 0$ if $\alpha_k^j > 0$ for all $j = 1,\ldots,m$. As before (see (9.9)), we denote by $I_{\bar{a}}^{\tilde{\alpha}}$ the integral operator in which in each variable x_l there are taken $\tilde{\alpha}_l$ times the indefinite integrals $\int_{a_l}^{x_l} \cdots d\xi_l$ and for $u(x) \in L_{p,\text{loc}}(\Omega,\Gamma_a)$ it is possible to set

$$I_{\bar{a}}^{\tilde{\alpha}} u(x) = v(x). \tag{10.5}$$

It is then obvious that $(I_{\bar{a}}^{\tilde{\alpha}})^{-1} v(x) = u(x) = D^{\tilde{\alpha}} v(x)$, and for all $v(x) \in D[(I_{\bar{a}}^{\tilde{\alpha}})^{-1}] = R(I_{\bar{a}}^{\tilde{\alpha}})$

$$
\begin{aligned}
v(x) &= \frac{\partial v}{\partial x_1} = \cdots = \frac{\partial^{|\tilde{\alpha}|-1} v(x)}{\partial x_1^{|\tilde{\alpha}_1|-1}} = 0, \quad x \in \partial\Omega: x_1 = a_1, \\
&\cdots\cdots\cdots\cdots\cdots\cdots\cdots\cdots\cdots\cdots\cdots\cdots \\
v(x) &= \frac{\partial v}{\partial x_n} = \cdots = \frac{\partial^{|\tilde{\alpha}|-1} v(x)}{\partial x_n^{|\tilde{\alpha}|-1}} = 0, \quad x \in \partial\Omega: x_n = a_n
\end{aligned}
\tag{10.6}
$$

(if $\tilde{\alpha}_k = 0$ for some $k \in [1,n]$, then for $x_k = a_k$ in (10.6) there are no boundary conditions, since there are no integrals on x_k in (10.5)).

We introduce the vectors $\bar{l}^k = (l_1^k,\ldots,l_n^k)$, $k = 1,\ldots,m$, with components

$$l_j^k = \alpha_j^k - \tilde{\alpha}_j \ge 0, \qquad j = 1,\ldots,n, \tag{10.7}$$

i.e.,

$$\bar{l}^k = \bar{\alpha}^k - \tilde{\alpha}, \qquad k = 1,\ldots,m. \tag{10.8}$$

For the functions $v(x)$ of (10.5) we define the seminorm

$$\|v|W_p^L(\Omega)\| = \left(\sum_{k=1}^{m} \|v|W_p^{\bar{l}^k}(\Omega)\|^p\right)^{1/p}, \tag{10.9}$$

where for the matrix L it is possible to write, with consideration of (10.7) or (10.8),

$$L = A - \tilde{A}, \qquad (10.10)$$

where $\tilde{A}$ is the matrix formed from a row $(\tilde{\alpha}_1, \ldots, \tilde{\alpha}_n)$ repeated m times.

In (10.4) or (10.9) we have so far not restricted the order of the maximal partial derivatives or required, as is usual, that the set of vectors $\tilde{l}^k$ of (10.8) in (10.9) form a convex set [195]; therefore, it is difficult to present substantial results for the spaces $W_p^A(\Omega)$ of (10.4) or $L_p^L(\Omega)$ of (10.9) so defined. We have presented these constructions in the general case in order to show how the operator $L_{\tilde{a}}^{\tilde{\alpha}}$ of (10.5) maps the space $W_p^A(\Omega)$ of (10.4) onto a Sobolev space $W_p^L(\Omega)$ (10.9). For example, in the case of Example 4 at the end of §10.1 the space $W_2^A(\Omega)$ is mapped by means of the operator $I_{a,b,-}^{(-1,-1,0)}$,

$$I_{a,b,-}^{(-1,-1,0)} u(x,y,t) = \int_a^x d\xi \int_b^y u(\xi, \eta, t)\, d\eta = v(x,y,t), \qquad (10.11)$$

into the space $W_2^L(\Omega_3, \Gamma)$ of functions with finite norm

$$\|v \,|\, \dot{W}_2^L(\Omega, \Gamma)\| = \left\{ \int_{\Omega_3} [v_{xy}^2 + v_t^2 + v_{xx}^2 + v_{yy}^2]\, dx\, dy\, dt \right\}^{1/2},$$

$$v(a,y,t) = \frac{\partial v}{\partial y}(a,y,t) = 0, \qquad b < y < B_0,\ 0 < t < T,$$

$$v(x,b,t) = \frac{\partial v}{\partial x}(x,b,t) = 0, \qquad a < x < A_0,\ 0 < t < T.$$

Here (see Example 4)

$$L = A - \tilde{A} = \begin{pmatrix} 0, & 0, & 0 \\ -1, & -1, & +1 \\ -1, & +1, & 0 \\ +1, & -1, & 0 \end{pmatrix}$$

$$= \begin{pmatrix} -1, & -1, & 0 \\ -1, & -1, & 0 \\ -1, & -1, & 0 \\ -1, & -1, & 0 \end{pmatrix} = \begin{pmatrix} 1, & 1, & 0 \\ 0, & 0, & 1 \\ 0, & 2, & 0 \\ 2, & 0, & 0 \end{pmatrix}.$$

The mapping of $W_p^A(\Omega)$ onto some Sobolev space $W_p^L(\Omega, \Gamma)$ in a number of cases makes it possible to investigate rather simply by means of Lemma 9.3 such important properties of $W_p^A(\Omega)$ as completeness, denseness of smooth functions, and stability of boundary values.

LEMMA 10.1. *The operator $I_{\tilde{a}}^{\tilde{\alpha}}$ defined by relation (9.9) for $\bar{l} = \tilde{\alpha}$ and*

$$\tilde{\alpha}_k = \min\left\{\min_{j=1,\ldots,m}(\alpha_k^j); 0\right\}, \qquad k = 1,\ldots,n,$$

establishes an isometric correspondence of the space $W_p^A(\Omega)$ of (10.4) and the space $\mathring{W}_p^L(\Omega,\Gamma)$ (of functions $v(x)$ satisfying (10.6) with a finite norm (10.9)), where the matrices L and A are related by (10.10).

Indeed, considering (10.5), (10.8), and (10.3), we get

$$\|v|\mathring{W}_p^L(\Omega,\Gamma)\| = \|I_{\tilde{a}}^{\tilde{\alpha}}u|\mathring{W}_p^L(\Omega,\Gamma)\| = \sum_{k=1}^m \|I_{\tilde{a}}^{\tilde{\alpha}}u|\mathring{W}_p^{\bar{l}^k}(\Omega,\Gamma)\|$$

$$= \sum_{k=1}^m \|u|W_p^{\bar{\alpha}^k}(\Omega)\| = \|u|W_p^A(\Omega)\| \qquad (10.12)$$

and hence the norm of the operator $I_{\tilde{a}}^{\tilde{\alpha}}: \mathring{W}_p^A(\Omega) \to W_p^L(\Omega,\Gamma)$ is equal to 1.
Conversely, considering (10.5), (10.8), and (10.6),

$$\|u|W_p^A(\Omega)\| = \|(I_{\tilde{a}}^{\tilde{\alpha}})^{-1}v|W_p^A(\Omega)\|$$

$$= \|D^{\tilde{a}}v|W_p^A(\Omega)\| = \sum_{k=1}^m \|D^{\tilde{\alpha}}v|W_p^{\bar{\alpha}^k}(\Omega)\|$$

$$= \sum_{k=1}^m \|v|W_p^{\bar{l}^k}(\Omega)\| = \|v|W_p^L(\Omega)\|, \qquad (10.13)$$

i.e., the norm of the inverse operator $(I_{\tilde{a}}^{\tilde{\alpha}})^{-1} \equiv D^{\tilde{\alpha}}$ mapping $\mathring{W}_p^L(\Omega,\Gamma)$ of (10.9) and (10.6) onto $W_p^A(\Omega)$ is also equal to 1. ∎
From Lemmas 9.3, 10.1 we obtain

COROLLARY 10.1. *The space $W_p^A(\Omega)$ of (10.4) is complete if and only if the space $\mathring{W}_p^L(\Omega,\Gamma)$ of (10.9) and (10.6) is complete.*

Since the boundary Γ of the domain Ω is piecewise smooth, from the density of the set $C^\infty(\bar{\Omega})$ in the Sobolev space $\mathring{W}_p^L(\Omega,\Gamma)$ with the help of Lemma 10.1 it is not hard to obtain

COROLLARY 10.2. *The set $C^\infty(\Omega)$ is dense in $W_p^A(\Omega)$.*

We shall establish a further relation of the space $W_p^A(\Omega)$ to a certain other Sobolev space; this relation shows that $W_p^A(\Omega)$ is larger than the Sobolev space (even a "weighted" space) in the norm of which there are derivatives of the same maximal order as in the norm of $W_p^A(\Omega)$ (10.4).

For the vectors $\bar{\alpha}^j$, $j = 1, \ldots, m$, defining the matrix

$$A = \left\{ \begin{array}{c} \bar{\alpha}^1 \\ \vdots \\ \bar{\alpha}^m \end{array} \right\},$$

we defined earlier a decomposition

$$\bar{\alpha}^j = \{\bar{\alpha}_+^j\} + \{\bar{\alpha}_-^j\}, \qquad j = 1, \ldots, m \qquad (10.14)$$

into two sets: $\{\bar{\alpha}_+^j\}$ is the set of positive components in $\bar{\alpha}^j$, $\alpha_{i_s}^j > 0$, $s = 1, \ldots, n$, $j = 1, \ldots, m$, and $\{\bar{\alpha}_-^j\}$ is the set of nonpositive components, $\bar{\alpha}_{k_p}^j \leq 0$, $p = 1, \ldots, n$, $j = 1, \ldots, m$. The decomposition (10.14) obviously generates a corresponding representation of the matrix

$$A = A_+ + A_-, \qquad (10.15)$$

where A_+ (A_-) is a matrix with positive (nonpositive) components.

Then the "weight" space $W_p^{A_+}(\Omega, x^{A_-})$, $1 < p < \infty$, can be defined as the set of functions for which the following seminorm exists and is infinite:

$$\|u|W_p^{A_+}(\Omega, x^{A_-})\| = \left(\sum_{j=1}^m x^{-\bar{\alpha}_-^j} \cdot D^{\bar{\alpha}_+^j} u(x)|L_p(\Omega)\|^p(\Omega)\|^p \right)^{1/p}$$

$$= \left(\sum_{j=1}^m \left\| x_{k_1}^{-\alpha_{k_1}^j} \cdots x_{k_n}^{-\alpha_{k_n}^j} \frac{\partial^{|\bar{\alpha}_+^j|} u(x)|}{\partial x_{i_1}^{+\alpha_{i_1}^j} \cdots \partial x_{i_n}^{+\alpha_{i_n}^j}} \right\|_{L_p(\Omega)}^p \right)^{1/p}. \qquad (10.16)$$

Since the decomposition (10.14) defines nonintersecting sets $\{i_1, \ldots, i_n\}_j$ and $\{k_1, \ldots, k_n\}_j$ for each $j \in [1, m]$, in each jth term of the sum in (10.16) the derivative is computed with respect to one group of variables $\{x_{i_1}, \ldots, x_{i_n}\}_j$, while the "weight" is computed with respect to the other group of remaining variables $\{x_{k_1}, \ldots, x_{k_n}\}_j$ (to simplify notation we have set $a_j = 0$, $j = 1, \ldots, n$, in the definition of Ω).

LEMMA 10.2. *Suppose A is a given integral matrix for which the decomposition (10.15) holds. Then*

$$W_p^{A_+}(\Omega, x^{A_-}) \to W_p^A(\Omega). \qquad (10.17)$$

The proof can be carried out by direct estimation of the norms, applying Hardy's inequality (9.42); but the lemma follows from Lemma 9.2 if the representation (10.3) for the terms in the norm (10.4) is considered. ∎

10.3. We mention some limitations of the approach presented here to establishing properties of spaces of the type $W_p^A(\Omega)$ by means of the isometric mapping (10.5). For example, consider the norm

$$\|u|W_2^{\tilde{A}}(\Omega)\| = \left\{ \int_\Omega \left[u^2 + \left(\int_0^x u_y(\xi,y,z)\,d\xi \right)^2 \right. \right.$$

$$\left. \left. + \left(\int_a^x u_z(\xi,y,z),d\xi \right)^2 \right] d\Omega \right\}^{1/2},$$

$$\Omega = \{x,y,z : 0 < x < a,\ 0 < y < b,\ 0 < z < d\},$$

$$\tilde{A} = \begin{pmatrix} 0, & 0, & 0 \\ -1, & 1, & 0 \\ -1, & 0, & 1 \end{pmatrix}. \tag{10.18}$$

By the changes

1) $\displaystyle\int_a^x u(\xi,y,z)\,d\xi = v,\quad u = v_x',\quad v(a,y,z) = 0;$

2) $\displaystyle\int_0^x u(\xi,y,z)\,d\xi = v,\quad u = v_x',\quad v(0,y,z) = 0;$

3) $\displaystyle\int_0^x d\theta \int_a^\theta u(\xi,y,z)\,d\xi = v,\quad u = v_{xx}'',\quad v(0,y,z) = v_x'(a,y,z) = 0;$

4) $\displaystyle\int_a^x d\theta \int_0^\theta u(\xi,y,z)\,d\xi = v,\quad u = v_{xx}'',\quad v(a,y,z) = v_x'(0,y,z) = 0$

the function space $W_2^{\tilde{A}}(\Omega)$ is mapped respectively onto the space of functions with finite norm

1′) $\displaystyle \|v\|_1 = \left(\int_\Omega [v_x^2 + [v_y - v_y(0,y,z)]^2 + v_z^2]\,d\Omega \right)^{1/2},$

$$(x,y,z) \in \partial\Omega : v(a,y,z) = 0;$$

2′) $\displaystyle \|v\|_2 = \left(\int_\Omega [v_x^2 + v_y^2 + (v_z - v_z(a,y,z))^2]\,d\Omega \right)^{1/2},$

$$(x,y,z) \in \partial\Omega : v(0,y,z) = 0;$$

3′) $\displaystyle \|v\|_3 = \left(\int_\Omega [v_{xx}^2 + [v_{xy} - v_{xy}(0,y,z)]^2 + v_{xz}^2]\,d\Omega \right)^{1/2},$

$$(x,y,z) \in \partial\Omega : v(0,y,z) = v_x'(a,y,z) = 0;$$

4′) $\displaystyle \|v\|_4 = \left(\int_\Omega \{v_{xx}^2 + v_{xy}^2 + [v_{xz} - v_{xz}(a,y,z)]^2\}\,d\Omega \right)^{1/2},$

$$(x,y,z) \in \partial\Omega;\, v(a,y,z) = v_x'(0,y,z) = 0.$$

In any case in each of the norms $1'-4'$ the square of the difference of a derivative defined in Ω and its boundary value is present, which makes it difficult to draw direct conclusions regarding properties of functions in spaces with such finite norms $1'-4'$.

Moreover, we remark that, in the theory of functions, in considering properties of elements of a concrete function space "translation of all facts obtained from the 'language' of one space to the 'language' of a space isometric to it is often not realizable in practice. In this case identification of isometric spaces is not practical" ([170], Chapter I, §2.4).

Commentary to Chapter II

To §6. Formulations of results of negative character for parabolic equations have sometimes been expounded abstractly in applications without reference to the class of functionals; as a result, as Yu. I. Nyashin has noted [245], imprecise assertions have been made regarding the impossibility of solving parabolic equations by variational methods [31], [113], [276].

With Corollary 6.1 in mind it may now be said that for any linear equation

$$Au = f \tag{1}$$

which is well posed in the Hadamard sense in some Hilbert space $H_A \subset H$ $(\overline{D(A)} = H, \overline{R(A)} = H)$ it is possible to formulate the variational problem of minimizing a quadratic functional with properties A–C of the Introduction analogous to the variational properties of the Dirichlet functional for the Laplace equation. Of course, such a generalization became possible as a result of condition C, which allows that from

$$\delta F[u] = 0 \tag{2}$$

for the corresponding functional $F[u]$ being minimized follows not equation (1) directly but the equivalent equation

$$B^*(Au - f) = 0 \tag{3}$$

(equivalent in the sense that $B^*\varphi = 0$ if and only if $\varphi = 0$).

This generalization, however, is not new. Copson [54] used the multiplier $B\varphi = \mu(x)\varphi(x)$; with the help of an integrating variational multiplier $B\varphi = \mu(\varphi)\varphi$ Bolza [40] solved the inverse problem of the calculus of variations for the equation $\varphi'' = f(x, \varphi(x), \varphi'(x))$; for the quasilinear differential equation (6.8) Balatoni [18] took $B^*\varphi = \mu(x, y, u(x, y), u_x, u_y)\varphi$. For some nonlinear equations Vaĭnberg [331] considered an operator factor B (see also [192] and [193]); for general nonlinear equations this method was developed by Nashed [235] and Lyashko [202] (see Chapter 4), while for linear equations (see Chapter 1) it was developed by Martynyuk, Petryshyn, and Shalov. We point out that we do not mention results with $B = A$ in

(3), since in this case the functional $F[u]$ for a differential operator A in the general case does not satisfy condition A of the Introduction.

From the results of Copson [54] and Balatoni [18] it follows that for broad classes of nonsymmetric linear and quasilinear equations of elliptic and hyperbolic types corresponding variational problems can also be constructed in the class of Euler functionals; this fact must be borne in mind when, for example, "nonvariational" nonsymmetric elliptic or hyperbolic equations are mentioned.

To §§6 and 7. Copson [54] emphasized the importance of solving the inverse problem of the calculus of variations for differential equations with the object obtaining various conservation laws with the help of theorems of E. Noether. As noted in [329], in theoretical physics in the last decade an extensive literature has appeared devoted to the search for a Lagrangian for a given differential equation (see, for example, some of the work in this direction [128], [78], and also [4] and [16], where there is an extensive bibliography on this question). The papers of Gel'fand and Dikiĭ [117] and Gel'fand and Dorfman [118], which are important for the development of the theoretical foundations of this approach, should be mentioned. However, within the framework of this direction the inverse problem of the calculus of variations is considered for a differential equation without boundary conditions, and boundedness above or below is not required of the equivalent functional—solutions of the corresponding equation are, generally speaking, only a stationary point for such a functional, which may not have an extremum. Constructions of corresponding variational problems go through also only in the class of Euler functionals.

To §7. We emphasize again that mainly results of Pomraning [267] and some generalizations of them are presented in this section.

To §§9 and 10. 1. Results of Nikol'skiĭ [24], which were used repeatedly in §9, were applied by Nikol'skiĭ [24] to the investigation by a variational method of rather general hypoelliptic equations and even equations beyond the limits of this class; the symmetric quadratic form corresponding to the equation may not have definite sign. These investigations were further developed by Kazaryan [154], [155].

2. In §§9 and 10 we have tried to limit ourselves to an elementary exposition of those results on the function spaces which we will use later on, obtaining these results mainly from the analysis of a mapping of spaces to known Lebesgue-Sobolev spaces and weight spaces. Of course, this makes it possible to present only simple facts in far from general form, leaving aside a number of properties of the spaces $W_p^A(\Omega)$ and their generalizations.

CHAPTER III

Construction and Investigation
of Variational Principles
for Linear Boundary Value Problems

In this chapter we present various techniques for the constructive approach to variational principles (analogues in variational properties to the Dirichlet principle for the Laplace equation) for boundary value problems for some equations and classes of partial differential equations (PDE), generally speaking, with nonsymmetric and nonpositive operators. To understand this chapter, a reader who is interested only in these questions (rather than in the mathematical aspects of the theory—properties of B-positive operators, the existence of symmetrizing operators, generalized solutions, and solutions of variational problems, extensions of operators, etc.) needs to be familiar only with the formulations of the basic facts presented in Chapters 1 and 2. For such a reader in §11.5 the general scheme of constructing and investigating variational principles for nonsymmetric linear equations, which follows from the results of Chapter 1, is presented.

§§12–16 are devoted to the construction of quasiclassical solutions of the inverse problem of the calculus of variations for various PDE, but first in §11 some other approaches to the solution of inverse problems of the calculus of variations are discussed.

§11. On various formulations of the inverse problem
of the calculus of variations and some solutions of it

11.1. As is known, the direct problem of the calculus of variations consists in investigating the problem of maximizing or minimizing some functional, possibly by obtaining and investigating the corresponding differential equation (the Euler equation) for this functional. Formulation of the inverse problem of the calculus of variations on the other hand, presupposes the definition of a differential equation on some set of functions with initial and boundary conditions. In a manner similar to the way "direct" problems in mathematical physics during recent decades have undergone essential modifications as a result of the introduction of various concepts of a generalized solution, so the problem of constructing variational principles for a given equation and the evolution of the formulation of inverse problems have been shaped by mathematical results on the nonexistence of "ordinary" solutions and by practical requirements of applications in extending variational methods to ever broader classes of differential equations.

We consider, for example, a general nonlinear PDE of second order

$$N(x, y, u, u_x, u_y, u_{xx}, u_{xy}, u_{yy}) = 0, \qquad u \in D(N) \qquad (11.1)$$

(to shorten the notation we have taken $u = u(x, y)$, $(x, y) \in \Omega \subset R_2$), where $N(\cdots)$ is a smooth function in all its variables on R_8; the operator N acts in a Hilbert space H with inner product $\langle \cdot, \cdot \rangle$. The following formulation of the inverse problem of the calculus of variations has long been known [138]: find a functional $\Phi(u)$ such that

$$\delta\Phi(u) = \int_\Omega N(u)\delta u\, dx\, dy, \qquad u \in D(N). \qquad (11.2)$$

Here until recently authors have restricted their attention mainly to the class of Euler functional

$$\Phi(u) = \int_\Omega \Phi(x, y, u(x, y), u_x, u_y)\, dx\, dy. \qquad (11.3)$$

Such an a priori restriction is natural, generally speaking, for differential equations but not integro-differential equations, and is a reflection of the local dependence in field theory of the Lagrangian, as a function, on a variable characteristic of the field and its derivatives [324]. In such a formulation([24]) the inverse problem of the calculus of variations has a solution [331] if and only if for the Gâteaux derivative N'_u the potential criterion is satisfied—the condition of symmetry

$$\langle N'_u\varphi, \psi \rangle = \langle \varphi, N'_u\psi \rangle \quad \forall u, \varphi, \psi \in D(N). \qquad (11.4)$$

([24])In this section we repeat from somewhat different positions some facts presented in §6: earlier we were interested in classes of functionals of variational problems for nonsymmetric equations; now we are interested in the formulation of inverse problems of the calculus of variations and constructive approaches to their solution. For more details regarding inverse problems for nonlinear equations, see §17.

Since for a linear operator N we have $N'_\varphi = N\varphi$ for all $\varphi \in D(N)$, in this case many simple differential equations of any type (for example, the elliptic equation $Nu \equiv \Delta u + \partial u/\partial x_i$, $D(N) = \overset{\circ}{C}^2(\Omega)$) turn out to be nonvariational.

Generalizing this formulation of the inverse problem of the calculus of variations for a general multidimensional linear PDE of second order with variable coefficients (§6.1)

$$\mathscr{L}_n u = 0, \quad u \in \overset{\circ}{C}^2(\Omega), \tag{11.5}$$

Copson [54] showed that for broad classes of equations of elliptic and parabolic types there exists a "variational multiplier"—a function $\mu(x)$ ($\mu(x) \neq 0$, $x \in \Omega$), and in the class of Euler functionals (6.2) there is a functional $F(u)$ such that the condition

$$\delta F(u) = 0, \quad u \in \overset{\circ}{C}^2(\Omega), \tag{11.6}$$

implies the equation

$$\mu(x) \cdot \mathscr{L}_n u = 0, \quad u \in \overset{\circ}{C}^2(\Omega), \tag{11.7}$$

equivalent to (11.5). If (11.5) is of parabolic type in Ω, then such a function $\mu(x)$ does not exist, and in the class of Euler functionals (6.2) there is no functional for which equation (11.5) or an equivalent parabolic equation (11.7) would follow from condition (11.6).

Generalizing Copson's variational multiplier $\mu = \mu(x) \rightarrow \mu = \mu(x, y, u(x,y), u_y, u_x)$, in 1960 Balatoni [18] obtained analogous results for the quasilinear equation (6.8). Here there were many classes of equations of parabolic, mixed, composite, and other types (see, for example, [1] and [226]) for which the question of the existence of a solution of the inverse problem of the calculus of variations either remained open or was solved negatively in the formulation of [18], [54], and [138].

In G. Minty's dissertation [227] the variational multiplier was taken in the form of a functional; for a given nonlinear operator N he established necessary and sufficient conditions for the existence of a positive functional $\varphi(u)$ and a functional $f(u)$ such that

$$\operatorname{grad} f(u) = \varphi(u)N(u), \quad u \in D(N) \tag{11.8}$$

(see also the discussion of closely related questions in [235]).

Martynyuk [209], Petryshyn [250], and Shalov [293] introduced the "variational multiplier" as an auxiliary symmetrizing operator. Thanks to this, it was possible to show that for any linear equation, for example, which is uniquely solvable in a Hilbert space H,

$$Au = f, \qquad \overline{D(A)} = H = \overline{R(A)} \qquad (11.9)$$

there exists an auxiliary operator B such that for the functional

$$D(u) = (Au, Bu) - 2(f, Bu) \qquad (11.10)$$

the condition

$$\delta D(u) = 0, \qquad u \in D(A, B), \qquad (11.11)$$

implies (11.9), and the functional (11.10) may possess properties A–C of the Introduction. For nonlinear equations solutions of such an inverse problem of the calculus of variations (using the operator "variational multiplier") were first investigated by Nashed [235] and Lyashko [202] (see §18).

On the other hand, the following general formulation of the inverse problem of the calculus of variations is known [60]. For a given equation

$$N(u) = 0, \qquad u \in D(N), \qquad (11.12)$$

find a functional whose set of critical points would coincide with the set of solutions of equation (11.12), i.e.,

$$N(\bar{u}) = 0 \Leftrightarrow \delta\Psi(\bar{u}) = 0. \qquad (11.13)$$

This formulation of the inverse problem of the calculus of variations admitted arbitrariness in the choice of classes of functionals $\Psi(u)$, and therefore made it possible to considerably extend the classes of equations for which it is possible to construct a variational principle. A rather general solution of this inverse problem of the calculus of variations for the nonlinear equation (11.12) was obtained in 1982 by Tonti [324]; in Chapter 4 we present this result and some of its generalization and concretizations.

To conclude this subsection we emphasize again the formulation of the inverse problem of the calculus of variations which is the object of investigation in the present monograph: we consider a quasiclassical solution of the inverse problem of the calculus of variations, i.e., the construction of a functional possessing properties A–C of the Introduction analogous to the variational properties of the Dirichlet functional for the Laplace equation.

11.2. The most attractive class of functionals for solving the inverse problem of the calculus of variations for differential equations is the class of Euler functionals (6.2). In this case the norm of the corresponding "energy" space generates certain Sobolev function spaces where by known

imbedding theorems it is possible to establish the smoothness of a generalized solution of the equation, while by results of the theory of approximation of functions it is also possible to find an a priori estimate of the
approximate solution of the variational problem (see §5.1).

We shall analyze in more detail Copson's approach [54], with the ob-
jective of a constructive approach to a classical solution of the inverse
problem of the calculus of variations for a general, multidimensional, linear, inhomogeneous PDE of second order

$$\mathcal{L}u \equiv \sum_{\beta,\gamma=1}^{n} p^{\beta\gamma} D_\beta D_\gamma u + \sum_{\beta=1}^{n} q^\beta D_\beta u + ru = f(x), \qquad x \in \Omega. \quad (11.14)$$

Here $D_i = \partial/\partial x_i$, $i = 1,\ldots,n$, $x \in \Omega \subset R^n$, the boundary $\Gamma = \partial\Omega$
of the bounded domain Ω is piecewise smooth, the functions $p^{\beta\gamma}(x) = p^{\gamma\beta}(x) \in C^2(\Omega) \cap C(\overline{\Omega})$, $q^\beta(x) \in C^1(\overline{\Omega})$, $\beta, \gamma = 1,\ldots,n$, $r(x) \in C(\overline{\Omega})$, and
$f(x) \in C(\overline{\Omega})$; and $u(x)$ is the unknown function, $u(x) \in M = C^2(\Omega) \cap C^1(\overline{\Omega}) \cap \overset{\circ}{C}(\Omega)$.

We introduce the class of functionals

$$V_\Phi[u] = \int_\Omega \mu(x) \left\{ \sum_{\beta,\gamma=1}^{n} a^{\beta\gamma}(x) \cdot D_\beta u \cdot D_\gamma u \right.$$

$$\left. +2c(x)u^2 + 2\sum_{\beta=1}^{n} b^\beta(x)uD_\beta u + 2g(x)u \right\} dx. \quad (11.15)$$

We consider the problem of determining for $V_\Phi[u]$ functions $\mu = \mu(x) \in C^1(\overline{\Omega})$, $a^{\beta\gamma} = a^{\gamma\beta}$, b^β $(\beta, \gamma = 1,\ldots,n)$, c, and g, smooth in $\overline{\Omega}$, so that the
condition

$$\delta V_\Phi[u] = 0 \quad (11.16)$$

implies (11.14) or

$$\mu(x)(\mathcal{L}u - f) = 0, \qquad \mu(x) \neq 0, x \in \Omega. \quad (11.17)$$

The condition of equivalence of equations (11.17) and (11.14) im-
poses the condition of strict positivity (or negativity) on the function
$\mu(x) \in C^1(\overline{\Omega})$; it may therefore be assumed that $\mu(x) = \exp\Phi(x)$, and
the function $\Phi(x) \in C^1(\overline{\Omega})$ and the unknown coefficients in (11.15) are to
be determined.

The first variation of the functional (11.15) on the set M has the form

$$\delta V_\Phi[u] = -2 \int_\Omega \delta u \exp \Phi(x) \left\{ \sum_{\beta,\gamma=1}^{n} a^{\beta\gamma} D_\beta D_\gamma u \right.$$

$$+ \sum_{\beta=1}^{n} \sum_{\gamma=1}^{n} (a^{\beta\gamma} D_\gamma \Phi + D_\gamma a^{\beta\gamma}) D_\beta u$$

$$\left. + \left[\sum_{\beta=1}^{n} (D_\beta b^\beta + D_\beta \Phi) - c \right] u - g(x) \right\} dx,$$

and from (11.16) we find that

$$\exp \Phi \{ \mathscr{L}_\Phi u - g \}$$

$$\equiv \Phi \left\{ \sum_{\beta,\gamma=1}^{n} a^{\beta\gamma} D_\beta D_\gamma u + \sum_{\beta=1}^{n} \sum_{\gamma=1}^{n} (a^{\beta\gamma} D_\gamma \Phi + D_\gamma a^{\beta\gamma}) D_\beta u \right.$$

$$\left. + \left[\sum_{\beta=1}^{n} (D_\beta b^\beta + b^\beta D_\beta \Phi) - c \right] u - g \right\} = 0 \quad (11.18)$$

or, since $\exp \Phi(x) \neq 0$,

$$\mathscr{L}_\Phi u - g(x) = 0, \qquad x \in \Omega. \qquad (11.19)$$

It is important to emphasize that the operator $\exp \Omega \cdot \mathscr{L}_\Phi$ of (11.18) is selfadjoint, while the operator $\mathscr{L}_\Phi$ is nonsymmetric.

In order that (11.14) should follow from condition (11.16) for the functional (11.15) it now obviously suffices to determine the functions $a^{\beta\gamma}(x)$, $b^\beta(x)$ $(\gamma, \beta = 1, \ldots, n)$, $c(x)$, $g(x)$, and $\Phi(x)$, so that

$$\mathscr{L}_\Phi u - g = \mathscr{L} u - f \quad \forall u \in M. \qquad (11.20)$$

From (11.20), (11.14), and (11.19), considering the symmetry conditions $p^{\beta\gamma} = p^{\gamma\beta}$ and $a^{\beta\gamma} = a^{\gamma\beta}$, $\beta, \gamma = 1, \ldots, n$, we obtain a system of equations for determining the unknown functions:

$$p^{\beta\gamma}(x) = a^{\beta\gamma}(x), \qquad \beta, \gamma = 1, \ldots, n, \qquad (11.21)$$

$$q^\beta(x) = \sum_{\gamma=1}^{n} [a^{\beta\gamma}(x) D_\gamma \Phi + D_\gamma a^{\beta\gamma}(x)], \qquad \beta = 1, \ldots, n, \qquad (11.22)$$

$$r(x) = \sum_{\beta=1}^{n} [D_\beta b^\beta(x) + b^\beta(x) D_\beta \Phi] - c(x), \qquad (11.23)$$

$$f(x) = g(x). \qquad (11.24)$$

The coefficients $a^{\beta\gamma}(x)$, $\beta,\gamma = 1,\ldots,n$, and the function $g(x)$ can be determined from (11.21) and (11.24) directly and uniquely. From (11.21)–(11.24) it is also evident that if there exists at least one function $\Phi(x) \in C^1(\overline{\Omega})$ satisfying (11.22), then there exists an infinite set of functions $b^1(x),\ldots,b^n(x)$, $c(x)$ satisfying (11.23): it suffices to substitute the function $\Phi(x)$ into (11.23), take arbitrary functions $b^\beta(x) \in C^1(\overline{\Omega})$, $\beta = 1,\ldots,n$, and satisfy (11.23) by an appropriate choice of $c(x)$.

For the existence of a function $\Phi(x)$ satisfying the system (11.22) it is necessary and sufficient that this system be compatible. To write the compatibility conditions in a more usual form ([57], vol. II) we suppose that $\Phi(x)$ is determined by a relation of the form

$$\tilde{f}(x_1,\ldots,x_n,\Phi(x)) = 0. \tag{11.25}$$

In this notation, (11.22) can be written in the form

$$\sum_{\gamma=1}^{n} p^{\beta\gamma} D_\gamma \tilde{f} + \left(q^\beta - \sum_{\gamma=1}^{n} D_\gamma p^{\beta\gamma} \right) D_0 \tilde{f} = 0,$$

$$\text{where } D_0 = \partial/\partial\Phi. \tag{11.26}$$

If $\det P \equiv \|P\|$ of the matrix of the coefficients $(p^{\beta\gamma})_{\beta,\gamma=1,\ldots,n}$ is nonzero in $\overline{\Omega}$, then system (11.26) can be solved for $D_\alpha \tilde{f}$:

$$D_\alpha \tilde{f} + \frac{1}{\|P\|} \sum_{\beta=1}^{n} P_{\alpha\beta} \left(q^\beta - \sum_{\gamma=1}^{n} D_\gamma p^{\beta\gamma} \right) D_0 \tilde{f} = 0, \tag{11.27}$$

where the $P_{\alpha\beta}$ are the cofactors of the elements $p^{\alpha\beta}$. We now obtain a condition for the existence of a function $\Phi(x) \in C^1(\overline{\Omega})$ satisfying (11.22) from the condition of completeness of the "Jacobi system" (11.27):

$$D_\alpha \left[\sum_{\gamma=1}^{n} P_{\beta\gamma} \left(q^\gamma - \sum_{\varepsilon=1}^{n} D_\varepsilon p^{\gamma\varepsilon} \right) \frac{1}{\|P\|} \right]$$

$$= D_\beta \left[\sum_{\gamma=1}^{n} P_{\alpha\gamma} \left(q^\gamma - \sum_{\varepsilon=1}^{n} D_\varepsilon p^{\gamma\varepsilon} \right) \frac{1}{\|P\|} \right]. \tag{11.28}$$

Thus, if $\det(p^{\beta\gamma})$ of the matrix of coefficients in (11.14) is nonzero in $\overline{\Omega}$ (i.e., the equation is elliptic or hyperbolic in $\overline{\Omega}$), then under condition (11.28) the system (11.21)–(11.24) is solvable, and the functional $V_\Phi[u]$ of (11.25) gives a solution of this inverse problem of the calculus of variations. If $\det(p^{\beta\gamma}(x)) = 0$, $x \in \Omega$, then for such a parabolic equation, as shown in [54], a corresponding functional (11.15) with properties (11.16) and (11.17) does not exist.

11.3. The functional $V_\Phi[u]$ constructed for equation (11.14) possesses properties A and C of the Introduction, but the question of boundedness above or below of such functionals requires further investigation. We shall show that in rather general cases of equations (11.14) of elliptic and hyperbolic types even with constant coefficients the functionals $V_\varphi[u]$ of (11.14) will not be bounded either above or below and are therefore not a quasiclassical solution of this inverse problem of the variational calculus.

We present at the same time a constructive approach to the functional (11.15) for an equation (11.14) where the $p^{\beta\gamma}$ and q^β are constants in $\overline{\Omega}$ $(\beta, \gamma = 1, \ldots, n)$ such that

$$\|P\| \equiv \det(p^{\beta\gamma}) \neq 0, \tag{11.29}$$

$$u|_{\partial\Omega} = 0. \tag{11.30}$$

In this case the system (11.21)–(11.24) has the form

$$\begin{cases} p^{\beta\gamma} = a^{\beta\gamma}, & \beta, \gamma = 1, \ldots, n, \\ q^\beta = \sum_{\gamma=1}^n a^{\beta\gamma} D_\gamma \Phi(x), & \beta = 1, \ldots, n, \\ r(x) = \sum_{\beta=1}^n b^\beta D_\beta \Phi(x) - c(x), \\ g(x) = f(x). \end{cases} \tag{11.31}$$

Setting $b^1 = \cdots = b^n = 0$ in the third relation, we find the functions sought in $V_\varphi[u]$:

$$a^{\beta\gamma} = p^{\beta\gamma}, \quad \gamma, \beta = 1, \ldots, n, \tag{11.32}$$
$$b^1 = b^2 = \cdots = b^n = 0, \quad c(x) = -r(x), \quad g(x) = f(x),$$

while the unknown function $\Phi(x)$ is determined from the system of equations with constant coefficients

$$q^\beta = \sum_{\gamma=1}^n p^{\beta\gamma} D_\gamma \Phi(x), \quad \beta = 1, \ldots, n.$$

It is obvious that a solution of the last system is

$$\Phi(x) = \sum_{i=1}^n k_i x_i, \quad k_i \equiv \text{const}, \ i = 1, \ldots, n. \tag{11.33}$$

and to find the constants $k_1, \ldots, k_n$ we obtain the system of linear algebraic equations

$$q^\beta = \sum_{\gamma=1}^n p^{\beta\gamma} k_\gamma, \quad \beta = 1, \ldots, n, \tag{11.34}$$

whence the k_j, $j = 1, \ldots, n$, are determined uniquely because of condition (11.29):

$$k_j = \det(P_j)/\|P\|, \qquad j = 1, \ldots, n. \tag{11.35}$$

Here the matrix P_j is obtained by replacing the jth column in the matrix $P = (p^{ij})$ by the column of free terms $(q^1, \ldots, q^n)$.

Thus, for the Dirichlet problem for an arbitrary linear PDE of second order with constant coefficients under condition (11.29) a functional $V_\Phi[u]$ of (11.15), whose coefficients $a^{\beta\gamma}$, b^γ ($\beta, \gamma = 1, \ldots, n$), and c, and functions $g(x)$ and $\Phi(x)$ are determined by (11.32) and (11.35), has been constructed in explicit form in the class of Euler functionals.

In a similar way it is not hard to construct a functional $V_\Phi[u]$ of (11.15) (i.e. to solve in explicit form the system (11.21)–(11.24) in the case of the Dirichlet problem for (11.14), where $p^{\alpha\beta} = p^{\beta\alpha} = 0$ for all $\alpha \neq \beta$, $\alpha, \beta = 1, \ldots, n$, $p^{\beta\beta}(x) = p^{\beta\beta}(x_\beta) \in C^1(\overline{\Omega})$, $q^\beta = q^\beta(x_\beta) \in C^1(\overline{\Omega})$, and $c(x) \in C(\overline{\Omega})$, $\beta = 1, \ldots, n$, and, in particular, for the ordinary differential equation $p(x)u''(x) + q(x)u'(x) + a(x)u(x) = f(x)$, $x \in (a, b) \subset R^1$, $u(a) = u(b) = 0$.

REMARK 11.1. If the linear PDE (11.14) is of elliptic type in Ω, then under condition (11.30) and for $c \neq \lambda_j$ (λ_j is an eigenvalue of the operator $\mathscr{L}$) there exists a unique solution of this boundary value problem $\bar{u} \in M$. Then by the connection we have established,

$$\delta V_\Phi[u] = 0 \Leftrightarrow \mathscr{L}u - f = 0, \ u \in M,$$

there exists a unique element $\bar{u} \in M$ which is a critical point of the functional $V_\Phi[u]$.

Conversely, if an element $\bar{u} \in M$ is a critical point of $V_\Phi[u]$ then this element is a classical solution of problem (11.14), (11.30).

For a general equation (11.14), as is known (see [30] and [63]), the Dirichlet problem is not well posed in an arbitrary domain Ω. In these cases it should be borne in mind that from the results of the preceding and present subsections it follows only that the set of solutions in M of problem (11.14), (11.30) coincides with the set of critical points of the functional $V_\Phi[u]$ in M.

To consider the question of the lower boundedness of the functionals (11.15) and (11.32)–(11.35), we remark that in the case of a general linear PDE with constant coefficients of elliptic type (11.24) this equation can be transformed ([57], Vol. 2, §3.1) to the form

$$-\Delta u + \lambda u = f(x), \qquad \lambda \equiv \text{const}, \ x \in \Omega, \tag{11.36}$$

while the wave equation (11.14) can be transformed to the form

$$u_{tt} - \Delta_{n-1}u + \lambda u = f(x), \qquad \lambda \equiv \text{const}, \ x \in \Omega, \tag{11.37}$$

where we have set

$$x_n = t, \quad u_{tt} = \frac{\partial^2 u}{\partial t^2}, \quad \Delta_{n-1} = \sum_{i=1}^{n-1} \left(\frac{\partial}{\partial x_i} \right)^2.$$

For equation (11.36), from the system (11.21)–(11.24) we find particular values of the coefficients of the corresponding functional (11.15):

$$V_e[u] = \int_\Omega \left\{ -\sum_{i=1}^{n} u_{x_i}^2 - \lambda u^2 + 2fu \right\} dx, \tag{11.38}$$

where the constants $k_1, \ldots, k_n$ are determined from (11.34) for $q_1 = \cdots = q_n = 0$:

$$k_1 = k_2 = \cdots = k_n = 0. \tag{11.39}$$

Similarly, for (11.37) we obtain the functional

$$V_h[u] = \int_\Omega \left\{ -\sum_{i=1}^{n-1} u_{x_i}^2 + u_t^2 - \lambda u^2 + 2fu \right\} dx, \tag{11.40}$$

where the constants k_i, $i = 1, \ldots, n$, are again determined from the special case of (11.34) for $q_1 = \cdots = q_n = 0$:

$$k_1 = k_2 = \cdots = -k_n = 0. \tag{11.41}$$

It can be shown (see the Appendix) that in the general case of equations (11.36) and (11.37) the functionals (11.38) and (11.40) are not bounded either above or below; the unboundedness on M of the functionals (11.36) and (11.37) is established (see [294] or [28], Chapter III, §3) even in the simplest cases: $f \equiv 0$ and $\lambda \le \lambda_0 < 0$ for (11.36) and the functional (11.38); $f \equiv 0$ and $\lambda = 0$ for (11.37) and the functional (11.40).

Thus, in the case of a general nonsymmetric equation (11.14) of elliptic, hyperbolic, or parabolic type in the class of Euler functionals of the form (11.15) there does not always exist a classical solution of the inverse problem of the calculus of variations in the formulation (11.16), (11.17). For special cases of equation (11.14) functionals (11.15) with properties A–C of the Introduction may, of course, exist (for example, the Dirichlet functional for the Laplace equation).

11.4. For the inverse problem of the calculus of variations in the rather general formulation [60] indicated in §11.1—on the basis of a given equation find a functional whose set of critical points coincides with the set of solutions of this equation—various approaches are known to the construction of the corresponding functionals. As noted in §6, for an arbitrary linear equation

$$Au = f \qquad (11.42)$$

these approaches are based (when some other weak conditions are satisfied; see Lemma 6.1) on the following assertion.

ASSERTION 11.1. *If there exist an auxiliary linear operator B and a nondegenerate symmetric bilinear form* $\langle \cdot, \cdot \rangle$ *such that*

$$\langle Au, Bv \rangle = \langle Bu, Av \rangle \quad \forall u, v \in D(A, B), \qquad (11.43)$$

then an element $\bar{u} \in D(A, B)$ *is a solution of equation* (11.42) *if and only if* $\delta D(u) = 0$, *where*

$$D(u) = \langle Au, Bu \rangle - 2\langle f, Bu \rangle. \qquad (11.44)$$

The following two general approaches, noted by Magri [205], of solving the inverse problem of the calculus of variations for linear equations are combined in this assertion.

First method. For a given linear equation (11.42) considered in some real Hilbert space H with inner product $(\cdot, \cdot)$ an auxiliary linear operator B is sought with $\overline{D(A, B)} = H$ and $\overline{R_A(B)} = H$ such that

$$(Au, Bv) = (Bu, Av) \quad \forall u, v \in D(A, B). \qquad (11.45)$$

According to Assertion 11.1, the desired functional then has the form

$$D_1(u) = (Au, Bu) - 2(f, Bu), \qquad u \in D(A, B). \qquad (11.46)$$

Second method. For a given linear equation (11.42), a nondegenerate symmetric bilinear form $\langle \cdot, \cdot \rangle$ is sought such that

$$\langle Au, v \rangle = \langle u, Av \rangle \quad \forall u, v \in D(A). \qquad (11.47)$$

The functional $D(u)$ of (11.44) then has the form

$$D_2(u) = \langle Au, u \rangle - 2\langle f, u \rangle. \qquad (11.48)$$

As follows from the results of Chapter 1, in the first method for any well-posed problem (11.42) in H there exists a set of solutions of the inverse problem of the calculus of variations (see Corollary 6.1); if A^{-1} exists on $R(A)$, $\overline{D(A)} = H$, and $\overline{R(A)} = H$, then there exists a set of linear operators B with property (11.45), and hence there exists a set of functionals (11.46).

Within the framework of the second method Magri [205] proposed the following simple procedure of constructing for a given linear operator A $(\overline{D(A)} = H, \overline{R(A)} = H)$ a functional (11.48):

a) An arbitrary nondegenerate symmetric bilinear form $(\cdot, \cdot)_0$ defined on $D(A)$ is chosen.

b) We set $\langle v, u \rangle = (v, Au)_0$, $u, v \in D(A)$. Then obviously $\langle Au, v \rangle = \langle u, Av \rangle$ for all $u, v \in D(A)$, i.e., condition (11.47) is satisfied.

c) The desired functional has the form

$$D(u) = \langle Au, u \rangle = 2\langle f, u \rangle. \tag{11.49}$$

In the special case where the nondegenerate symmetric form $(\cdot, \cdot)_0$ is an inner product in the Hilbert space H, $\|u\| = (u, u)_0^{1/2}$, then the functional (11.49) differs only by a constant from the functional of the method of least squares in H:

$$D(u) = \langle Au, u \rangle - 2\langle f, u \rangle = (Au, Au)_0 - 2(f, Au)_0$$
$$= (Au - f, Au - f)_0 - (f, f)_0 = \|Au - f\|_H^2 - \|f\|_H^2. \tag{11.50}$$

EXAMPLE 1. Suppose there is given the simple nonsymmetric differential equation

$$Au \equiv \frac{du(t)}{dt} = f(t), \qquad 0 < t < T, \tag{11.51}$$
$$u(0) = 0,$$
$$D(A) = \tilde{C}_0^1 = \{u(t) \in C^1[0, T], \ u(0) = 1\}. \tag{11.52}$$

As a symmetric, bilinear form it is possible to take

$$(v_1, v_2)_0 = \int_0^T v_1(t) \int_0^{T-t} (T - t - \tau)v_2(\tau)\,d\tau\,dt. \tag{11.53}$$

Setting $\langle v, u \rangle = (v, Au)_0$, we obtain

$$\langle v, u \rangle \equiv \left(v, \frac{du}{dt}\right)_0 = \int_0^T v(t) \int_0^{T-t} (T - t - \tau)\frac{du(\tau)}{d\tau}\,dt$$
$$= \int_0^T v(t) \int_0^{T-t} u(\tau)\,d\tau\,dt.$$

It is not hard to verify that here, for all $u, v \in \tilde{C}_0^1$,

$$\langle Av, u \rangle = \left(\frac{dv}{dt}, \frac{du}{dt}\right)_0 = \left(\frac{du}{dt}, \frac{dv}{dt}\right)_0 = \langle v, Au \rangle.$$

The functional (11.49) then has the form

$$D(u) = \int_0^T \frac{du(t)}{dt} \int_0^T u(\tau)\,d\tau\,dt - 2\int_0^T f(t) \int_0^{T-t} u(\tau)\,d\tau\,dt$$
$$= \int_0^T u(t)u(T - t)\,dt - 2\int_0^T f(t)\int_0^{T-t} u(\tau)\,d\tau\,dt. \tag{11.54}$$

The functional $D(u)$ of (11.54) constructed for the differential equation (11.51) does not contain derivatives, and the element $\tilde{u}$ is a solution of (11.51) if and only if $\delta D(\tilde{u}) = 0$.

Functionals for various initial boundary value problems are constructed in a similar manner in [129], [134], [205], [322], and [323]. The functionals $D_1(u)$ of (11.46) and $D_2(u)$ of (11.48) may contain derivatives of lower order than the original differential equation, and, of course, they possess the property $A\bar{u} = f$ if and only if $\delta D(\bar{u}) = 0$, $\bar{u} \in D(A)$. However, in the general case such functionals are not bounded either above or below.

REMARK 11.2. The approaches presented here to the investigation of linear equations and, in particular, the procedure of Magri [205] have known analogues. For example, in setting forth a general functional method of investigating linear PDE, Sobolev and Vishik [312] considered the equation

$$\begin{cases} \mathscr{L}u \equiv -\Delta u + \sum_{i=1}^{n} b_i \dfrac{\partial u}{\partial x_i} + cu = f(x), & x \in \Omega \subset R_n, \\ u(x) = 0, & x \in \partial\Omega. \end{cases} \quad (11.55)$$

Here the b_i, $i = 1, \ldots, n$ are constants in $\overline{\Omega}$, $\sum_1^n b_i^2 \neq 0$, $D(\mathscr{L}) = C_0^2(\Omega) = \{u \in C^2(\overline{\Omega}): u(x) = 0, x \in 0, x \in \partial\Omega\}$, and $c(x) \leq 0$, $x \in \overline{\Omega}$.

The operator $\mathscr{L}$ is obviously not symmetric in $L_2(\Omega)$:

$$(\mathscr{L}u, v) \neq (u, \mathscr{L}v) \quad \text{on } C_0^2(\Omega),$$

but

$$(\mathscr{L}u, u) = \int_\Omega \left\{ \sum_{i=1}^{n} \left(\frac{\partial u}{\partial x_i} \right)^2 + cu^2 \right\} dx \geq 0 \quad \forall u \in C_0^2(\Omega),$$

and $(\mathscr{L}u, u) = 0$ if and only if $u = 0$ almost everywhere in Ω. Further, in [312] a new bilinear form

$$\langle u, v \rangle = (u, \mathscr{L}v), \qquad u, v \in D(\mathscr{L}),$$

was introduced relative to which the operator $\mathscr{L}$ is symmetric, and the function space $H_\mathscr{L}$ was defined as the completion of $D(\mathscr{L})$ in the norm $\|u\|_\mathscr{L} = \langle u, u \rangle^{1/2}$.

EXAMPLE 2. The operator of a boundary value problem for the wave equation may be symmetric but not positive, and the corresponding Hamiltonian functional is not bounded either above or below (see [28] or [293]). For the problem

$$\begin{cases} Au \equiv u_{tt} - \Delta u(x, t) = f(x, t), & (x, t) \in \Omega \times (0, T), \\ u = \partial u/\partial t = 0, & t = 0, \ x \in \Omega, \\ u = 0, & x \in \partial\Omega, \ 0 \leq t \leq T \end{cases} \quad (11.56)$$

Vishik [339] used the positive-definiteness of the form

$$\langle Au, u\rangle \equiv \left(Au, (T-t)\frac{\partial u}{\partial t}\right) = \frac{1}{2}\int_0^T \int_\Omega \left\{ u_t^2 + \sum_{i=1}^n u_{x_i}^2 \right\} dx\, dt,$$

where the functions $u(x,t) \in C^2(\Omega \times (0,T))$ satisfy the homogeneous boundary conditions in (11.56).

11.5. The following scheme of using a variational method to study the linear equation (11.42) with, generally speaking, nonsymmetric vectorial operator([25]) A acting in a real Hilbert space H, $\|u\|_H \equiv \|u\| = (u,u)^{1/2}$, $\overline{D(A)} = H$, can be obtained from the results of Chapter 1.

a) Some linear operator B is constructed, $D(B) \supseteq D(A)$, $\overline{R_A(B)} = H$, such that the following conditions are satisfied: the property of B-symmetry

$$(Au, Bv) = (Bu, Av) \quad \forall u, v \in D(A); \tag{11.57}$$

and the property of B-positivity

$$(Au, Bu) \geq \alpha^2 \|u\|^2 \quad \forall u \in D(A) \tag{11.58}$$

or B-positive-definiteness

$$(Au, Bu) \geq \alpha^2 \|u\|^2 \quad \forall u \in D(A), \tag{11.59}$$

$$(Au, Bu) \geq \beta^2 \|Bu\|^2 \quad \forall u \in D(A), \tag{11.60}$$

where the positive constants α and β do not depend on u.

The operator B should be such that (Au, Bu) after integration by parts on the set $D(A)$ contains derivatives of the unknown function u of lower order than the original equation (11.42) (we therefore assume that $D(B) \supseteq D(A)$; see also Remark 4.1).

We note that in the case of a B-symmetric, B-positive operator A its weak closability is required: for $\{u_n\} \subset D(A, B)$, $\|u_n\| \to 0$ implies $(Au_n, Bv) \to 0$ for all $v \in D(A, B)$. For a B-symmetric, B-positive-definite operator A the condition of weak closability of A is not required.

The set of conditions (1.2), (1.3) is equivalent to (11.59). It is useful also to bear in mind (Lemma 4.2) that condition (11.60) is equivalent to (11.64), where f is a fixed but arbitrary element of H.

b) The inner product and norm of the "energy" space

$$[u, v] \equiv [u, v]_{AB} = (Au, Bv), \tag{11.61}$$

$$\|u\|_{AB} = [u, u]^{1/2}, \tag{11.62}$$

are defined first on $D(A)$.

([25]) Whose components in the case of a boundary value problem may be the operators of the initial and boundary conditions.

The Friedrichs space H_{AB} is (§3) the completion of $D(A)$ in the norm (11.62).

c) The following functional is introduced:

$$D[u] = \|u\|^2_{AB} - 2(f, Bu), \qquad u \in D(A). \tag{11.63}$$

If

$$|(f, Bu|) \le C\|f\|\,\|u\|_{AB} \quad \forall u \in D(A),\ f \in H, \tag{11.64}$$

where the constant $C > 0$ does not depend on f or u, or if (Lemma 4.2) the operator A is B-positive-definite, then the bounded linear functional (f, Bu) on H_{AB} can be extended by continuity (in u) in H_{AB} to a bounded linear functional $l(u) = (f, B_0 u)$ on H_{AB}, $l(u) = (f, Bu)$ for all $u \in D(A)$ (Theorems 4.2 and 4.3), so that the functional

$$D[u] = \|u\|^2_{AB} - 2(f, B_0 u) \tag{11.63'}$$

is defined on all of H_{AB}.

d) A generalized solution of (11.42) is defined (Lemma 4.1) as an element $u_0 \in H_{AB}$ such that

$$[u_0, v] = (f, Bv) \quad \forall v \in D(A), \tag{11.65}$$

If $u_0 \in D(A)$, then, by (11.61),

$$(Au_0 - f, Bv) = 0 \quad \forall v \in D(A), \tag{11.66}$$

and from the condition $\overline{R_A(B)} = H$ it follows that

$$Au_0 - f = 0 \tag{11.67}$$

in the sense of the original Hilbert space H. In the general case the properties of the generalized solution u_0 of (11.65) are characterized in Lemma 11.1, proved below.

e) The variational problem of minimizing the functional $D[u]$ of (11.63') is considered, i.e., the problem of determining an element $u_0 \in H_{AB}$ such that

$$D[u_0] = \min_{H_{AB}} D|u| \equiv d. \tag{11.68}$$

If (11.64) is satisfied or the operator A is B-positive-definite, then (Theorem 4.2, Lemma 4.2, and their corollaries) in the space H_{AB} there exists a unique element u_0 minimizing the functional $D[u]$ of (11.63'). In §4 it was shown also (Theorems 4.1 and 4.2, and Corollaries 4.2–4.4) that this element u_0 coincides with a generalized solution of the original equation, but we shall present here an independent proof of this basic result in the variational scheme.

THEOREM 11.1. *An element $u_0 \in H_{AB}$ is a generalized solution of equation* (11.42) *if and only if u_0 minimizes the functional $D[u]$ of* (11.63') *in H_{AB}.*

PROOF. *Necessity.* Suppose u_0 is a generalized solution of (11.42), i.e., (11.65) holds. Then, since $(f, Bu) = [u_0, u]$ for all $u \in D(A)$ and by continuity $(f, B_0 u) = [u_0, u]$ for all $u \in H_{AB}$, from (11.63') we find that

$$D[u] = [u, u] - 2[u_0, u] = ||u - u_0||_{AB}^2 - ||u_0||_{AB}^2$$

and this element u_0 actually realizes a minimum of $D[u]$ in H_{AB}.

Sufficiency. Suppose u_0 minimizes the functional $D[u]$ of (11.63') in H_{AB}, i.e.,

$$D[u] \geq D[u_0] \quad \forall u \in H_{AB}. \tag{11.69}$$

Setting $u = u_0 + tv$, where v is any element of $D(A)$ and t is an arbitrary real parameter, we obtain from (11.69)

$$D[u_0 + tv] - D[u_0] = ||u_0 + tv||_{AB}^2 - ||u_0||_{AB}^2 - 2t(f, Bv) \geq 0,$$

whence

$$t^2||v||_{AB}^2 - 2t\{(f, Bv) - [u_0, v]\} \geq 0,$$

which is possible only if

$$(f, Bv) = [u_0, v] \quad \forall v \in D(A),$$

i.e., u_0 is a generalized solution of (11.42). ∎

From part e) of the scheme and Theorem 11.1 we obtain

COROLLARY 11.1. *If inequality* (11.64) *holds or the operator A is B-positive-definite, then for all $f \in H$ there exists a unique generalized solution u_0 of equation* (11.42), *and this generalized solution coincides with the unique element in H_{AB} which minimizes the functional $D[u]$ of* (11.63'). *Moreover, the element u_0 depends continuously on the function $f \in H$.*

The last assertion for a B-positive-definite operator A follows from Theorem 4.3 and Lemma 4.3, but it can easily be obtained independently with consideration of (11.64) and (11.65):

$$||u_0||_{AB}^2 = [u_0, u_0] = (f, Bu_0) \leq C||f|| \, ||u_0||_{AB}^2, \tag{11.70}$$

whence

$$||u_0||_{AB} \leq C||f||_H. \quad \blacksquare \tag{11.71}$$

The next lemma establishes properties of a generalized solution which are usual for the variational method (see [295] and [311]).

LEMMA 11.1. *An element $u_0 \in H_{AB}$ is a generalized solution of problem* (11.42) *if and only if there exists a sequence* $\{u_k\}$, $u_k \in D(A)$, $k = 1, 2, \ldots$, *such that the relations*

$$\|u_k - u_0\|_{AB} \to 0 \qquad (k \to \infty), \tag{11.72}$$

$$(Au_k - f, Bv) \to 0 \qquad (k \to \infty) \tag{11.73}$$

are satisfied simultaneously, where the last condition is satisfied for all $v \in D(A)$.

PROOF. *Necessity.* Suppose $u_0 \in H_{AB}$ is a generalized solution of problem (11.42), i.e., (11.65) holds. By definition of the space H_{AB} it is possible to select in it a sequence $\{u_k\}$ of "smooth functions $u_k \in D(A)$, $k = 1, 2, \ldots$, which converges in H_{AB} to the element u_0, i.e., (11.72) holds.

Since for these "smooth" elements $u_k \in D(A)$. $k = 1, 2, \ldots$, we have on the basis of (11.61)

$$(Au_k, Bv) = [u_k, v], \tag{11.74}$$

on subtracting (11.65) from (11.74) we find that

$$|(Au_k - f, Bv)| = |[u_k - u_0, v]| \le \|u_k - u_0\|_{AB}\|v\|_{AB},$$

whence by the convergence $u_k \to u$ $(k \to \infty)$ in H_{AB} we obtain (11.73).

Sufficiency. Suppose it is given that for some $u_0 \in H_{AB}$ there exists a sequence $\{u_k\}$ of functions $u_k \in D(A)$, $k = 1, 2, \ldots$, such that (11.72) and (11.73) are satisfied. We show that then u_0 is a generalized solution of problem (11.42). For this we consider the obvious relation

$$|(f, Bv) - [u_0, v]|$$
$$\le |(-Au_k + f, Bv)| + |(Au_k, Bv) - [u_k, v]| + |[u_k - u_0, v]|,$$

where v is an arbitrary element of $D(A)$. From (11.72)–(11.74) we conclude that the right side of this equality can be arbitrarily small as $k \to \infty$, whence it follows that $(f, Bv) = [u_0, v]$ for all $v \in D(A)$.

§12. A variational method for solving parabolic problems

In applications, various approaches to the solution of boundary value problems for the heat equation by a variational method are known (see [9], [129], [139], [245], [246], [276], [322], and [323]). However, the functionals constructed in these works for parabolic equations do not satisfy conditions A–C of the Introduction which we have demanded of a quasiclassical solution of the inverse problem of the calculus of variations: either the functionals contain derivatives of the same order as the original differential equation [9], or the functional does not contain an integral with respect to time and the variational principle is a differential rather than an integral principle [34], [131], [246], [276]; or the solution of the original boundary value problem is only a stationary point—the functional is not bounded below—which impedes direct investigation of the variational problem and precludes the application of minimization methods. The difficulty in extending variational methods to parabolic equations consists in the fact, as already noted in [1], [18], [54], and [245], that for parabolic equations there does not exist a solution of the inverse problem of the calculus of variations in the class of Euler functionals.

12.1. For a local boundary value problem for the heat equation with Dirichlet and Neumann boundary conditions a functional with properties A–C of the Introduction was constructed in [100] and [101] in the class of functionals (6.18). We shall not reproduce these results here, but rather present closely related constructions of a quasiclassical solution of the inverse problem of the calculus of variations for the following boundary value problem for the heat equation with the condition of periodicity in the time variable:

$$c(x)\frac{\partial u}{\partial t} - \frac{\partial}{\partial x}\left(k(x,t)\frac{\partial u}{\partial x}\right) = g(x,t), \qquad (x,t) \in \Omega \subset R_2, \quad (12.1)$$

$$u(x,0) = u(x,T), \qquad a \leq x \leq \gamma(0) = \gamma(T), \quad (12.2)$$

$$k\frac{\partial u}{\partial n} = \psi(t), \qquad (x,t) \in \Gamma_l = \{x, t: x = a, 0 \leq t \leq T\}, \quad (12.3)$$

$$u(x,t) = 0, \qquad (x,t) \in \Gamma_r = \{x, t: x = \gamma(t),\ 0 \leq t \leq T\}, \quad (12.4)$$

where the bounded domain $\Omega = \{x, t: a < x < \gamma(t),\ 0 < t < T\}$ is such that the boundary $\Gamma = \partial\Omega$ is piecewise smooth, and the lines $t = \tau$, $0 < \tau < T$, are not tangent to Γ and intersect Γ_r only at one point;

$$-k(x,t) \in C^1(\Omega) \cap C(\overline{\Omega}),$$

$$0 < \kappa \leq k(x,t) \leq K < \infty, \ (x,t) \in \overline{\Omega}, \ K, \kappa = \text{const}; \quad (12.5)$$

$$c(x) \in C\left[a, \max_{[0,T]} \gamma(t)\right], \qquad c(x) \neq 0 \quad \text{for all } x \in \left[a, \max_{[0,T]} \gamma(t)\right]; \quad (12.6)$$

n is the outer normal to the boundary Γ of the domain Ω.

To apply the scheme expounded in §11.5 for constructing and investigating a variational principle for problem (12.1)–(12.4) we present some preliminary constructions.

Suppose first that

$$u, v \in \mathring{C}^{2,1}_{x,t}(\overline{\Omega}; \Gamma_r, T) = \{u(x,t) \in C^{2,1}_{x,t}(\overline{\Omega}): u(x,t) = 0,$$
$$(x,t) \in \Gamma_r; u(x,0) = u(x,T),\ a \leq x \leq \gamma(T)x\}.$$

An operator A defined by the left sides of (12.1)–(12.4) is defined on this set of functions:

$$Au = \left\{ c\frac{\partial u}{\partial t} - \frac{\partial}{\partial x}\left(k\frac{\partial u}{\partial x}\right); k\frac{\partial u}{\partial n} \right\} \qquad (12.7)$$

(conditions (12.2) and (12.4) are satisfied, since $u \in \overset{\circ}{C}{}^{2,1}(\overline{\Omega}; \Gamma_r, T)$).

We introduce the auxiliary operator

$$Bv = \{Rv; Rv\}^{(26)}, \qquad (12.8)$$

$$Rv = v(x,t) - \int_{\gamma(t)}^{x} \frac{d\theta}{k(\theta,t)} \int_{0}^{0} c(\xi)v_t(\xi,t)d\xi. \qquad (12.9)$$

The components of the operators A and B are defined on Ω and Γ_l respectively.

Integration by parts for functions $u(x,t)$, $v(x,t) \in \overset{\circ}{C}{}^{2,1}(\overline{\Omega}; \Gamma_r, T)$ establishes the identity

$$\begin{aligned}
(Au, Bv) &= \int_{\Omega} \{cu_t - (ku_x)_x\}Rv\,d\Omega + \int_{\Gamma_l} k\frac{\partial u}{\partial n}Rv\,d\Gamma_l \\
&= \int_{\Omega} \left\{ ku_xv_x + \frac{1}{k}\int_{a}^{x} cu_t\,d\xi \int_{a}^{x} cv_t\,d\xi \right\} d\Omega \qquad (12.10)
\end{aligned}$$

(the bracket $(\cdot, \cdot)$ denotes the inner product of vector-valued functions of the form (12.11) in $L_2(\Omega, \Gamma_l) \equiv L_2(\Omega) \times (L_2(\Gamma_l))$.

Equality (12.10) shows that the operator A is B-symmetric and B-positive in the sense of the definitions of §1.

We introduce the space $\widetilde{S}_{a,-}^{-1,0}W_2(\Omega, \Gamma_l)$ (§9) of vector-valued functions of the form

$$f(x,t) = \{g(x,t); \psi(t)\} \qquad (12.11)$$

(with components defined on Ω and Γ_l respectively) for which the following norm exists and is finite:

$$\|f|\widetilde{S}_{a,-}^{-1,0}W_2(\Omega, \Gamma_l)\| = \left[\int_{\Omega}\left(\int_{a}^{x} g(\xi,t)d\xi\right)^2 d\Omega + \int_{0}^{T} \psi^2(t)dt \right]^{1/2}. \qquad (12.12)$$

$\widetilde{S}_{a,-}^{-1,0}W_2(\Omega, \Gamma_l)$ is the space of right sides of equations (12.1) and (12.3).

We define also the space $\overset{\circ}{W}{}_2^{\widetilde{A}}(\Omega; \Gamma_l, T)$ (in which a generalized solution of

---

[26] In the example of §8 we discussed heuristic considerations for the choice of the operator R of (12.9) for equation (12.1) with constant coefficients.

the problem will be determined) as the closure of the set $\overset{\circ}{C}{}^{2,1}(\overline{\Omega}; \Gamma_r, T)$ in the norm generated by the inner product (the right side of (12.10))

$$[u, v]_{\widetilde{A}} = \int_{\Omega} \left\{ k u_x v_x + \frac{1}{k} \int_a^x d\xi \int_a^x c v_t \, d\xi \right\} d\Omega, \qquad (12.13)$$

$$\|u | \overset{\circ}{W}{}^{\widetilde{A}}_2(\Omega; \Gamma_r, T)\| = [u, u]^{1/2}_{\widetilde{A}}, \qquad (12.14)$$

$$\widetilde{A} = \begin{pmatrix} 0, & 0 \\ 1, & 0 \\ -1, & +1 \end{pmatrix} \qquad (12.15)$$

(see §10).

According to the general scheme of §11.5, the original boundary problem can now be formulated in a well-posed manner.

DEFINITION 12.1. A *generalized solution* of problem (12.1)–(12.4) is a function $u \in \overset{\circ}{W}{}^{\widetilde{A}}_2(\Omega; \Gamma_r, T)$ such that for all $v \in C^{2,1}(\overline{\Omega}; \Gamma_r, T)$

$$[u, v]_{\widetilde{A}} = (f, Bv).^{(27)} \qquad (12.16)$$

It is obvious that for $u(x, t) \in \overset{\circ}{C}{}^{2,1}(\overline{\Omega}; \Gamma_r, T)$ from (12.16) by (12.10) and (12.13)

$$(Au - f, Bv) = 0 \quad \forall v \in \overset{\circ}{C}{}^{2,1}(\overline{\Omega}; \Gamma_r, T). \qquad (12.17)$$

The set $\overset{\circ}{C}{}^{\infty}(\Omega)$ is dense in $L_2(\Omega)$ and in $\widetilde{S}^{-1,0}_{a,-}W_2(\Omega)$ (§9), and it is not hard to verify that the sets $D(A)$ and $R(B)$ are also dense in the spaces $L_2(\Omega, \Gamma_l)$ and $\widetilde{S}^{-1,0}_{a,-}W_2(\Omega, \Gamma_l)$ in the corresponding metrics.

In the general case the essence of the generalized solution is clarified by the following concretization of Lemma 11.1.

LEMMA 12.1. *An element* $u \in \overset{\circ}{W}{}^{\widetilde{A}}_2(\Omega; \Gamma_r, T)$ *is a generalized solution of problem* (12.1)–(12.4) *if and only if there exists a sequence* $\{u_m\}$ *of functions* $u_m \in \overset{\circ}{C}{}^{2,1}(\overline{\Omega}; \Gamma_r, T)$, $m = 1, 2, \ldots$, *such that the relations*

$$\|u_m - u | \overset{\circ}{W}{}^{\widetilde{A}}_2(\Omega; \Gamma_r, T)\| \to 0 \quad (m \to \infty), \qquad (12.18)$$

$$(Au_m - f, Bv) \to 0 \quad (m \to \infty) \qquad (12.19)$$

hold simultaneously, and (12.19) *is satisfied for all* $v \in \overset{\circ}{C}{}^{2,1}(\overline{\Omega}; \Gamma_r, T)$.

The properties of B-symmetry and B-positivity of the operator A make it possible to consider for problem (12.1)–(12.4) the variational problem

---

(27)The definition of a generalized solution and the proof of its existence can be carried out with a weaker condition (12.5); for example, it suffices to require that $k(x, t) \in C(\overline{\Omega}$ and $0 < \kappa \le k(x, t) \le K$, where $K, \kappa = \text{const}$.

of minimization in $\overset{\circ}{W}{}_2^{\tilde{A}}(\Omega;\Gamma_r, T)$ of the functional

$$D_1[u] = \|u|\overset{\circ}{W}{}_2^{\tilde{A}}(\Omega;\Gamma_r, T)\|^2 - 2(f, Bu). \tag{12.20}$$

It is obvious that on the set of smooth functions $\overset{\circ}{C}{}^{2,1}(\overline{\Omega};\Gamma_r, T)$ we have, by (12.10) and (12.11),

$$D_1[u] = (Au, Bu) - 2(f, Bu). \tag{12.21}$$

DEFINITION 12.2. A *solution* of the variational problem of minimizing the functional (12.20)

$$\inf_{\overset{\circ}{W}{}_2^{\tilde{A}}} D_1[u] = d > -\infty \tag{12.22}$$

is an element $u_0 \in \overset{\circ}{W}{}_2^{\tilde{A}}(\Omega, \Gamma_r, T)$ for which there exists a sequence $\{u_n\}$ of elements $u_n \in \overset{\circ}{C}{}^{2,1}(\overline{\Omega};\Gamma_r, T)$ such that

$$\begin{aligned}\|u_n - u_0|\overset{\circ}{W}{}_2^{\tilde{A}}(\Omega;\Gamma_r, T)\| &\to 0 \quad (n \to \infty), \\ D_1[u)n] &\to d \quad (n \to \infty).\end{aligned} \tag{12.23}$$

Moreover, the following concretization of Theorem 11.1 holds.

THEOREM 12.1. *An element $u_0 \in \overset{\circ}{W}{}_2^{\tilde{A}}(\Omega;\Gamma_r, T)$ is a generalized solution of problem* (12.1)–(12.4) *if and only if u_0 is a solution of the variational problem of minimization of the functional* (12.20).

To establish the existence of an element minimizing the functional (12.20) it remains to demonstrate that the linear functional (f, Bu) is bounded in u in $\overset{\circ}{W}{}_2^{\tilde{A}}(\Omega;\Gamma_r, T)$.

LEMMA 12.2. *For an arbitrary $f \in \widetilde{S}_{a,-}^{-1,0}W_2(\Omega, \Gamma_l)$ and for all $u \in \overset{\circ}{W}{}_2^{\tilde{A}}(\Omega;\Gamma_r, T)$*

$$|(f, Bu)| \le C_2\|f|\widetilde{S}_{a,-}^{-1,0}W_2(\Omega, \Gamma_l)\|\,\|u|\overset{\circ}{W}{}_2^{\tilde{A}}(\Omega;\Gamma_r, T)\|, \tag{12.24}$$

where the constant $C_2 > 0$ does not depend on u or f (nor do the constants $C_n > 0$ below, $n = 3, 4, 5, \dots$).

PROOF. We establish (12.24) first for an arbitrary $u(x, t) \in \overset{\circ}{C}{}^{2,1}(\overline{\Omega};\Gamma_r, T)$. From (12.8) and (12.11) we get

$$\begin{aligned}(f, Bu) = &\int_\Omega g(x, t)\left\{u - \int_{\gamma(t)}^x \frac{d\theta}{k(\theta, t)}\int_a^0 cu_t(\xi, t)\,d\xi\right\}dt\,dx \\ &+ \int_0^T \psi(t)\left\{u(a, t) - \int_{\gamma(t)}^a \frac{d\theta}{k(\theta, t)}\int_a^\theta c(\xi, t)u_t(\xi, t)\,d\xi\right\}dt \\ \equiv\; &I_1 + I_2.\end{aligned} \tag{12.25}$$

By integration by parts with the help of the Cauchy-Schwarz-Bunyakovskiĭ inequality we obtain

$$|I_1| = \left| \int_\Omega gRu \, d\Omega \right|$$

$$\leq C_3 \left[\int_\Omega \left(\int_a^x g(x \subset, t) \, d\xi \right)^2 d\Omega \right]^{1/2}$$

$$\times \left[\int_\Omega \left\{ u_x^2 + \frac{1}{k} \left(\int_a^x cu_t \, d\xi \right)^2 \right\} c\Omega \right]^{1/2}$$

$$\leq C_4 \| f |\widetilde{S}_{a,-}^{-1,0} W_2(\Omega, \Gamma_l) \| \, \| u | \mathring{W}_2^{\widetilde{A}}(\Omega; \Gamma_r, T) \|. \qquad (12.26)$$

$$|I_2| = \left| \int_0^T \psi(t) \left\{ u(a, t) - \int_{\gamma(t)}^a \frac{d\theta}{k(\theta, t)} \int_a^\theta c(\xi) u_t(\xi, t) \, d\xi \right\} dt \right|$$

$$\leq 2 \left(\int_0^T \psi^2(t) \, dt \right)^{1/2} \left\{ \int_0^T \left[u^2(a, t) \right. \right.$$

$$\left. \left. + \left(\int_{\gamma(t)}^a \frac{d\theta}{k(\theta, t)} \int_a^0 c(\xi) u_t(\xi, t) \, d\xi \right)^2 \right] dt \right\}^{1/2}$$

$$\leq C_5 \| \psi | L_2(0, T) \| \left\{ \int_\Omega \left[u_x^2 + \frac{1}{k} \left(\int_1^x cu_t \, d\xi \right)^2 \right] d\Omega \right\}^{1/2}$$

$$\leq C_6 \| f \int_\Omega |\widetilde{S}_{a,-}^{-1,0} W_2(\Omega, \Gamma_l) \| \, \| u | \mathring{W}_2^{\widetilde{A}}(\Omega; \Gamma_r, T) \|. \qquad (12.27)$$

The estimate (12.24) for the functional (f, Bu) of (12.25) follows from (12.26) and (12.27) first of all $u(x, t) \in \mathring{C}^{2,1}(\overline{\Omega}; \Gamma_r, T)$; using the operation of closure in the norm of $\mathring{W}_2^{\widetilde{A}}(\Omega; \Gamma_r, T)$, we obtain the complete assertion of the lemma.[28] ∎

From what was proved in part e) of §11.5 we obtain

THEOREM 12.2. *For an arbitrary $f \in \widetilde{S}_{a,-}^{-1,0} W_2(\Omega, \Gamma_l)$ there exists a unique element $u_0 \in \mathring{W}_2^{\widetilde{A}}(\Omega, \Gamma_r, T)$ which is a solution of the variational problem of minimizing the functional $D_1[u]$ of (12.20). Moreover, the element u_0 minimizing the functional $D_1[u]$ coincides with a generalized*

_____
[28]Here we assume (§4) that (f, Bu) has been extended by continuity to all of $\mathring{W}_2^{\widetilde{A}}(\Omega; \Gamma_r, T)$.

solution in $\overset{\circ}{W}{}_2^{\tilde{A}}(\Omega; \Gamma_r, T)$ *of the original boundary value problem and depends continuously on the vector-valued function* f *(12.11) of right sides of equations* (12.1)–(12.4):

$$u_0|\overset{\circ}{W}{}_2^{\tilde{A}}(\Omega; \Gamma_r, T)\| \le C_7\|f|\tilde{S}_{a,-}^{-1,0}W_2(\Omega, \Gamma_l)\|.$$

This theorem completes the construction and investigation of the variational problem with properties A–C of the Introduction for the boundary value problem for the heat equation (12.1)–(12.4).

REMARK 12.1. From the variational problem of minimizing the functional $D_1[u]$ of (12.20) we shall obtain some other known variational problems for the heat equation.

Following [34] and [131], we write the functional $D_1[u]$ of (12.20) for a fixed time $t \in (0, T)$ (i.e., without the integral $\int_0^T$ in (12.20)) on a set of functions satisfying the boundary conditions (12.2)–(12.4), assuming the variation $\delta\{cu_t\} \equiv 0$ ([131], Chapter VI, §1). Assuming for simplicity that $\gamma(t) \equiv b = \mathrm{const} > a$, $t \in (0, T)$, we obtain Fourier's variational principle for the heat equation ([131], formula (6.20) familiar in theoretical physics, from the necessary condition for a minimum of the functional (12.20):

$$\delta D_t[u] = \delta\left\{\int_a^b [2cuu_t + ku_x^2 - 2gu](x, t)\,dx + C^*[u_t]\right\}$$
$$= 0 \quad \forall t \in (0, T),$$

where the quantity

$$C^*[u_t] = \int_a^b \frac{1}{k}\left(\int_a^x cu_t\,d\xi\right)^2 dx - 2\int_a^b g(x, t)\int_b^x \frac{d\theta}{k(\theta, t)}\int_a^\theta cu_t\,d\xi\,dx$$

contains terms of the functional $D_1[u]$ of (12.20) which are constant relative to the variation $\delta\{\cdots\}$.

For the boundary value problem

$$\frac{\partial u}{\partial t} - \frac{\partial^2 u}{\partial x^2} = 0, \qquad (x, t) \in \Omega' = (a, b) \times (0, T), \qquad (12.1')$$

$$u(x, t) = 0, \qquad (x, t) \in \{x = a, 0 < t < T\} \cup \{x = b, 0 < t < T\},$$
$$(12.2')$$

$$u(x, 0) = u_0(x), \qquad a < x < b, \qquad (12.3')$$

with the help of the nonlinear theory of semigroups it is established in [43] and [44] that a generalized solution in a certain sense of this problem coincides with the unique element minimizing the functional

$$E(u) = \int_0^T \left(\||u|\overset{\circ}{W}_2^1(a,b)\|^2 + \left\|\frac{\partial u(x,t)}{\partial t}\Big|W_2^{-1}(a,b)\right\|^2 \right) dt$$

$$+ \int_a^b |u(x,T)|^2 \, dx. \tag{12.20'}$$

Setting $c(x) = k(x,t) \equiv 1$, $\psi(t) = 0$, $(x,t) \in \Gamma_l$, and $g(x,t) = 0$, $(x,t) \in \Omega = \Omega'$, in (12.1)–(12.4) and (12.20), and considering in virtue of (9.35) the relation

$$\||u|W_2^{-1,0}(a,b;a)\|^2 = \int_0^T \int_a^b \left(\int_a^x u(\xi,t)\,d\xi \right)^2 dx$$

$$= \int_0^T \||u|W_2^{-1}(a,b;a)\|^2 \, dt,$$

we write the functional (12.20) in the form

$$E_1[u] = \int_0^T \left(\||u|\overset{\circ}{W}_2^1(a,b;b)\|^2 + \left\|\frac{\partial u}{\partial t}(x,t)|W_2^{-1}(a,b;a)\right\|^2 \right) dt, \tag{12.20''}$$

which is close to (12.20')—the differences are connected only with the different boundary conditions in (12.2)–(12.4) and (12.2'), (12.3').

12.2. For parabolic equations we shall indicate below (§§14 and 16) other approaches to the construction of symmetrizing operators B in classes of functionals distinct from (6.2) and (6.18), but now we present closely related modifications of the constructions of §12.1.

The nonlinear heat equation encountered in applications frequently has the form [167]

$$\frac{\partial}{\partial t}[c(u)u] - \frac{\partial}{\partial x}\left(k(u)\frac{\partial u}{\partial x} \right) = f(u), \tag{12.28}$$

i.e., the coefficients of heat capacity $c(u)$ and thermal conductivity $k(u)$ are determined experimentally as functions of the temperature $u(x,t)$, and not of the space-time variables x and t. However, in linearizing (12.28) in some manner (when an initial approximation $u_0(x,t)$ is given and the $N+1$ approximations $u_{N+1}(x,t)$, $N = 0,1,2,\ldots$, are determined successively) the necessity arises of investigating the equation

$$\frac{\partial}{\partial t}[c(u_N)u_{N+1}] - \frac{\partial}{\partial x}\left(k(u_N)\frac{\partial u_{N+1}}{\partial x} \right) = f(u_N). \tag{12.29}$$

For an equation of the form (12.29)

$$\mathscr{L}_1 u \equiv \frac{\partial}{\partial t}[c(x,t)u(x,t)] - \frac{\partial}{\partial x}\left(k(x,t)\frac{\partial u}{\partial x} \right) = g(x,t), \tag{12.\tilde{1}}$$

considered on the set of functions

$D(\mathscr{L}_1) = \{u(x,t) \in C^{2,1}(\overline{\Omega})$ satisfying (12.4) and the condition $u(x,t) = 0,$
$(x,t) \in \Gamma_0 = \{(x,t) : t = 0,\ a \le x \le \gamma(0)\}\}.$ (12.$\tilde{2}$)

analogous results can be obtained by the scheme presented; it is only necessary to require that $c(x,t) > 0$ for all $(x,t) \in \overline{\Omega}$,

$$0 \le \partial c/\partial t \le C_0 = \text{const}, \qquad (x,t) \in \overline{\Omega}.$$

Indeed, for any functions $u(x,t)$ and $v(x,t)$ in $D(\mathscr{L}_1)$ by integration by parts we establish the identity

$$(A_2 u, \tilde{B} v) \equiv \int_\Omega \{(cu)_t - (ku_x)_x\} \left\{ v - \int_{\gamma(t)}^x \frac{d\theta}{k(\theta,t)} \right.$$
$$\left. \times \int_a^\theta [c(\xi,t)u(\xi,t)]_t\, d\xi \right\} d\Omega$$
$$+ \int_{\Gamma_l} k \frac{\partial u}{\partial n} \tilde{R} v\, dt \tag{12.30}$$
$$= \int_\Omega \left\{ c'_t u v + k u_x v_x + \frac{1}{k} \int_a^x (cu)'_t\, d\xi \int_a^x (cv)'_t\, d\xi \right\} d\Omega$$
$$+ \int_a^{\gamma(t)} c(\xi,T)u(\xi,T)v(\xi,T)\, d\xi,$$

where

$$A_2 u = \left\{ \mathscr{L}_1 u; k \frac{\partial u}{\partial n} \right\}, \tag{12.31}$$
$$\tilde{B} v = \{\tilde{R} v; \tilde{R} v\},$$
$$\tilde{R} v = v(x,t) - \int_{\gamma(t)}^x \frac{d\theta}{k(\theta,t)} \int_a^0 [c(\xi,t)v(\xi,t)]'_t\, d\xi. \tag{12.32}$$

The identity (12.30) demonstrates the $\tilde{B}$-symmetry and $\tilde{B}$-positivity of the operator A_2 of problem (12.$\tilde{1}$), (12.$\tilde{2}$), (12.3), (12.4).

For a parabolic equation of order $2m$ (in the space variable), $m \ge 1$,

$$\mathscr{L}_2 u \equiv c \frac{\partial u}{\partial t} - \frac{\partial^{2m} u}{\partial x^{2m}} = g(x,t), \qquad (x,t) \in \Omega \tag{12.33}$$

it is possible to consider [279] the mixed boundary value problem

$$u(x,0) = u(x,T), \qquad a \le x \le \gamma(0) = \gamma(0) = \gamma(T), \tag{12.34}$$
$$\frac{\partial^{2m-k-1} u}{\partial n^{2m-k-1}} = h_k(t), \qquad (x,t) \in \Gamma_l, k = 0,\dots,m-1, \tag{12.35}$$
$$\frac{\partial^k u}{\partial n^k} = 0, \qquad (x,t) \in \Gamma_r, k = 0,\dots,m-1. \tag{12.36}$$

On the set $\overset{\circ}{C}{}^{2m,1}(\overline{\Omega};\Gamma_r,T)$ of functions in $C_{x,t}^{2m,1}(\overline{\Omega})$ which satisfy the boundary conditions (12.34) and (12.36) we define the boundary operator A_3 with boundary values of problem (12.33)–(12.36):

$$A_3 u = \left\{ \mathscr{L}_2 u; \frac{\partial^m u}{\partial n^m}, \ldots, \frac{\partial^{2m-1}u}{\partial n^{2m-1}} \right\}, \tag{12.37}$$

and the auxiliary operator B_3,

$$B_3 v = \{\tilde{\tilde{R}}v; R_1 v; \ldots; R_m v\}, \tag{12.38}$$

where

$$\tilde{\tilde{R}}v = (-1)^{m+1}(v - I_{\gamma,a}^{2m}[cv_t]),$$

$$I_{\gamma,a}^{2m}[cv_t] \equiv I_\gamma^m I_a^m C v_t$$
$$= \int_{\gamma(t)}^x d\theta_1 \cdots \int_{\gamma(t)}^{\theta_{m-1}} d\theta_m \int_a^{\theta_m} d\theta_{m+1} \cdots \int_a^{\theta_{2m-1}} c(\theta_{2m})v_t(\theta_{2m},t)d\theta_{2m},$$

$$R_k v = (-1)^{2m-k}\frac{\partial}{\partial x^k}(v - I_\gamma^m i_a^m(cv_t)).$$

The first components of the operators A_3 and B_3 are defined on Ω, while the remaining components are defined on Γ_l. By direct integration by parts it is not hard to verify that for all $u, v \in \overset{\circ}{C}{}^{2m,1}(\overline{\Omega};\Gamma_r,T)$

$$(A_3 u, B_3 v) \equiv \int_\Omega \mathscr{L}_2 u \tilde{\tilde{R}}v d\Omega + \sum_{k=0}^{m-1}\int_{\Gamma_l}\frac{\partial^{2m-k-1}u}{\partial n^{2m-k-1}}R_{k+1}v\,dt$$
$$= \int_\Omega \left\{ \frac{\partial^m u}{\partial x^m}\frac{\partial^m v}{\partial x^m} + \int_a^x d\theta_1 \cdots \int_a^{\theta_{m-1}} cv_t\,d\theta_m \right.$$
$$\left. \times \int_a^x d\theta_1 \cdots \int_a^{\theta_{m-1}cu_td\theta_m} \right\} d\Omega. \tag{12.39}$$

From this it is obvious that the operator A_3 is B_3-symmetric; it is also B_3-positive, since in this case

$$\|u|H_{A_3B_3}\| = \left\| u|\overset{\circ}{W}_2\begin{pmatrix} m, & 0 \\ -m, & 1 \end{pmatrix}(\Omega;\Gamma_r,T) \right\|$$
$$= \left\{ \int_\Omega \left[\left(\frac{\partial^m u}{\partial x^m}\right)^2 + \left(\int_a^x d\theta_1 \cdots \int_a^{\theta_{m-1}} cu_t\,d\theta_m\right)^2 \right] d\Omega \right\}^{1/2}. \tag{12.40}$$

$H_{A_3B_3}$ is the Hilbert space obtained as the closure of the original set $\overset{\circ}{C}{}^{2m,1}(\overline{\Omega}; \Gamma_r, T)$ in the norm (12.40). In this space we define, according to the scheme of §11.5, a generalized solution of problem (12.33)–(12.36) and an element minimizing the functional

$$D_3[u] = \||u|H\int_{A_3B_3}\||^2 - 2(f_3, B_3 u), \qquad (12.41)$$

where $f_3 = \{g(x,t); h_0(t), \dots, h_{m-1}(t)\}$ is the given vector-valued function of right sides of equations (12.33) and (12.35),

$$\||f_3|\widetilde{S}_{a,-}^{-m,0} W_2(\Omega, \Gamma_l)\||$$

$$= \left\{ \int_\Omega \left(\int_a^x d\theta_1 \int_a^{\theta_1} d\theta_2 \cdots \int_a^{\theta_{m-1}} g(\theta_m, t) \, d\theta_m \right)^2 d\Omega \right.$$

$$\left. + \sum_{k=0}^{m-1} \int_0^T h_k^2(t) \, dt \right\}^{1/2}. \qquad (12.42)$$

We shall show that for all $f_3 \in \widetilde{S}_{a,-}^{-m,0} W_2(\Omega, \Gamma_l)$ and for all $u \in H_{A_3B_3}$

$$|(f_3, B_3 u)| \le C_9 \||f_3|\widetilde{S}_{a,-}^{-m,0} W_2(\Omega, \Gamma_l)\|| \cdot \||u|H_{A_3B_3}\||. \qquad (12.43)$$

Suppose first that $u \in \overset{\circ}{C}{}^{2m,1}(\overline{\Omega}; \Gamma_r, T)$ and $f_3 \in \widetilde{S}_{a,-}^{-m,0} W_2(\Omega, \Gamma_l)$; then

$$|(f_3, B_3 u)| = \left| \int_\Omega g\widetilde{R}u \, d\Omega + \sum_{k=0}^{m-1} \int_0^T h_k(t) R_{k+1} u(a,t) \, dt \right|$$

$$= I_1 + I_2; \qquad (12.44)$$

$$|I_1| = \left| \int_\Omega g[u(x,t) - I_\gamma^m I_a^m(cv_t) \, dx \, dt \right|$$

$$\le 2 \left[\int_\Omega (I_a^m g(x,t))^2 \, d\Omega \right]^{1/2}$$

$$\times \left[\int_\Omega \left\{ \left(\frac{\partial^m u}{\partial x^m} \right)^2 + (I_a^m(cv_t))^2 \right\} d\Omega \right]^{1/2}$$

$$\le 2\||f_3|\widetilde{S}_{a,-}^{-m,0} W_2(\Omega, \Gamma_l)\|| \cdot \||u|H_{A_3B_3}\||; \qquad (12.45)$$

$$|I_2| \equiv \left| \sum_{k=0}^{m-1} \int_0^T h_k(t)(R_{k+1}u)(a,t)\,dt \right|$$

$$\leq C_{10} \| f_3 | \widetilde{S}_{a,-}^{-m,0} W_2(\Omega,\Gamma_l) \|$$

$$\times \sum_{k=0}^{m-1} \left\{ \int_0^T \left[\left(\frac{\partial^k u}{\partial x^k} \right)^2 + (I_{\gamma,a}^{2m-k-1} cu_t)^2 \right] dt \right\}^{1/2}$$

$$\leq C_{11} \| f_3 | \widetilde{S}_{a,-}^{-m,0} W_2(\Omega,\Gamma_l) \| \, \| u | H_{A_3 B_3} \|. \tag{12.46}$$

The validity of (12.43) for all $u \in \overset{\circ}{C}{}^{2m,1}(\overline{\Omega};\Gamma_r,T)$ follows from (12.44)–(12.46), while it is possible to extend the linear functional $(f_3, B_3 u)$ by continuity to all of $H_{A_3 B_3}$ and obtain (12.43) for all $u \in H_{A_3 B_3}$.

According to Lemma 4.2, the estimate (12.43) is equivalent in the present case to

$$\|u\|_{A_3 B_3} \geq C_{12} \| B_3 u | \widetilde{L}(\Omega,\Gamma_l) \|,$$

$$\widetilde{L}_2(\Omega,\Gamma_l) \equiv L_2(\Omega) \times \overbrace{L_2(\Gamma_l) \times \cdots \times L_2(\Gamma_l)}^{m},$$

and the operator A_3 of problem (12.33)–(12.36) is B_3-positive-definite in the sense of Definition 1.3. From the general scheme of §11.5 we also obtain

ASSERTION 12.1. *For any $f_3 \in \widetilde{S}_{a,-}^{-m,0} W_2(\Omega,\Gamma_l)$ there exists a unique element $u_0 \in H_{A_3 B_3}$ which in $H_{A_3 B_3}$ minimizes the functional $D_3[u]$ of (12.41); this element u_0 coincides with a generalized solution (in the sense of the definition of §11.5 d)) in $H_{A_3 B_3}$ of the boundary value problem (12.33)–(12.36), and*

$$\|u_0 | H_{A_3 B_3}\| \leq C_8 \| f_3 | \widetilde{S}_{a,-}^{-m,0} W_2(\Omega,\Gamma_l).$$

12.3. For completeness of the exposition we present some properties of the function spaces

$$W_2 \begin{pmatrix} 0, & 0 \\ 1, & 0 \\ -1, & 1 \end{pmatrix} (\Omega), \qquad W_2 \begin{pmatrix} m, & 0 \\ -m, & 1 \end{pmatrix} (\Omega),$$

introduced in §§12.1 and 12.2 in which generalized solutions of the corresponding boundary value problems were determined, and of the spaces $\widetilde{S}_{a,-}^{-m,0}(\Omega,\Gamma_l)$ of given functions—the right sides of the equations. To shorten computations we restrict our attention to equivalent spaces of functions with the finite norms

$$\|u|W_2^{\widetilde{A}}(\Omega)\| = \left\|u\Big|W_2\begin{pmatrix} 0, & 0 \\ 1, & 0 \\ -1, & 0 \end{pmatrix}(\Omega)\right\|$$

$$= \left\{\int_\Omega \left[u^2 + u_x^2 + \left(\int_a^x u_t(\xi,t)\,d\xi\right)^2\right]d\Omega\right\}^{1/2}, \quad (12.47)$$

$$\|g|\widetilde{S}_{a,-}^{-1,0}W_2(\Omega)\| = \left[\int_\Omega\left(\int_a^x g(\xi,t)\,d\xi\right)^2 dx\,dt\right]^{1/2}. \quad (12.48)$$

Making the change

$$v(x,t) = Iu \equiv \int_a^x u(\xi,t)\,d\xi, \qquad v(a,t) = 0,\ 0 < t < T, \quad (12.49)$$

$$f(x,t) = Ig \equiv \int_a^x g(\xi,t)\,d\xi, \qquad f(a,t) = 0,\ 0 < t < T, \quad (12.50)$$

we consider the Hilbert space $\overset{\circ}{W}_2^{2,1}(\Omega,a)$ of functions $v(x,t)$,

$$\|v|\overset{\circ}{W}_2^{2,1}(\Omega,a)\| = \left(\int_\Omega[v_x^2 + v_{xx}^2 + v_t^2]dx\,dt\right)^{1/2} < \infty, \quad (12.51)$$

where $v(a,t) = 0$ in the sense of $L_2(0,T)$, and the space $L_2(\Omega,a)$ of functions $f(x,t)$,

$$\|f|L_2(\Omega,a)\| = \left(\int_\Omega f^2 dx\,dt\right)^{1/2}, \qquad f(a,t) = 0. \quad (12.52)$$

The latter space is not complete, since the condition $f(a,t) = 0$ is an unstable boundary condition for functions in $L_2(\Omega)$.

It is obvious that the operators I and I^{-1} in (12.49) and (12.50) realize an isometric mapping of the spaces

$$W_2\begin{pmatrix} 0, & 0 \\ 1, & 0 \\ -1, & +1 \end{pmatrix}(\Omega) \leftrightarrow W_2^{2,1}(\Omega,a), \qquad \widetilde{S}_{a,-}^{-1,0}(\Omega) \leftrightarrow L_2(\Omega,a),$$

and then by Lemma 9.1 we obtain

ASSERTION 12.2. *The space*

$$W_2\begin{pmatrix} 0, & 0 \\ 1, & 0 \\ -1, & 1 \end{pmatrix}(\Omega)$$

is complete; the space $\widetilde{S}_{a,-}^{-1,0}W_2(\Omega)$ *is not complete.*

We shall show that the closure of $\widetilde{S}_{0,-}^{-1,0}W_2(\Omega)$ is the space $\overset{\circ}{W}{}_2^{-1,0}(\Omega, a)$ of generalized functions in the Schwartz sense with finite norm

$$\|u|W_2^{-1,0}(\Omega, a)\| = \sup_{v \neq 0} \frac{|(u, v)|}{\|v|\overset{\circ}{W}{}_2^{1,0}(\Omega, \Gamma_r)|}, \tag{12.53}$$

where $\overset{\circ}{W}{}_2^{1,0}(\Omega, \Gamma_r)$—a Sobolev space—is the set of functions $v(x, t)$ with $v(x, t) = 0$ on Γ_r in the sense of $L_2(\Gamma_r)$ for which the following seminorm exists and is finite:

$$\|v|\overset{\circ}{W}{}_2^{1,0}(\Omega, \Gamma_r)\| = \left(\int_\Omega \left| \frac{\partial v}{\partial x} \right|^2 dx\, dt \right)^{1/2}. \tag{12.54}$$

LEMMA 12.3. *For all* $u \in \widetilde{S}_{a,-}^{-1,0}W_2(\Omega)$,

$$\|u|\widetilde{S}_{a,-}^{-1,0}W_2(\Omega)\| = \|u|\overset{\circ}{W}{}_2^{1,0}(\Omega, a)\|. \tag{12.55}$$

PROOF. For $u \in \widetilde{S}_{a,-}^{-1,0}W_2(\Omega)$, $v \in \overset{\circ}{W}{}_2^{1,0}(\Omega, \Gamma_r)$, integrating by parts and applying the Cauchy-Schwarz-Bunyakovskii inequality, we obtain

$$\left| \int_\Omega uv\, dx\, dt \right| = \left| \int_\Omega \int_a^x u(\xi, t)\, d\xi\, v_x\, dx\, dt \right|$$

$$\leq \left[\int_\Omega \left(\int_a^x u(\xi, t)\, d\xi \right)^2 dx\, dt \right]^{1/2} \left[\int_\Omega v_x^2\, dx\, dt \right]^{1/2} \tag{12.56}$$

$$= \|u|\widetilde{S}_{a,-}^{-1,0}W_2(\Omega)\| \, \|v|\overset{\circ}{W}{}_2^{1,0}(\Omega, \Gamma_r)\|,$$

whence

$$\|u|\overset{\circ}{W}{}_2^{-1,0}(\Omega, a) = \sup_{v \neq 0} \frac{|(u, c)|}{\|v|\overset{\circ}{W}{}_2^{-1,0}(\Omega, \Gamma_r)\|}$$

$$\leq \|u|\widetilde{S}_{a,-}^{-1,0}W_2(\Omega, a)\|. \tag{12.57}$$

Since for the function

$$v(x, t) = \int_{\gamma(t)}^x \left(\int_a^\theta u(\xi, t)\, d\xi \right) d\theta \in \overset{\circ}{W}{}_2^{1,0}(\Omega, \Gamma_r)$$

in (12.56), and hence also in (12.57), equality is achieved, the validity of (12.55) follows from this. ∎

From the lemma we immediately obtain

COROLLARY 12.1.

$$\widetilde{S}_{a,-}^{-1,0} W_2(\Omega, a) \to \overset{\circ}{W}{}_{2}^{-1,0}(\Omega, a),$$

$$\overline{\widetilde{S}_{a}^{-1,0} W_2(\Omega, a)} = \overset{\circ}{W}{}_{2}^{-1,0}(\Omega, a). \tag{12.58}$$

Equality (12.55), in particular, makes it possible to constructively find the norm (12.53) which is difficult to compute.

If $\overset{\circ}{W}_2 \begin{pmatrix} 1, & 0 \\ -1, & +1 \end{pmatrix}(\Omega, \Gamma_r)$ is the subspace of functions of

$$W_2 \begin{pmatrix} 0, & 0 \\ 1, & 0 \\ -1, & +1 \end{pmatrix}(\Omega)$$

of (12.47) which vanish on Γ_r in the sense of $L_2(\Gamma_r)$, then by inequalities established for $u \in W_2^{\widetilde{A}}(\Omega)$ in the usual manner [242]

$$\left(\int_0^T u^2(\gamma(t), t)\, dt \right)^{1/2} \le C_{13} \left\| u \middle| W_2 \begin{pmatrix} 0, & 0 \\ 1, & 0 \\ -1, & +1 \end{pmatrix}(\Omega) \right\|, \tag{12.59}$$

$$\left(\int_\Omega u^2\, d\Omega \right)^{1/2} \le C_{14} \left[\int_\Omega \left\{ u_x^2 + \left(\int_a^x u_t\, d\xi \right)^2 \right\} d\Omega + \int_{\Gamma_r} u^2\, d\Gamma_r \right]^{1/2} \tag{12.60}$$

we conclude that for all $u \in \overset{\circ}{W}_2 \begin{pmatrix} +1; & 0 \\ -1, & +1 \end{pmatrix}(\Omega)$ the norm

$$\left\| u \middle| \overset{\circ}{W}_2 \begin{pmatrix} 1, & 0 \\ -1, & 1 \end{pmatrix}(\Omega) \right\| = \left\{ \int_\Omega \left[u_x^2 + \left(\int_a^x u_t\, d\xi \right)^2 \right] d\Omega \right\}^{1/2} \tag{12.61}$$

is equivalent to (12.47) and that the boundary condition $u = 0$ (understood in the sense of $L_2(\Gamma_r)$) is a stable boundary condition for all functions in $\overset{\circ}{W}_2 \begin{pmatrix} 1, & 0 \\ -1, & +1 \end{pmatrix}(\Omega)$.

Similarly, considering (12.55), we obtain

$$\| u(x, T) - u(x, 0) | W_2^{-1}(a, \gamma(0)) \|$$

$$\le C_{14} \left\| u \middle| \overset{\circ}{W}_2 \begin{pmatrix} 1, & 0 \\ -1, & +1 \end{pmatrix}(\Omega) \right\|;$$

hence, the condition $u(x, 0) = u(x, T)$, $a < x < \gamma(0)$ (understood in the sense of $W_2^{-1}(a, \gamma(0))$) is a stable boundary condition for functions in $\overset{\circ}{W}_2 \begin{pmatrix} 1, & 0 \\ -1, & +1 \end{pmatrix}(\Omega)$ with finite norm (12.61).

Thus, the boundary conditions (12.32) and (12.34) will be stable in the sense indicated above on minimization of the functional $D_1[u]$ of (12.20) in

$$\overset{\circ}{W}{}_2^{\widetilde{A}}(\Omega, \Gamma_r, T) \sim \overset{\circ}{W}_2 \begin{pmatrix} 1, & 0 \\ -1, & +1 \end{pmatrix} (\Omega).$$

Having established by means of Lemma 12.3 that the space $\widetilde{S}_{a,-}^{-1,0} W_2(\Omega)$ is a certain "regular restriction" of $\overset{\circ}{W}_2^{-1,0}(\Omega, a)$, we now consider the relation of the space $W_2^{\widetilde{A}}(\Omega)$ of (12.47) to the familiar Sobolev, Nikol'skiĭ, and Besov function spaces $W_2^1(\Omega)$, $H_2^1(\Omega)$, and $B_2^1(\Omega)$ respectively (see [242] and [33]).

The Sobolev space $W_2^1(\Omega)$ can be defined as the closure of $C^\infty(\overline{\Omega})$ in the norm

$$\|u|W_2^1(\Omega)\| = \left[\int_\Omega \{u^2 + u_x^2 + u_t^2\} \, dx \, dt \right]^{1/2}. \tag{12.62}$$

It is known that [242]

$$W_2^1(\Omega) \to H2^1(\Omega), \tag{12.63}$$
$$B_2^1(\Omega) = W_2^1(\Omega), \tag{12.64}$$
$$H_2^{1+\varepsilon}(\Omega) \to W_2^1(\Omega) \quad \forall \varepsilon > 0, \tag{12.65}$$

where the imbeddings (12.63) and (12.65) are not reversible.

LEMMA 12.4. *For the space* $W_2^{\widetilde{A}}(\Omega)$ *of* (12.47) *and* $W_2^1(\Omega)$ *of* (12.62) *there is the irreversible imbedding*

$$W_2^1(\Omega) \to W_2^{\widetilde{A}}(\Omega). \tag{12.66}$$

For the proof it obviously suffices to establish the inequality

$$\|u|W_2^{\widetilde{A}}(\Omega)\| \le \text{const} \, \|u|W_2^1(\Omega)\| \quad \forall u \in W_2^1(\Omega) \tag{12.67}$$

and its irreversibility. Inequality (12.67) follows from the obvious estimates

$$\int_\Omega u^2 \, dx \, dt \le \|u|W_2^1(\Omega)\|^2, \tag{12.68}$$

$$\int_\Omega u_x^2 \, dx \, dt \le \|u|W_2^1(\Omega)\|^2, \tag{12.69}$$

$$\int_\Omega \left(\int_a^x u_t d\xi \right)^2 dx \, dt \le C_{16} \|u|W_2^1(\Omega)\|^2. \tag{12.70}$$

We demonstrate the irreversibility of the imbedding (12.66). Considering the norms (12.62), (12.47) on the functions

$$u_n(x, t) = n^{-3/4} \sin(nt) \sin(n^{1/3}x), \quad n = 1, 2, \ldots, \tag{12.71}$$

by direct computations we find that

$$I_1 \equiv \int_\Omega u_n^2 \, d\Omega \leq C_{17} n^{-3/2}, \tag{12.72}$$

$$I_2 \equiv \int_\Omega \left(\frac{\partial u_n}{\partial x}\right)^2 d\Omega \leq C_{18} n^{-5/6}, \tag{12.73}$$

$$I_3 \equiv \int_\Omega \left(\int_a^x \frac{\partial u_n}{\partial t} \, d\xi\right)^2 d\Omega \leq C_{19} n^{-1/6}, \tag{12.74}$$

$$I_4 \equiv \int_\Omega \left(\frac{\partial u_n}{\partial t}\right)^2 d\Omega = \int_\Omega n^{1/2} \cos^2(nt) \sin^2(nt) \, d\Omega$$

$$\geq \frac{n^{1/2}}{4} \, (\text{area } \Omega - \varepsilon) \quad \forall \varepsilon > 0, \ n \geq N(\varepsilon), \ \frac{\partial C_k}{\partial n} = 0, \ k = 17, \ 18, \ 19. \tag{12.75}$$

From the definition of the norms (12.47) and (12.62) and the estimates (12.72)–(12.75) we get

$$\|u_n | W_2^{\widetilde{A}}(\Omega)\| = (I_1 + I_2 + I_3)^{1/2} < \varepsilon \quad \forall \varepsilon > 0, \ n \geq N_0(\varepsilon),$$

$$\|u_n | W_2^1(\Omega)\| = (I_1 + I_2 + I_3)^{1/2} \geq C_{20} n^{1/4}, \quad n \geq N_0(\varepsilon).$$

The last two relations show that an inequality the reverse of (12.67)

$$\|u_n | W_2^1(\Omega)\| \leq C_{21} \|u_n | W_2^{\widetilde{A}}(\Omega)\| \tag{12.76}$$

is violated for any constant $C_{21} > 0$ not depending on the index n. Hence, the norms of the spaces $W_2^1(\Omega)$, $W_2^{\widetilde{A}}(\Omega)$ are not equivalent, and the imbedding $W_2^1(\Omega) \to W_2^{\widetilde{A}}(\Omega)$ is not reversible. ∎

We shall now show that in the space $W_2^{\widetilde{A}}(\Omega)$ there exists at least one function not belonging to $W_2^1(\Omega)$. This assertion is obtained from the following theorem of Nikol'skiĭ ([242], §7.6, Theorem 2)

THEOREM. *If for Banach spaces E_1 and E_2 their intersection $E_1 \cdot E_2$ is a Banach space and if there does not exist a constant $C > 0$ such that the inequality*

$$\|u | E_2\| \leq C \|u | C_1\| \tag{12.77}$$

holds for all $u \in E_1 \cdot E_2$, then in E_1 there exists an element not belonging to E_2.

COROLLARY 12.2. *In the space $W_2^{\widetilde{A}}(\Omega)$ there exists at least one function not belonging to $W_2^1(\Omega)$.*

Indeed, setting $W_2^{\widetilde{A}} = E_1$ and $E_2 = W_2^1(\Omega)$, from the imbedding (12.66) we find that $E_1 \cdot E_2 = W_2^1(\Omega)$, and $W_2^1(\Omega)$ is a Banach space. As shown in the proof of Lemma 12.4, there does not exist a constant $C > 0$ not depending on the function $u_n(x, t)$ of (12.71) ($u_n \in E_1 \cdot E_2 = W_2^1(\Omega)$) such that (12.76) holds, i.e., all conditions of Nikol'skiǐ's theorem are satisfied. Therefore in $E_1 = W_2^{\widetilde{A}}(\Omega)$ there exists a function not belonging to $E_2 = W_2^1(\Omega)$. ∎

COROLLARY 12.3. *The imbeddings* (12.64)–(12.6) *entail the irreversible imbeddings*

$$B_2^1(\Omega) \to W_2^{\widetilde{A}}(\Omega), \tag{12.78}$$

$$H_2^{1+\varepsilon}(\Omega) \to W_2^{\widetilde{A}}(\Omega) \quad \forall \varepsilon > 0. \tag{12.79}$$

§13. The principle of a minimum of a quadratic functional for the wave equation

13.1. The equation of a vibrating string

$$u_{tt} - u_{xx} = g(x, t), \qquad (x, t) \in Q = \{x, t : 0 << x < l, 0 < t < T\}, \tag{13.1}$$

as we know, follows from the condition that the following Hamiltonian functional be stationary:

$$H[u] = \int\!\!\int_Q \{u_x^2 - u_t^2 - 2gu\}\, dx\, dt. \tag{13.2}$$

This functional belongs to the class of Euler functionals but does not belong to the class of functionals with properties A–C of the Introduction we are considering: the functional $H[u]$ even on $C^2(Q)$ is not bounded either above or below (see [28] and [294]).

As shown in §11, even for the more general hyperbolic equation with a nonsymmetric operator

$$A_n u = \frac{\partial^2 u}{\partial x_n^2} - \sum_{i=1}^{n-1} a_i \frac{\partial^2 u}{\partial x_i^2} + \sum_{i=1}^{n} q_i \frac{\partial u}{\partial x_i} + ru = g(x),$$

$$a_i, q_i = \text{const}, \quad i = 1, \ldots, n, \quad x \in \Omega \subset R_n \tag{13.3}$$

it is possible to constructively find a function $\Phi(x)$ and construct the quadratic functional (11.19) in the class of Euler functionals

$$F_\Phi[u] = \int_\Omega F(x, \Phi(x), u(x), u'_{x_1}, \ldots, u'_{x_n})\, dx, \tag{13.4}$$

so that the condition

$$\delta F_\Phi[u] = 0 \tag{13.5}$$

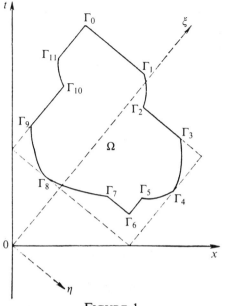

FIGURE 1

implies the equation

$$\exp \Phi(x)\{A_n u - g(x)\} = 0, \tag{13.6}$$

equivalent to (13.3). In the general case, however, these functionals are not bounded either above or below even on $\overset{\circ}{C}{}^2(\Omega)$, i.e., they do not satisfy condition B of the Introduction.

13.2. For the wave equation

$$\frac{\partial}{\partial \xi}\left(k(\xi,\eta)\frac{\partial u}{\partial \eta}\right) = g(\xi,\eta), \tag{13.7}$$

written in characteristic variables we shall construct a functional analogous in its variational properties A–C of the Introduction to the Dirichlet functional for the Laplace equation.

Suppose in (13.7) $(\xi,\eta) \in \Omega$ and Ω is a domain with piecewise smooth boundary Γ such that (Figure 1) each of the lines $\xi = x + t$ and $\eta = x - t$ either intersects Γ in no more than two points or coalesces with some part of Γ. The $\Gamma_i = \{\xi_i, \eta_i\}$, $i = 0, \dots, 11$, are points of contact of Γ with the corresponding lines ξ or η.

We set $\eta_2 = \Gamma_0\Gamma_1$, $l_2^2 = \Gamma_1\Gamma_2$, $\eta_1 = \Gamma_2\Gamma_3$, $l_2^1 = \Gamma_3\Gamma_4$, $\gamma'' = \Gamma_4\Gamma_5$, $\xi_0 = \Gamma_5\Gamma_6$, $\eta_0 = \Gamma_6\Gamma_7$, $\gamma' = \Gamma_7\Gamma_8$, $l_1^1 = \Gamma_8\Gamma_9$, $\xi_1 = \Gamma_9\Gamma_{10}$, $l_1^2 = \Gamma_{10}\Gamma_{11}$, $\xi_2 = \Gamma_{11}\Gamma_0$, $\gamma_H = \gamma' + \gamma''$, $l_1 = l_1^1 + l_1^2$, $l_2 = l_2^1 + l_2^2$, $\gamma_1 = l_2 + \eta_1 + \eta_2$, and

$\gamma_2 = l_1 + \xi_1 + \xi_2$. Suppose the equation of the curve $\Gamma_8\Gamma_9\Gamma_{10}\Gamma_{11}\Gamma_0$ has the form $\xi = \gamma_1(\eta)$ $(\eta_0 \leq \eta \leq \eta_8)$, while the curve $\Gamma_0\Gamma_1\Gamma_2\Gamma_3\Gamma_4$ has the form $\eta = \gamma_2(\xi)$ $(\xi_4 \leq \xi \leq \xi_0)$.

In the domain Ω we consider the following problem for equation (13.7):

$$u|_{\gamma_H} = \varphi(s), \tag{13.8}$$

$$(\partial/\partial s)(u|_{\gamma_H}) = \partial\varphi/\partial s, \tag{13.9}$$

$$\partial u/\partial n|_{\gamma_H} = \psi(s), \tag{13.10}$$

$$u|_{\xi_0} = \theta(s), \tag{13.11}$$

$$\partial u/\partial s|_{\eta_0} = \alpha(s), \tag{13.12}$$

$$\partial u/\partial \eta|_{l_1} = \chi(s), \tag{13.13}$$

$$u|_{l_2} = \omega(s). \tag{13.14}$$

We first assume that the function $k(\xi, \eta)$ satisfies the conditions

$$k(\xi, \eta) \in C^1(\Omega) \cap C(\overline{\Omega}), \tag{13.15}$$

$$0 < \kappa \leq k(\xi, \eta) \leq K < \infty, \qquad \kappa, K = \text{const}, \ (\xi, \eta) \in \overline{\Omega}. \tag{13.16}$$

The use of (13.15) and (13.16) as well as a well-posed formulation of problem (13.7)–(13.14) will be specified below. We remark that from the general problem (13.7)–(13.14) there follow familiar special problems for the wave equation (13.7), for example:

1) the Cauchy problem if the segments ξ_0, η_0, and l^i_j, i, $j = 1, 2$, are absent;

2) a Goursat problem if we have only the segments ξ_0, η_2, and l^1_1 which form a closed region in the form of a right triangle with a diagonal curve;

3) a Darboux problem if only the segments ξ_0, η_0, ξ_2, and η_2, which form a rectangle, are present; and

4) a mixed problem if η_k and ξ_k, $k = 0, 1, 2$, are absent, the initial conditions (13.8)–(13.10) are given on the portion γ_H, and on the segments l_1 and l_2 conditions (13.13) and (13.14) of second and first kind, respectively, are given.

Suppose first that $u(\xi, \eta)$, $v(\xi, \eta) \in C^2(\Omega) \cap C^1(\overline{\Omega})$. We introduce the operator A defined by the left sides of (13.7)–(13.14),

$$Au = \left\{ \frac{\partial}{\partial \xi}\left(k(\xi, \eta)\frac{\partial u}{\partial \eta}\right); u; \frac{\partial u}{\partial \xi}; \frac{u}{\partial n}; u; \frac{\partial u}{\partial s}; \frac{\partial u}{\partial \eta}; u \right\}, \tag{13.17}$$

and the auxiliary operator B defined by

$$Bv = \{Kv; K_Hv; K_{sH}v; K_{HH}v; T_\xi v; T_\eta v; T_{l_1}v; T_{l_2}v\}, \tag{13.18}$$

with components given, respectively, on $\Omega, \gamma_H, \gamma_H, \gamma_H, \xi_0, \eta_0, l_1$ and l_2, where

$$Kv = \int_{\gamma_1(\eta)}^{\xi} \frac{d\zeta}{k(\zeta,\eta)} \int_{\gamma_2(\zeta)}^{\eta} v(\zeta,\tau), d\tau, \tag{13.19}$$

$$K_Hv = \cos(\widehat{n,\eta}) \int_{\gamma_2(\xi)}^{\eta} v(\xi,\tau)\, d\tau = T_\xi v = T_{l_2} v, \tag{13.20}$$

$$K_{HH}v = -Kv \sin[2(\widehat{n,\eta})]\frac{k}{2}, \tag{13.21}$$

$$T_{l_1}v = -Kv \cdot k \cdot \sin(\widehat{n,\eta}), \tag{13.22}$$

$$T_\eta v = -Kv \cdot k \cdot \sin^2(\widehat{n,\eta}) = -K_{sH}v. \tag{13.23}$$

We define the Hilbert space $L_2(\Omega, \Gamma)$ of vector-valued functions of the form

$$f = \left\{ g(\xi,\eta); \varphi(s); \frac{\partial \varphi}{\partial s}; \psi(s); 0(s); \alpha(s); \chi(s); \omega(s) \right\} \tag{13.24}$$

(whose components are given on $\Omega, \gamma_H, \gamma_H, \gamma_H, \xi_0, \eta_0, l_1, l_2$ respectively) with inner product

$$(f_1, f_2) = \int_\Omega g_1 g_2 \, d\xi \, d\eta + \int_{\gamma_H} \varphi_1 \varphi_2 \, ds + \int_{\gamma_H} \psi_1 \psi_2 \, ds$$
$$+ \int_{\xi_0} \theta_1 \theta_2 \, ds + \int_{\gamma_H} \frac{\partial \varphi_1}{\partial s} \frac{\partial \varphi_2}{\partial s} \, ds + \int_{\eta_0} \alpha_1 \alpha_2 \, ds$$
$$+ \int_{l_1} \chi_1 \chi_2 \, ds + \int_{l_2} \omega_1 \omega_2 \, ds. \tag{13.25}$$

The following identity can be established by integration by parts:

$$(Au, Bv) = \int_\Omega uv \, d\eta \, d\xi \quad \forall u, v \in C^2(\Omega) \cap C^1(\overline{\Omega}), \tag{13.26}$$

i.e., B-symmetry and B-positivity of the operator A in the sense of Definitions 1.1 and 1.2. According to the general scheme of §11.5, following from results of Chapter 1, these properties make it possible to consider for problem (13.7)–(13.14) the corresponding variational problem of minimizing the functional

$$D[u] = \|u|L_2(\Omega)\|^2 - 2(f, Bu). \tag{13.27}$$

It is obvious that on $C^2(\Omega) \cap C^1(\overline{\Omega})$

$$\|u|L_2(\Omega)\| = (Au, Bu)^{1/2}, \tag{13.28}$$

$$D[u] = (Au, Bu) - 2(f, Bu). \tag{13.29}$$

DEFINITION 13.1. A *generalized solution* of problem (13.7)–(13.14) is a function $u \in L_2(\Omega)$ such that for all $v \in C^2(\Omega) \cap C^1(\overline{\Omega})$

$$(f, Bv) = \int_{\Omega} uv \, d\Omega. \tag{13.30}$$

In the definition of a generalized solution it suffices to assume that the function $k(\xi, \eta)$ satisfies only condition (13.16).

If $u \in C^2(\Omega) \cap C^1(\overline{\Omega})$, then (13.30) is equivalent by (13.26) to

$$(Au - f, Bv) = 0 \quad \forall v \in C^2(\Omega) \cap C^1(\overline{\Omega}). \tag{13.31}$$

In the general case properties clarifying the essence of the generalized solution are established in the following special case of Lemma 11.1.

LEMMA 13.1. *An element* $u_0 \in L_2(\Omega)$ *is a generalized solution of problem* (13.7)–(13.14) *if and only if there exists a sequence* $\{u_n\}$ *of functions* $u_n \in C^2(\Omega) \cap C^1(\overline{\Omega})$, $n = 1, 2, \ldots$, *such that the relations*

$$\|u_n - u_0 | L_2(\Omega)\| \to 0 \quad (n \to \infty),$$
$$(Au_n - f, Bv) \to 0 \quad (n \to \infty),$$

are satisfied simultaneously, and the latter relation is true for all $v \in C^2(\Omega) \cap C^1(\overline{\Omega})$.

It is not hard to see that the set $D(A) = C^2(\Omega) \cap C^1(\overline{\Omega})$ is dense in $L_2(\Omega)$, while $R_A(B)$ is dense in $L_2(\Omega, \Gamma)$.

To apply the scheme of §11.5 it now suffices to establish boundedness in $L_2(\Omega)$ of the linear functional (f, Bu) in (13.27).

LEMMA 13.2. *If the measurable function* $k(\xi, \eta)$ *satisfies condition* (13.16), *then the linear functional* (f, Bu) *satisfies*

$$|(f, Bu)| \leq C \|f| L_2(\Omega, \Gamma)\| \, \|u| L_2(\Omega)\| \tag{13.32}$$

for all $f \in L_2(\Omega, \Gamma)$ *and* $u \in L_2(\Omega)$; *the constant* $C > 0$ *does not depend on* f *or* u.

PROOF. We first establish (13.32) for an arbitrary $f \in L_2(\Omega, \Gamma)$ and $u \in C^2(\Omega) \cap C^1(\overline{\Omega})$. We have

$$(f, Bu) = \int_{\Omega} gKu \, d\Omega + \int_{\gamma_H} \varphi K_H u \, ds + \int_{\gamma_H} \frac{\partial \varphi}{\partial s} K_{sH} u \, ds$$
$$+ \int_{\gamma_H} \psi K_{HH} u \, ds + \int_{\xi_0} \theta(s) T_\xi u \, ds + \int_{\eta_0} \alpha(s) T_\eta u \, ds$$
$$+ \int_{l_1} \chi(s) T_{l_1}, u \, ds + \int_{l_2} \omega(s) T_{l_2} u \, ds. \tag{13.33}$$

Considering (13.16), (13.19)–(13.23), and (13.25), and using the Cauchy-Schwarz-Bunyakovskiĭ inequality, we obtain

$$\left| \int_{\Omega} g K u \, d\Omega \right| \le C_1 \left(\int_{\Omega} g^2 \, d\Omega \right)^{1/2} \left(\int_{\Omega} u^2 \, d\Omega \right)^{1/2}, \qquad (13.34)$$

$$\left| \int_{\gamma_H} \varphi \cos(\widehat{n,\eta}) \int_{\gamma_2}^{\eta} v(\xi,\tau) d\tau \, ds \right|$$

$$\le C_2 \left(\int_{\gamma_H} \varphi^2 \, ds \right)^{1/2} \left| \int_{\gamma_H} \int_{\gamma_2}^{\eta} u^2(\xi,\tau) \, d\tau \, ds \right|^{1/2}$$

$$\le C_2 \left(\int_{\gamma_H} \varphi^2 \, ds \right)^{1/2} \left| \int_{\xi_8}^{\xi_7} d\xi \int_{\gamma_2}^{\gamma_H} u^2(\xi\tau) \, d\tau + \int_{\xi_5}^{\xi_4} u^2(\xi,\tau) \, d\tau \right|^{1/2}$$

$$\le C_3 \left(\int_{\gamma_H} \varphi^2 \, ds \right)^{1/2} \left(\int_{\Omega} u^2(\xi,\tau) \, d\xi \, d\tau \right)^{1/2}, \qquad (13.35)$$

where $\eta = \gamma_H(\xi)$ is the equation of the curve γ_H. Analogously

$$\left| \int_{\gamma_H} \frac{\partial \varphi}{\partial s} K_{sH} u \, ds \right| \le \left[\int_{\gamma_H} \left(\frac{\partial \varphi}{\partial s} \right)^2 ds \right]^{1/2}$$

$$\times \left| \int_{\gamma_H} \left(\int_{\gamma_1}^{\xi} \frac{d\zeta}{k(\zeta,\eta)} \int_{\gamma_2(\zeta)}^{\eta} u(\zeta,\tau) \, d\tau \right)^2 ds \right|^{1/2}$$

$$\le C_5 \left[\int_{\gamma_H} \left(\frac{\partial \varphi}{\partial s} \right)^2 ds \right]^{1/2} \left(\int_{\Omega} u^2(\zeta,\tau) \, d\tau \right)^{1/2}, \qquad (13.36)$$

$$\left| \int_{\gamma_H} \psi K_{HH} u \, ds \right| \le C_6 \left(\int_{\gamma_H} \psi^2 \, ds \right)^{1/2}$$

$$\times \left(\int_{\gamma_H} \left(\int_{\gamma_1(\eta)}^{\xi} \frac{d\zeta}{k(\zeta,\eta)} \int_{\gamma_2}^{\eta} u(\zeta,\tau) \, d\tau \right)^2 ds \right)^{1/2}$$

$$\le C_7 \left(\int_{\gamma_H} \psi^2 \, ds \right)^{1/2} \left(\int_{\Omega} u^2(\zeta,\tau) \, d\zeta \, d\tau \right)^{1/2}, \qquad (13.37)$$

$$\left|\int_{\xi_0} \theta(s) T_\xi u\, ds\right| \le C_8 \left(\int_{\xi_0} \theta^2(s)\, ds\right)^{1/2} \left[\int_{\xi_0} \left(\int_{\gamma_2}^\eta u(\xi,\tau)\, d\tau\right)^2\right]^{1/2}$$

$$\le C_9 \left[\int_{\xi_0} \theta^2(s)\, ds\right]^{1/2} \left|\int_{\xi_6}^{\xi_5} d\xi \int_{\gamma_2(\xi)}^{\eta_5} u^2(\xi,\tau)\, d\tau\right|^{1/2}$$

$$\le C_9 \left(\int_{\xi_0} \theta^2(s)\, ds\right)^{1/2} \left(\int_\Omega u^2(\xi,\tau)\, d\tau\, d\xi\right)^{1/2}, \qquad (13.38)$$

$$\left|\int_{\eta_0} \alpha(s) T_\eta u\, ds\right| \le C_{10} \left(\int_{\eta_0} \alpha^2\, ds\right)^{1/2} \left(\int_\Omega u^2\, d\Omega\right)^{1/2}, \qquad (13.39)$$

$$\left|\int_{l_1} \chi T_{l_1} u\, ds\right| \le C_{11} \left(\int_{l_1} \chi^2(s)\, ds\right)^{1/2} \left|\int_{l_1} \left(\int_{\gamma_1(\eta)}^\xi \frac{d\zeta}{k(\zeta,\eta)}\right.\right.$$

$$\left.\left.\times \int_{\gamma_2(\zeta)}^\eta u(\zeta,\tau)\, d\tau\right)^2 ds\right|^{1/2}$$

$$= C_{11} \left(\int_{l_1} \chi^2\, ds\right)^{1/2} \left|\int_{\eta_9}^{\eta_8} d\eta \int_{\gamma_2^{-1}(\eta)}^{\gamma_1(\eta)} u^2(\zeta,\eta)\, d\zeta\right.$$

$$\left.+ \int_{\eta_{11}}^{\eta_{10}} d\eta \int_{\gamma_2^{-1}(\eta)}^{\gamma_1(\eta)} u^2(\zeta,\eta)\, d\zeta\right|^{1/2}$$

$$\le C_{11} \left(\int_{l_1} \chi^2\, ds\right)^{1/2} \left(\int_\Omega u^2\, d\Omega\right)^{1/2}, \qquad (13.40)$$

$$\left|\int_{l_2} \omega(s) T_{l_2} u\, ds\right|$$

$$\le C_{12} \left(\int_{l_2} \omega^2(s)\, ds\right)^{1/2} \left|\int_{l_2} \int_{\gamma_2(\xi)}^\eta u^2(\xi,\tau)\, d\tau\, ds\right|^{1/2}$$

$$\le C_{13} \left(\int_{l_2} \omega^2(s)\, ds\right)^{1/2} \left|\int_{\xi_2}^{\xi_3} \int_{\gamma_2(\xi)}^{\gamma_1^{-1}(\xi)} u^2(\xi,\tau)\, d\tau\right.$$

$$\left.+ \int_{\xi_2}^{\xi_1} d\xi \int_{\gamma_2(\xi)}^{\gamma_1^{-1}(\xi)} u^2(\xi,\tau)\, d\tau\right|^{1/2}$$

$$\le C_{13} \left(\int_{l_2} \omega^2(s)\, ds\right)^{1/2} \left(\int_\Omega u^2\, d\Omega\right)^{1/2}. \qquad (13.41)$$

From (13.34)–(13.41) we obtain (13.32) for an arbitrary $f \in L_2(\Omega, \Gamma)$ and for all $u \in D(A)$, and, using the operation of closure by continuity in $L_2(\Omega)$, we establish (13.32) for all $u \in L_2(\Omega)$. ∎

REMARK 13.1. Because of (13.32) it may be assumed that the functional (f, Bu) has been extended by continuity (in u) to the entire space $L_2(\Omega)$. From the general scheme of §11.5 we now have

ASSERTION 13.1. *For any $f \in L_2(\Omega, \Gamma)$ in the space $L_2(\Omega)$ there exists a unique element u_0 minimizing the functional $D[u]$ of* (13.37). *The element u_0 coincides in $L_2(\Omega)$ with a generalized solution of the original boundary value problem, and*

$$\|u_0|L_2(\Omega)\| \leq C\|f|L_2(\Omega, \Gamma)\|.$$

REMARK 13.2. As is evident from (13.18)–(13.24), the quadratic functional (13.27) does not contain derivatives of the unknown function $u(\xi, \eta)$. Use of such a functional in practical computations raises the stability of numerical methods and reduces the volume of computations, since it allows one to use a shorter Ritz series (for example, in the finite element method). We note further that for the complex boundary value problem (13.7)–(13.14) minimization of the functional $D[u]$ in $L_2(\Omega)$ is carried out on a set of functions not satisfying any boundary conditions, which is also essential in a numerical realization.

REMARK 13.3. Shalov [294] constructed for the operator A_g of boundary value problem (13.7)–(13.14) (but with the essential restriction $k(\xi, \eta) \equiv$ const, $(\xi, \eta) \in \overline{\Omega}$) an auxiliary operator B_g so that

$$(A_g u, B_g v) = [u, v]_{W_2^1(\Omega)} \quad \forall u, v \in C^2(\overline{\Omega}),$$

$$[u, v]_{W_2^1(\Omega)} = \int_\Omega \left\{ \frac{\partial u}{\partial \eta} \frac{\partial v}{\partial \eta} + \frac{\partial u}{\partial \xi} \frac{\partial v}{\partial \xi} \right\} d\Omega + \int_{\gamma_H} uv \, ds.$$

The corresponding quadratic functional

$$D_g[u] = [u, u]_{W_2^1(\Omega)} - 2(f, B_g u)$$

contained derivatives of first order, belonged to a class of functionals of the form (6.18), and possessed the necessary properties A–C of the Introduction.

REMARK 13.4. From boundary value problem (13.7)–(13.14) there follows [294] a problem for the wave equation (13.7) with data on the entire boundary of Ω. This problem corresponds to the case where the segments of characteristics $\Gamma_0\Gamma_1$, $\Gamma_3\Gamma_2$, $\Gamma_0\Gamma_{11}$, and $\Gamma_9\Gamma_{10}$, on which there are no boundary conditions, are not present and the domain Ω is situated strictly within the angle formed by the characteristics ξ and η and "suspended from above" on Ω, i.e., Ω is situated within the characteristics ξ and η tangent to Ω only at one point, namely, at the point Γ_0 at which ξ and η intersect.

On the other hand, a generalization of boundary value problem (13.8)–(13.14) is possible: in a manner similar to that presented above a problem is considered in a domain Ω whose boundary Γ can consist of any finite number of segments of characteristic lines of the type $\Gamma_0\Gamma_1$, $\Gamma_2\Gamma_4$, $\Gamma_0\Gamma_{11}$, and $\Gamma_9\Gamma_{10}$.

The results of this subsection extend also to the "n-wave" equation

$$\frac{\partial^n}{\partial \xi^n}\left(k(\xi,\eta)\frac{\partial^n u}{\partial \eta^n}\right) = g(\xi,\eta), \qquad k \in C^n(\Omega), \qquad (13.42)$$

considered with boundary conditions

$$\mathscr{L}_k u(\xi,\eta) = g_k(\xi,\eta), \qquad (\xi,\eta) \in \Gamma_k \subset \Gamma$$

of the type (13.8)–(13.14) but with a corresponding increase in the number of them. Special cases of such a problem were considered, for example, in [166] and [307].

The basic identity of the form (13.26), demonstrating B-symmetry and B-positivity of the operator of the boundary value problem for equation (13.42),

$$\int_\Omega \frac{\partial^n}{\partial \xi^n}\left(k(\xi,\eta)\frac{\partial^n u}{\partial \eta^n}\right)\left[\int_{\gamma_1(\eta)}^\xi d\theta_1 \int_{\gamma_1(\eta)}^{\theta_1} d\theta_2 \cdots \int_{\gamma_1(\eta)}^{\theta_{n-1}} \frac{d\theta_n}{k(\theta_n,\eta)}\right.$$
$$\times \int_{\gamma_2(\theta_n)}^\eta d\tau_1 \int_{\gamma_2(\theta_n)}^{\tau_1} d\tau_2 \cdots \int_{\gamma_2(\theta_n)}^{\tau_{n-1}} v(\theta_n,\tau_n)\,d\tau_n \Bigg] d\xi\, d\eta$$
$$+ \sum_k \int_{\Gamma_k} \mathscr{L}_k u R_k v\, ds$$
$$= \int_\Omega v u\, d\xi\, d\eta,$$

can be obtained by direct integration by parts on the set of functions $C^{2n}(\overline{\Omega})$.

It can be shown also([29]) that Assertion 13.1 remains valid for f in $S^{-1,-1}\overset{\circ}{W}_2(\Omega,\gamma_l,\gamma_r)$, i.e., in a space dual to some Sobolev space with a dominant mixed derivative.

In conclusion we note Shalov's solution [295] by a variational method of a boundary value problem for a linearized wave equation, in which the auxiliary symmetrizing operator B was constructed on the basis of the Riemann function for the original equation. However, in the general case

---

([29])See [99].

the construction of symmetrizing operators B by means of a Riemann function is obviously nonconstructive.

§14. A variational principle for hypoelliptic PDE with constant coefficients

14.1. In this section we consider using a variational method to find a solution of the following boundary value problem for a linear, inhomogeneous, hypoelliptic PDE:

$$\mathscr{L}u \equiv \sum_{|\overline{\alpha}| \le m} a_{\overline{\alpha}} D^{\overline{\alpha}} u(x) = g(x), \qquad x = (x_1, \dots, x_n) \in \Omega \subset R_n, \tag{14.1}$$

$$\frac{\partial^j u}{\partial n^j} = 0, \qquad j = 0, 1, 2, \dots, k, \ x \in \partial\Omega. \tag{14.2}$$

Here $\overline{\alpha} = (\alpha_1, \dots, \alpha_n)$, $\alpha_s \ge 0$, $s = 1, \dots, n$, is an integral vector;

$$D^{\overline{\alpha}} = \frac{\partial^{|\overline{\alpha}|}}{\partial x_1^{\alpha_1} \dots \partial \alpha_n^{\alpha_n}};$$

the $a_{\overline{\alpha}}$ are real constant coefficients in $\overline{\Omega}$ for all $\overline{\alpha}$, $|\overline{\alpha}| \le m$, $\partial\Omega$ is the sufficiently smooth boundary of a bounded domain Ω, and $\overline{n}$ is the outer normal to $\partial\Omega$. The integer $k = k(\mathscr{L}, n) \ge m$ will depend on the operator $\mathscr{L}$ and the dimension n of the space R_n. We assume that the following condition is satisfied.

CONDITION 1. The function $g(x)$ belongs to a class G, so that for all $g \in G$ there exists a unique $u_0(x) \in \overset{\circ}{C}{}^k(\Omega)$ which is a classical solution of problem (14.1), (14.2).

Suppose first that $D(\mathscr{L}) = \overset{\circ}{C}{}^k(\Omega)$, $k \ge m$; we consider the case of hypoelliptic operators $\mathscr{L}$ which are characterized by the relation

$$\lim_{\xi \to \infty} \frac{\mathscr{L}^{(\overline{\alpha})}(\xi)}{\mathscr{L}(\xi)} = 0, \qquad \xi = (\xi_1, \dots, \xi_n),$$

$$\xi_k = \frac{-1}{i} \frac{\partial}{\partial x_k}, \quad i^2 = -1, \ \forall \overline{\alpha}, |\overline{\alpha}| \ge 1. \tag{14.3}$$

Here it is not hard to verify that the adjoint operator $\mathscr{L}^*$ is also hypoelliptic. Hörmander [140] showed that precisely hypoelliptic equations are characterized by the fact that their fundamental solution is infinitely differentiable away from the origin, so that in this case it is possible to assume that the singularity of the fundamental solution $E^*(x)$ is concentrated at the point $\{x = 0\} \in \partial\Omega: \mathscr{L}^* E^* = \delta(x)$.

We shall assume that equation (14.1), generally speaking, is anisotropic and of order $\overline{m} = (m_1, \dots, m_n)$, where m_j is the order of the leading derivative with respect to the variable x_j in (14.1). Setting $\partial/\partial x_k = \xi_k/i$ ($i^2 = -1$) in the expression for $\mathscr{L}^*$, we write the characteristic polynomial of

$\mathscr{L}^*$ in the form

$$\mathscr{L}(\xi) = \sum_{|\overline{\alpha}| \le m} b_{\overline{\alpha}} \xi^{\overline{\alpha}}, \qquad \xi(\xi_1, \dots, \xi_n), \tag{14.4}$$

where the $b_{\overline{\alpha}}$ are complex numbers. We denote by C_n the n-dimensional complex space of points $\xi = \sigma + i\tau$, $\sigma = (\sigma_1, \dots, \sigma_n)$, $\tau = (\tau_1, \dots, \tau_n)$, $\xi_j = \sigma_j + i\tau_j$, $j = 1, \dots, n$; C_n^k is the subset of points in C_n for which all coordinates ξ_j for $j \ne k$ are real:

$$C_n^k = \{\xi \in C_n : \operatorname{Im} \xi_j = 0 \ \forall j \ne k\}.$$

Let $N(\mathscr{L})$ be the manifold of all complex zeros of the polynomial $\mathscr{L}(\xi)$:

$$N(\mathscr{L}) = \{\xi \in C_n : \mathscr{L}(\xi) = 0\},$$
$$N^k(\mathscr{L}) = N(\mathscr{L}) \cap C_n^k. \tag{14.5}$$

It is known [123] that in the case of a hypoelliptic operator $\mathscr{L}^*$ for the roots of its characteristic polynomial $\mathscr{L}(\xi)$ (14.4) on the manifold $N^k(\mathscr{L})$ the inequality

$$|\tau| = |\tau_k| \ge a|\sigma_p|^\gamma - C \tag{14.6}$$

is satisfied for some constants $a > 0$, $C > 0$, and $\gamma > 0$. The largest of the possible numbers γ in (14.6) is called the *exponent of hypoellipticity* and is denoted by γ_p^k. Gorin [123] proved that among all such γ there exists a largest one, and it is necessarily a rational number; for more details on the computation of γ see [123], [126], and [297].

Defining for a given vector $\overline{\alpha} = (\alpha_1, \dots, \alpha_n)$ the quantity

$$\delta_s = \sum_{j=1}^n \frac{\alpha_j + 1}{\gamma_j^s} - m_s, \qquad s = 1, \dots, n, \tag{14.7}$$

we present the estimates established by Grushin [126] of the fundamental solution of a hypoelliptic equation: there exist constants $C_2, C_3 > 0$, not depending on x, such that for $\delta_s \ne 0$

$$|D^{\overline{\alpha}} E^*(x)| \le C_2 |x_s|^{-\delta_s}, \qquad x \in \Omega, \tag{14.8}$$

while for $\delta_s = 0$

$$|D^{\overline{\alpha}} E^*(x)| \le C_3 |\ln|x||, \qquad x \in \Omega. \tag{14.9}$$

14.2. We define the auxiliary operator B by the relation

$$Bv = (-1)^{\widetilde{m}} E^* * \Delta^{\widetilde{m}} v, \qquad D(B) = \overset{\circ}{C}^k(\Omega) = D(\mathscr{L}). \tag{14.10}$$

Here $\tilde{m} = m/2$; the operator $*$ denotes the convolution of two functions,

$$(\varphi * \psi)(x) = \int \varphi(\xi)\psi(x - \xi)\,d\xi;$$

and we assume functions $\psi \in \overset{\circ}{C}(\Omega)$ to be extended by zero outside Ω.

LEMMA 14.1. *The operator $\mathscr{L}$ is B-symmetric and B-positive on $\overset{\circ}{C}^k(\Omega)$, where*

$$k \geq \max\left\{ m;\ \max_{1\leq s\leq n} \sum_{j=1}^{n} \frac{1}{\gamma_j^s} - 1;\ \max_{1\leq s\leq n} \left(\sum_{j=1}^{n} \frac{m_j + 1}{\gamma_j^s} - m_s - 1 \right) \right\}. \quad (14.11)$$

Indeed, by direct integration by parts we establish the identity

$$\int_\Omega \mathscr{L} u E^* * \tilde{\Delta}^m v\,d\Omega = \int_\Omega u(\mathscr{L}^* E^* * \tilde{\Delta}^m v)\,d\Omega \quad (14.12)$$

for all $u, v \in \overset{\circ}{C}^k(\Omega)$, where k satisfies (14.11), since the functions $u, v \in \overset{\circ}{C}^k(\Omega)$ vanish together with their derivatives to the required order on $\partial\Omega$; the singularity of $D^{\bar{\alpha}}E^*(x)$, $|\bar{\alpha}| \leq m - 1$, on $\partial\Omega$ is thus neutralized; and all boundary integrals of the form

$$\int_{\partial\Omega} D^{\bar{\beta}}u(x) \int_\Omega E^{\bar{\alpha}}E^*(x - \xi)\tilde{\Delta}^m v(\xi)\,d\xi\,dx$$

exist and vanish for all $\bar{\alpha}, \bar{\beta} \geq 0$ with $|\bar{\alpha}|, |\bar{\beta}| \leq m - 1$ and $|\bar{\alpha} + \bar{\beta}| \leq m - 1$.

Considering (14.10) and (14.12), we then find that for all $u, v \in \overset{\circ}{C}^k(\Omega)$

$$\begin{aligned}
(\mathscr{L} u, Bv) &\equiv \int_\Omega \mathscr{L} u\{(-1)^{\tilde{m}} E^* * \tilde{\Delta}^m v\}\,d\Omega \\
&= \int_\Omega u(=1)^{\tilde{m}} \mathscr{L}^* E^* * \tilde{\Delta}^m v\,d\Omega \\
&= (-1)^{\tilde{m}} \int_\Omega u\tilde{\Delta}^m v\,d\Omega = \int_\Omega \nabla^{\tilde{m}}u\nabla^{\tilde{m}}v\,d\Omega, \quad (14.13)
\end{aligned}$$

i.e.,

$$(\mathscr{L} u, Bv) = \int_\Omega \nabla^{\tilde{m}}u\nabla^{\tilde{m}}v\,d\Omega \quad \forall u, v \in \overset{\circ}{C}^k(\Omega). \quad (14.14)$$

Completion of $\overset{\circ}{C}^k(\Omega)$ in the norm

$$\|u\|_{\mathscr{L}B} = \|u\|\overset{\circ}{W}_2^{\tilde{m}}(\Omega)\| \quad (14.15)$$

gives the energy Hilbert space $H_{\mathscr{L}B} = \overset{\circ}{\widetilde{W}}{}_2^m(\Omega)$ with inner product

$$[u,v] = \int_\Omega \nabla^{\widetilde{m}} u \nabla^{\widetilde{m}} v \, d\Omega, \tag{14.16}$$

$$[u,v] = (\mathscr{L}u, Bv) \quad \forall u, v \in \overset{\circ}{C}{}^k(\Omega), \tag{14.17}$$

$$\|u\|_{\mathscr{L}B} = [u,u]^{1/2}. \quad \blacksquare \tag{14.18}$$

LEMMA 14.2. *If $s > k + n/2$, where k satisfies (14.11), then for any $g \in \overset{\circ}{W}{}_2^s(\Omega)$*

$$|(g, Bu)| \leq C_4(\Omega, E^*) \|u|\overset{\circ}{\widetilde{W}}{}_2^m(\Omega)\| \, \||g|\overset{\circ}{W}{}_2^s(\Omega)\|. \tag{14.19}$$

Indeed, since the singularity of the fundamental solution $E^*(x)$ is of finite order (see (14.8) and (14.9)) and is concentrated at the point $\{x = 0\} \in \Gamma$, it follows that $|\nabla^{\widetilde{m}} E^*(x)| \in \overset{\circ}{W}{}_2^{-s}(\Omega)$. Therefore, applying the Schwarz inequality twice, we obtain

$$|(g, Bu) \equiv \left|\int_\Omega g(x) \int_\Omega \nabla^{\widetilde{m}} E^*(x - \xi) \nabla^{\widetilde{m}} u(\xi) \, d\xi \, d\Omega\right|$$

$$= \left|\int_\Omega \nabla^{\widetilde{m}} u(\xi) \int_\Omega \nabla^{\widetilde{m}} E^*(x - \xi) g(x) \, dx \, d\xi\right|$$

$$\leq C_4(\Omega, E^*) \|u|W_2^{\widetilde{m}}(\Omega)\| \, \||g|W_2^s(\Omega)\|. \quad \blacksquare$$

14.3. We define a generalized solution of problem (14.1), (14.2) as an element $u_0 \in \overset{\circ}{\widetilde{W}}{}_2^m(\Omega)$ such that

$$[u_0, v] = (g, Bv) \quad \forall v \in C^k(\Omega). \tag{14.20}$$

COROLLARY 14.1. *If there exists a generalized solution u_0, then it is unique in $\overset{\circ}{\widetilde{W}}{}_2^m(\Omega)$ for any $g \in \overset{\circ}{W}{}_2^s(\Omega)$, $s > k + n/2$.*

This assertion follows directly from the identity

$$[u_0 - u_0', v] = 0 \quad \forall v \in C^k(\Omega)$$

for two arbitrary generalized solutions u_0, u_0' of problem (14.1), (14.2). $\blacksquare$

In the general case properties of such a generalized solution are characterized by the following special case of Lemma 11.1.

LEMMA 14.3. *An element $u_0 \in \overset{\circ}{\widetilde{W}}{}_2^m(\Omega)$ is a generalized solution of problem (14.1), (14.2) if and only if there exists a sequence $\{u_n\} \subset \overset{\circ}{C}{}^k(\Omega)$ such* *that the relations*

$$\|u_n - u_0 | \overset{\circ}{W}{}^{\widetilde{m}}_2(\Omega)\| \to 0 \qquad (n \to \infty),$$

$$(\mathscr{L}u_n - g, Bv) \to 0 \quad \forall v \in \overset{\circ}{C}{}^k(\Omega) \qquad (n \to \infty)$$

are satisfied simultaneously.

REMARK 14.1. The generalized solution introduced will clearly be a "bad" generalized solution in the sense of [30] and [64] that if $u_0 \in \overset{\circ}{W}{}^{\widetilde{m}}_2(\Omega) \cap C^k(\overline{\Omega})$, then the function $u_0(x)$ may not satisfy conditions (14.2) for $j \geq m/2$—a "loss of boundary conditions" occurs. However, for the overdetermined problem (14.1), (14.2) this definition of a generalized solution can be used, since if for special functions $g(x) \in G \cap \overset{\circ}{W}{}^s_2(\Omega)$ it is possible to establish that $u_0 \in \overset{\circ}{C}{}^k(\Omega)$, then by (14.17) and (14.20)

$$(\mathscr{L}u_0 - g, Bv) = 0 \quad \forall v \in \overset{\circ}{C}{}^k(\Omega). \tag{14.21}$$

On $\overset{\circ}{C}{}^k(\Omega)$ we now define the functional

$$D[u] = \|u | \overset{\circ}{W}{}^{\widetilde{m}}_2(\Omega)\|^2 - 2(g, Bu). \tag{14.22}$$

Because of (14.19) the linear functional (g, Bu) can be extended by continuity (in u) to a bounded linear functional $l_g(u)$, $D(l_g) = \overset{\circ}{W}{}^{\widetilde{m}}_2(\Omega)$, $l_g(u) = (g, Bu)$, $u \in D(B)$, and for the functional $l_g(u)$ the estimate

$$|l_g(u)| \leq C_4(\Omega, E^*)\|u | \overset{\circ}{W}{}^{\widetilde{m}}_2(\Omega)\| \, \|g | \overset{\circ}{W}{}^s_2(\Omega)\| \tag{14.23}$$

remains valid for all $u \in \overset{\circ}{W}{}^{\widetilde{m}}_2(\Omega)$.

From the scheme of §11.5 we then immediately obtain

ASSERTION 14.1. *For any* $g \in \overset{\circ}{W}{}^s_2(\Omega)$, $s > k + n/2$, *where* k *satisfies* (14.11), *there exists a unique element* $u_0 \in \overset{\circ}{W}{}^{\widetilde{m}}_2(\Omega)$ *which is a generalized solution of problem* (14.1), (14.2), *and it coincides with the element in* $\overset{\circ}{W}{}^{\widetilde{m}}_2(\Omega)$ *minimizing the functional*

$$D_0[u] = \|u | \overset{\circ}{W}{}^{\widetilde{m}}_2(\Omega)\|^2 - 2l_g(u). \tag{14.24}$$

Since any classical solution is a generalized solution, considering uniqueness of generalized and classical solutions, we obtain

COROLLARY 14.2. *For any* $g \in G \cap \overset{\circ}{W}{}_2^s(\Omega)$, $s > k + n/2$, *where k satisfies* (14.11), *there exists a unique element* $u_0 \in \overset{\circ}{C}{}^k(\Omega)$—*a solution of problem* (14.1), (14.2)—*which coincides with the unique element in* $\overset{\circ}{W}{}_2^{\widetilde{m}}(\Omega)$ *minimizing the functional* (14.22).

14.4. The results presented above can be sharpened (in the direction of lowering the values of s and the lower bound of k in (14.11)), in particular, for those classes of hypoelliptic equations for which estimates of the fundamental solution sharper than (14.8) and (14.9) are known. For individual types of equations it is possible to use knowledge of the exact fundamental solution $E^*(x)$ or the order of its singularity.

EXAMPLE 1. $\mathcal{L} = -\Delta$ and $x \in \Omega \subset R_n$, $n \geq 3$. In this case

$$E^*(x) = \text{const} \cdot r^{2-n}, \qquad r = \left(\sum_1^n x_i^2\right)^{1/2}, \qquad (14.25)$$

i.e., the singularity of the fundamental solution $E^*(x)$ at $x = 0$ is of order $|x|^{2-n}$, $n \geq 3$. For all functions $u, v \in \overset{\circ}{C}{}^k(\Omega)$, where

$$k \geq \begin{cases} 2, & n = 3, \\ n - 2, & n > 3, \end{cases} \qquad (14.26)$$

the validity of (14.13) can be established by integration by parts. For $s > k + n/2$ we have $\overset{\circ}{W}{}_2^s(\Omega) \to \overset{\circ}{C}{}^k(\Omega)$, and, moreover,

$$\partial E^* / \partial x_i \in W_2^{-s}(\Omega), \qquad i = 1, \dots, n,$$

i.e., relations (14.14) and (14.21) are established. Therefore, it is possible to formulate

COROLLARY 14.3. *If in problem* (14.1), (14.2) $\mathcal{L} = -\Delta = \mathcal{L}^*$ *and k satisfies* (14.26), *then for all* $g \in G \cap \overset{\circ}{W}{}_2^s(\Omega)$, $s > k + n/2$, *the function* $u_0(x)$—*a classical solution of problem* (14.1), (14.2)—*coincides with the unique element in* $\overset{\circ}{W}{}_2^1$ *minimizing the functional* (14.22).

Since in this case

$$Bu = -\Delta E^* * u = \delta * u = u, \qquad u \in C^2(\Omega),$$

the functional $D[u]$ of (14.22) has the form

$$D[u] = \int_\Omega \{|\nabla u|^2 - 2gu\} \, dx,$$

i.e., it coincides with the familiar Dirichlet functional for the Laplace-Poisson equation $-\Delta u = g$.

EXAMPLE 2. Since for the operator $\mathscr{L}^* = \mathscr{L}$ of the equation

$$\mathscr{L}^* u \equiv -\Delta u + \lambda u = g(x), \qquad D(\mathscr{L}^*) = \overset{\circ}{C}{}^k(\Omega) \qquad (14.27)$$

the singularity of the fundamental solution is also characterized by (14.25), in complete analogy to Example 1 we obtain

COROLLARY 14.4. *If for problem* (14.1), (14.2) *the operator* $\mathscr{L}^*$ *is defined by* (14.27) *while* k *satisfies* (14.26), *then for all* $g(x) \in G \cap \overset{\circ}{W}{}^s_2(\Omega)$, $s > k + n/2$, *there exists a unique element* u_0 *which is a classical solution of this problem, and it coincides with the unique element in* $\overset{\circ}{W}{}^1_2(\Omega)$ *minimizing the functional* $D[u]$ (14.22), *where* $\tilde{m} = 1$.

We observe that a general elliptic PDE (14.1) of second order in Ω can be transformed ([57], vol. II) to the form (14.27).

EXAMPLE 3. If $\mathscr{L}^*$ is the operator of the heat equation, then, as is known [140], $E^*(x) \in C^\infty(R^n \backslash \{0\})$ and $\lim_{|x| \to 0} D^{\bar{\alpha}} E^*(x) = 0$ for all $|\bar{\alpha}| = 0, 1, 2 \ldots$. Then (14.14) and (14.21) are satisfied for any $u, v \in \overset{\circ}{C}{}^k(\Omega)$, $k \geq 2$, and $g(x) \in \overset{\circ}{W}{}^s_2(\Omega)$, $s \geq 2$. We obtain

COROLLARY 14.5. *For any* $g(x) \in G \cap \overset{\circ}{W}{}^s_2(\Omega)$, $s > k+n/2$, $k \geq 2$, $n \geq 2$, *the function* $u_0(x)$—*a classical solution of problem* (14.1), (14.2) *with the reverse heat operator*—*minimizes in* $\overset{\circ}{W}{}^1_2(\Omega)$ *the functional* $D[u]$ *of* (14.22), *where* $\tilde{m} = 1$.

§15. Symmetrization of some nonclassical PDE

In this section we continue examples of the constructive approach to functionals of the form (6.18) with variational properties A–C of the Introduction for some nonclassical PDE with nonsymmetric operators. Although the specific form of the equations will be used, we emphasize that there is something general in the constructions presented here—it will allow us in the next section to expound still another general approach to the constructive treatment of auxiliary symmetrizing operators B.

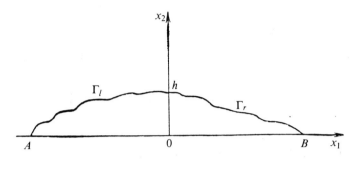

FIGURE 2

15.1. We consider a boundary value problem for an equation of composite type [127]

$$\mathscr{L}_c u \equiv \frac{\partial}{\partial x_1}\left(\sum_{i,j=1}^{2} \frac{\partial}{\partial x_i}\left(a_{ij}(x)\frac{\partial u}{\partial x_j}\right) + b(x)u\right) = g(x), \qquad x \in \Omega, \quad (15.1)$$

$$u(x) = 0, \qquad x \in \partial\Omega, \qquad\qquad\qquad (15.2)$$

$$\int_{\gamma_l(x_2)}^{\gamma_r(x_2)} u(\xi, x_2)d\xi = 0, \qquad 0 \le x_2 \le h, \qquad\qquad (15.3)$$

where

a) Ω is a domain (Figure 2) with piecewise smooth boundary $\partial\Omega$ such that any line parallel to the axis Ox_1 or Ox_2 intersects Γ_l and Γ_r at only one point each, $\Omega = \{(x_1, x_2): \gamma_l(x_2) < x_1 < \gamma_r(x_2), \ 0 < x_2 < h\}$, $\Gamma_l = \{(x_1, x_2): x_1 = \gamma_l(x_2), \ 0 \le x_2 \le h\}$, and $\Gamma_r = \{(x_1, x_2): x_1 = \gamma_r(x_2), \ 0 \le x_2 \le h\}$;

b) $a_{ij} \in C^2(\overline{\Omega})$, $i, j = 1, \ldots, n$; and $b(x) \in C^1(\overline{\Omega})$ (these conditions will be relaxed below); and

c) $a_{ij} = a_{ji}$, $i, j = 1, 2$; $b(x) \in C(\overline{\Omega})$, $b(x) \le b_0 = \text{const} < 0$, $x \in \Omega$; $\exists \mu = \text{const} > 0$; $\sum_{i,j=1}^{2} a_{ij}\xi_i\xi_j \ge \mu\ (\xi_1^2 + \xi_2^2)\ \forall(\xi_1, \xi_2) \in \overline{\Omega}$.

On the set $\widetilde{C}_0^3(\Omega, \Gamma) = D(\mathscr{L}_c)$ of functions $u(x_1, x_2) \in C^3(\overline{\Omega})$ satisfying conditions (15.2) and (15.3) the following identity can be established by integration by parts:

$$(\mathscr{L}_c u, Bv) \equiv \int_\Omega \frac{\partial}{\partial x_1} \left[\sum_{i,j=1}^{2} \frac{\partial}{\partial x_i} \left(a_{ij} \frac{\partial u}{\partial x_j} \right) + bu \right] \int_{\gamma_l(x_2)}^{x_1} v(\xi, x_2) \, d\xi \, d\Omega$$

$$= \int_\Omega \left(\sum_{i,j=1}^{2} a_{ij} \frac{\partial u}{\partial x_j} \frac{\partial v}{\partial x_j} - buv \right) dx_1 \, dx_2; \tag{15.4}$$

thus, the operator $\mathscr{L}_c$ of problem (15.1)–(15.3) is B-symmetric and B-positive for

$$Bv = \int_{\gamma_l(x_2)}^{x_1} v(\xi, x_2) \, d\xi, \qquad D(B) = D(\mathscr{L}_c).(^{30})$$

The "energy" space $\overset{\circ}{\widetilde{W}}{}^1_2(\Omega, \Gamma)$ is defined as the closure of the original set $\tilde{C}_0^3(\Omega, \Gamma)$ in the norm

$$\|u|\overset{\circ}{\widetilde{W}}{}^1_2(\Omega, \Gamma)\| = [u, u]^{1/2}, \tag{15.5}$$

$$[u, v] = \int_\Omega \left(\sum_{i,j=1}^{2} a_{ij} \frac{\partial u}{\partial x_i} \frac{\partial v}{\partial x_j} - buv \right) d\Omega, \tag{15.6}$$

while the space of functions $g(x)$ (the right sides of the equations) is the set of functions for which the following norm exists and is finite:

$$\|g|\tilde{S}_{\gamma_r,-}^{-1,0} W_2(\Omega)\| = \left(\int_\Omega \left(\int_{\gamma_r}^{x_1} g(\xi, x_2) \, d\xi \right)^2 d\Omega \right)^{1/2}. \tag{15.7}$$

It is obvious that $\tilde{S}_{\gamma_r,-}^{-1,0} W_2(\Omega) \supset L_2(\Omega)$, and conditions (15.2) and (15.3) are stable (in the mean-square sense) under passage to the limit in the norm of $\overset{\circ}{\widetilde{W}}{}^1_2(\Omega)$.

It is now possible to present a well-posed formulation of the initial boundary value problem.

DEFINITION 15.1. An element $u_0 \in \overset{\circ}{\widetilde{W}}{}^1_2(\Omega, \Gamma)$ is a *generalized solution*(^{31}) of problem (15.1)–(15.3) if

$$[u_0, v] = (g, Bv) \quad \forall v \in D(B). \tag{15.8}$$

(^{30}) It is not hard to verify that in this case the operator $\mathscr{L}_c$ will also be B-positive-definite in the sense of Petryshyn's Definition 1.3.

(^{31})In the definition of a generalized solution, and below, it may be assumed that the functions $b(x)$ and $a_{ij}(x), i, j = 1, 2$, are only bounded measurable functions in Ω which satisfy condition c) (understanding the integrals in the Lebesgue sense).

If it turns out that $u_0 \in D(\mathcal{L}_c)$, then from (15.4), (15.6), and (15.8) we get $(\mathcal{L}_c u_0 - g, Bv) = 0$ for all $v \in D(B)$, and it is easy to verify that the set $R(B)$ is dense in $L_2(\Omega)$.

In the general case properties of a generalized solution are characterized by a lemma analogous to Lemma 11.1.

According to the general scheme of §11.5, we obtain the functional

$$D_c[u] = \|u\|\overset{\circ}{\widetilde{W}}{}^1_2(\Omega, \Gamma)\|^2 - 2(g, Bu)$$

$$= \int_\Omega \left(\sum_{i,j=1}^2 a_{ij} \frac{\partial u}{\partial x_i} \frac{\partial u}{\partial x_j} - bu^2 - 2g(x) \int_{\gamma_l(x_2)}^{x_1} u(\xi, x_2)\, d\xi \right) d\Omega. \tag{15.9}$$

By an easily established estimate of the linear functional (g, Bu) in (15.9), for all $g \in \widetilde{S}^{-1,0}_{\gamma_r,-} W_2(\Omega)$ and for all $u \in \overset{\circ}{\widetilde{W}}{}^1_2(\Omega)$ with

$$|(g, Bu)| \le C_1(\Omega)\|g\|\widetilde{S}^{-1,0}_{\gamma_r,-} W_2(\Omega)\| \cdot \|u\|\overset{\circ}{\widetilde{W}}{}^1_2(\Omega, \Gamma)\|$$

we have (see Assertion 11.1)

ASSERTION 15.1. *For all $g \in \widetilde{S}^{-1,0}_{\gamma_r,-} W_2(\Omega)$ there exists a unique element $u_0 \in \overset{\circ}{\widetilde{W}}{}^1_2(\Omega, \Gamma)$ which in $\overset{\circ}{\widetilde{W}}{}^1_2(\Omega, \Gamma)$ minimizes the functional $D_c[u]$ of (15.9). The element u_0 coincides with a generalized solution in the sense of Definition 15.1 of the original boundary value problem (15.1)–(15.3) and depends continuously on the function g :*

$$\|u_0\|\overset{\circ}{\widetilde{W}}{}^1_2(\Omega, \Gamma) \le C_1\|g\|\widetilde{S}^{-1,0}_{\gamma_r,-} W_2(\Omega)\|.$$

15.2. For a general equation of composite type

$$A\mathcal{L}u(x) = g(x), \tag{15.10}$$

where A and $\mathcal{L}$ are linear integro-differential operators, the idea of symmetrization presented in the preceding subsection can be generalized as follows.

If there exists a closable linear operator B such that for all $u, v \in D(A, B)$

$$(Au, Bv) = (Bu, Av), \qquad (Au, Bv) \ge \|u\|^2_H \tag{15.11}$$

and there exists an operator C such that for all $u, v \in D(\mathscr{L}, C)$

$$[\mathscr{L}u, Cv] = [Cu, \mathscr{L}v], \qquad [\mathscr{L}u, Cu] \geq [u, u], \qquad (15.12)$$

where $[u, v] = (Au, Bv)$, then the operator $A\mathscr{L}$ will be BC-symmetric and BC-positive, since

$$(A\mathscr{L}u, BCv) = [\mathscr{L}u, Cv] = [Cu, \mathscr{L}v] = (BCu, A\mathscr{L}v).$$

In analogy to Theorem 1.1 it is possible to prove the existence of operators B and C with properties (15.11) and (15.12) (under analogous assumptions regarding A^{-1} and $\mathscr{L}^{-1}$); hence, it is possible to construct for equation (15.10) a functional $D[u]$ of (11.63) with variational properties A–C of the Introduction and to obtain assertions analogous to Assertion 11.1 according to the scheme of §11.5.

The construction of symmetrizing operators for equation (15.10) simplifies considerably if one of the operators A or $\mathscr{L}$ is symmetric and positive or negative, while for the other a symmetrizing operator can be constructively treated. For example, in the preceding subsection we consider the operator $A = \partial/\partial x_1$ for which it is easy to select a symmetrizing operator B,

$$Bv = \int_{\gamma_l(x_2)}^{x_1} v(\xi, x_2)\, d\xi,$$

and the operator $\mathscr{L}$,

$$\mathscr{L}u = \sum_{i,j=1}^{2} \frac{\partial}{\partial x_i}\left(a_{ij}\frac{\partial u}{\partial x_j}\right) + bu$$

is symmetric and negative-definite.

The idea presented here goes back to one of the first papers on the theoretical justification of the variational method of solving nonselfadjoint equations of (Martynyuk [209]). In it the linear equation

$$Au = f \qquad (15.13)$$

with an operator

$$A = A_0 B \qquad (15.14)$$

was considered, where A_0 is a selfadjoint operator and B is closed. Because of the selfadjointness of A_0 the property of B-symmetry is obvious for A:

$$(Au, Bv) = (A_0Bu, Bv) = (Bu, A_0Bv) = (Bu, Av).$$

The definition of generalized positivity of such an operator is given in Definition 1.4. Martynyuk [209] assumed that the representation (15.14) was known a priori; in practice this is satisfied only in rare cases (see also Theorem 1.2). In the next section we shall present a construction of a symmetrizing operator B which in a certain sense is close to the representation (15.14).

15.3. We present still another example of symmetrization of an operator of a boundary value problem for a linear PDE; this example clarifies obtaining a possible representation (15.14) and, moreover, is of independent interest, since for equation (15.15) considered below the inverse problem of the calculus of variations has no solution (see [1], [18], and [54]) in the class of Euler functionals.

We consider a nonlocal boundary value problem (with the condition of periodicity in the n-dimensional time $t = (t_1, \ldots, t_n)$) for an ultraparabolic equation

$$\mathscr{L}_p u \equiv \sum_{i=1}^{n} a_i(x,t) \frac{\partial u}{\partial t_i} - \frac{\partial}{\partial x} \left(k(x,t) \frac{\partial u}{\partial x} \right) = g(x,t), \qquad (x,t) \in \Omega,$$
$$(15.15)$$

$$u(x,t)|_{t_i=0} = u(x,t)|_{t_i=T},$$
$$a \leq x \leq b, \ 0 \leq t_s \leq T_s, \ \forall s \neq i, \ i = 1, \ldots, n; \quad (15.16)$$

$$\frac{\partial u}{\partial n} = \psi(t), \qquad (x,t) \in \Gamma_l = \{x, t : x = a, \ 0 \leq t_i \leq T_i, \ i = 1, \ldots, n\},$$
$$(15.17)$$

$$u(x,0) = 0, \qquad (x,t) \in \Gamma_r = \{x, t : x = b, \ 0 \leq t_i \leq T_i, \ i = 1, \ldots, n\}.$$
$$(15.18)$$

Here the bounded domain $\Omega = \{x, t : a < x < b, \ 0 < t_i < T_i, \ i = 1, \ldots, n\}$, and also

$$k(x,t) \in C^1(\Omega) \cap C(\overline{\Omega}), \quad 0 < \kappa \leq k(x,t), \ (x,t) \leq \Omega, \ \kappa \equiv \text{const}; \quad (15.19)$$

$$a_i(x,t) \in C(\overline{\Omega}); \qquad a_i(x,t)|_{t_i=0} = a_i(x,t)|_{t_i=T_i}, \qquad a \leq x \leq b,$$
$$0 \leq t_s \leq T_s, \quad \forall s \neq i; \ s, i = 1, \ldots, n;$$
$$a_i(x,t) \neq 0, \qquad (x,t) \in \Omega, \ i = 1, \ldots, n. \quad (15.20)$$

For functions $u, v \in \overset{\circ}{C}{}^{2,1}(\overline{\Omega}; \Gamma_r, T) = \{u(x,t) \in C^{2,1}_{x,t}(\overline{\Omega})$ satisfying conditions (15.16) and (15.18)$\}$ the following relation can be established by direct integration by parts:

$$
(\mathscr{L}_p u, B_p v) \equiv \int_\Omega \left[\sum_{i=1}^n a_i \frac{\partial u}{\partial t_i} - \frac{\partial}{\partial x} \left(k(x,t) \frac{\partial u}{\partial x} \right) \right]
$$

$$
\times \left[v(x,t) - \sum_{i=1}^n \int_b^x \frac{d\theta}{k(\theta,t)} \int_a^0 a_i(\xi,t) \frac{\partial v(\xi,t)}{\partial t_i} d\xi \right] d\Omega
$$

$$
+ \int_0^{T_1} \cdots \int_0^{T_1} \frac{\partial u(a,t)}{\partial n} k(a,t)
$$

$$
\times \left[v(a,t) - \sum_{i=1}^n \int_b^a \frac{d\theta}{k(\theta,t)} \int_a^b a_i(\xi,t) \frac{\partial v}{\partial t_i} d\xi \right] dt_1 \ldots dt_n
$$

$$
= \int_\Omega \left\{ k u_x v_x + \sum_{i=1}^n a_i \frac{\partial(uv)}{\partial t_i} + \frac{1}{k} \left(\sum_{i=1}^n \int_a^x a_i(\xi,t) \frac{\partial u(\xi,t)}{\partial t_i} d\xi \right) \right.
$$

$$
\left. \times \left(\sum_{i=1}^n \int_a^x a_i(\xi,t) \frac{\partial v}{\partial t_i} (\xi,t) d\xi \right) \right\} d\Omega.
$$

$$(15.21)$$

Thus, the operator $\mathscr{L}_p$ of boundary value problem (15.15)–(15.18) is B_p-symmetric on the set $\overset{\circ}{C}{}^{2,1}(\overline{\Omega}; \Gamma_r, T)$, and it will be B_p-positive if, for example, there exist the derivatives

$$
\frac{\partial a_i}{\partial t} \in C(\overline{\Omega}), \quad i = 1, \ldots, n \quad \text{and} \quad \sum_1^n (a_i)'_{t_i} \leq 0, \quad (x,t) \in \Omega.
$$

Therefore, according to the scheme of §11.5, for the problem in question it is possible to construct a functional of the type (11.63) with properties A–C of the Introduction and establish by the variational method results altogether analogous to those obtained in §12 for a special case ($n = 1$) of (15.15). In the present case, however, we note that the left side of (15.21) can be transformed for $v = u \in \overset{\circ}{C}{}^{2,1}(\overline{\Omega}; \Gamma_r, T)$, $\partial u/\partial n = 0$, $(x,t) \in \Gamma_l$, to the form

$$(\mathscr{L}_p, B_p u) = \int_\Omega \left[\sum_1^n a_i \frac{\partial u}{\partial t_i} - (ku_x)_x \right]$$

$$\times \left[u - \sum_1^n \int_b^x \frac{d\theta}{k(\theta, t)} \int_a^\theta a_i \frac{\partial u}{\partial t_i} d\xi \right] d\Omega$$

$$= \int_\Omega \left[\sum_1^n a_i \frac{\partial u}{\partial t_i} - (ku_x)_x \right] \int_a^x \frac{d\theta}{k(\theta, t)}$$

$$\times \int_a^\theta \left[(ku_\xi)_\xi - \sum_1^n a_i \frac{\partial u}{\partial t_i} \right] (\xi, t) \, d\xi \, d\Omega$$

$$\left(\equiv \int_\Omega \mathscr{L}_p u J \mathscr{L}_p u \, d\Omega, \quad \text{where } Ju \equiv -\int_b^x \frac{d\theta}{k(\theta, t)} \int_a^0 u(\xi, t) \, d\xi \right)$$

$$= \int_\Omega \frac{1}{k(x, t)} \left\{ \int_a^x \left[\sum_1^n a_i(\xi, t) \frac{\partial u(\xi, t)}{\partial t_i} \right. \right.$$

$$\left. \left. - \frac{\partial}{\partial \xi} \left(k(\xi, t) \frac{\partial u}{\partial \xi} \right) \right] d\xi \right\}^2 d\Omega$$

$$\sim \|\mathscr{L}_u u | \tilde{S}_{a,-}^{-1,0} W_2(\Omega) \|^2 - I_\Omega. \qquad (15.22)$$

The representation $I_\Omega = (\mathscr{L}_p u, I \mathscr{L}_p u)$, where I is a selfadjoint positive integral operator is evidence of the connection of the symmetrization carried out in (15.21) with Martynyuk's representation (15.14), while the equivalence $I_\Omega \sim \|\mathscr{L}_p u | \tilde{S}_{a,-}^{-1,0} W_2(\Omega) \|^2$ indicates the similarity in the present case of our variational approach to the method of least squares in some negative space $\overset{\circ}{W}_2^{-1,0}(\Omega, \Gamma)$. From (15.22) it is obvious that an a priori estimate (see §16) of the type $\|\mathscr{L} u | \overset{\circ}{W}_2^{-1,0} \| \geq C \| u \|$ must play an important role in this case.

The example presented here, however, can be considered in the plan of solving the inverse problem of the calculus of variations only as a heuristic exposition of a variational method which will be rigorously justified in the next section.

§16. A variational principle constructed on the basis of a priori estimates in negative spaces

V. P. Didenko [75] proposed a method of constructing symmetrizing operators B and functionals $D[u] = \langle Au, Bu \rangle - 2\langle f, Bu \rangle$ with properties

A–C of the Introduction for a broad class of systems of PDE regardless of their type. This approach is based on obtaining a priori estimates of the form $\|Au'\|W_{bdr^+}^-\| \geq \|u'|L_2\|$, and we shall call it the Didenko variational method.

We outline the contents of the section. In §16.1, for completeness of the exposition, following [30] and [76], we present (without proofs) some familiar facts from the theory of boundary value problems considered in a triple of rigged Hilbert spaces $W_{bdr}^- \supset L_2 \supset W_{bdr}^+$; Didenko's variational method is presented in a rather general form in §16.2. The connection of the Didenko functional with the functional of the method of least squares (MLS) is then established. A constructive modification of Didenko's method in application to a hyperbolic equation with the help of the function spaces $\widetilde{S}^{-l}\overset{\circ}{W}_2(\Omega)$ introduced in §9 is presented in §§16.4–16.6.

16.1. Suppose there is given a system of linear PDE

$$Au \equiv a_{ij}u_{x_ix_j} + a_iu_{x_i} + a_0u = f(x), \qquad x \in \Omega \subset R_n, \qquad (16.1)$$

$$A_1u = 0, \qquad x \in \partial\Omega, \qquad (16.2)$$

where $u(x)$ is an unknown and $f(x)$ a given m-dimensional vector-valued function, $m \geq 1$, $a_{ij}(x)$, $a_i(x)$ $(i, j = 1, \ldots, n)$, and $a_0(x)$ are given $m \times m$ matrices whose components we at first suppose belong respectively to $C^2(\overline{\Omega})$, $C^1(\overline{\Omega})$, and $C(\overline{\Omega})$ respectively. In (16.1) and below summation over repeated indices is understood; A_1 in (16.2) is a local linear boundary differential operator of order no more than one.([32])

Assuming the operator A to be defined first on the set of functions in $C^2(\overline{\Omega})$ satisfying conditions (16.2), $D(A) \equiv C_{bdr}^2(\overline{\Omega}) \cup \overset{\circ}{C}^2(\Omega)$, we introduce the space W_{bdr}^+ as the completion of $D(A)$ in the norm $\| \cdot |W_2^1(\Omega)\|$. It is obvious that $\overset{\circ}{W}_2^1(\Omega) \subseteq W_{bdr}^+ \subseteq W_2^1(\Omega)$.

The space $W_{bdr^+}^+$ is defined as the closure in the norm of $W_2^1(\Omega)$ of the set of functions $C_{bdr^+}^2(\Omega) = D(A^+)$, i.e., of functions in $C^2(\overline{\Omega})$ satisfying for any (fixed) $u \in C_{bdr}^2(\overline{\Omega})$ the equality

$$(Au, v) = (u, A^+v), \qquad (16.3)$$

where $A^+v = (a_{ij}v)_{x_ix_j} - (av)_{x_i} + a_0v$.

([32])In theoretical expositions of the method in the general case Didenko in [73] and [76] required of the boundary conditions that "membership of $u \in W_{bdr}^+$ meant that the boundary conditions were satisfied in mean at least", which actually eliminated operations of differentiation in (16.2). In concrete applications of the method (see [75] and [76]), boundary conditions of second kind were admitted.

It is now possible to present a well-posed formulation of the original boundary value problem, assuming henceforth that $\partial a_{ij}/\partial x_i$, $a_i(x)$ $(i, j = 1, \ldots, n)$, and $a_0(x)$ are bounded measurable functions in Ω.

DEFINITION 16.1. An element $u \in W_{\text{bdr}}^+$ is called a *strong solution* of problem (16.1), (16.2) if for all $v \in C_{\text{bdr}^+}^2(\Omega)$

$$\int_\Omega \{-(a_{ij}v)_{x_i}u_{x_j} + a_i v u_{x_i} + a_0 uv\}\, dx = \int_\Omega fv\, dx. \qquad (16.4)$$

It is convenient to investigate generalized solvability of the original problem

$$Au = f, \qquad u \in W_{\text{bdr}}^+ \qquad (16.5)$$

with the help of corresponding properties of the adjoint problem

$$A^+v = g, \qquad v \in W_{\text{bdr}^+}^+ \qquad (16.6)$$

and use of the negative spaces W_{bdr}^- and $W_{\text{bdr}^+}^-$ dual to W_{bdr}^+ and $W_{\text{bdr}^+}^+$ respectively. The space W_{bdr}^-, for example, can be defined [30] as the closure of $C_0^\infty(\Omega)$ in the norm

$$\|u|W_{\text{bdr}}^-\| = \sup_{\substack{v \neq 0 \\ v \in W_{\text{bdr}}^+}} \frac{|(u, v)|}{\|v|W_{\text{bdr}}^+\|}. \qquad (16.7)$$

The following result is also known (see [30] and [76]).

LEMMA 16.1. *An element* $u_0 \in W_{\text{bdr}}^+$ *is a strong solution of problem* (16.1), (16.2) *for* $f \in W_{\text{bdr}^+}^-$ *if and only if there exists a sequence* $\{u_m\}$ *of smooth functions* $u_m \in C_{\text{bdr}}^2(\overline{\Omega})$, $m = 1, 2, \ldots$, *such that as* $m \to \infty$

$$\|u_m - u_0|W_{\text{bdr}}^+\| \to 0, \qquad \|Au_m - f|W_{\text{bdr}^+}^-\| \to 0.$$

It is obvious that if a strong solution turns out to be in $C_{\text{bdr}}^2(\overline{\Omega})$, then it is a classical solution of the original problem.

From the general theory of spaces with negative norm [30] it is known that for $u \in W_{\text{bdr}}^+$ and $v \in W_{\text{bdr}}^-$ (or $u \in W_{\text{bdr}^+}^+$, and $v \in W_{\text{bdr}^+}^-$) a bilinear form $\langle \cdot, \cdot \rangle$ is defined which is an extension by continuity of the form $(\cdot, \cdot)_{L_2}$, and for all $u \in W_{\text{bdr}}^+$ and $v \in W_{\text{bdr}}^-$ a generalized Schwarz inequality holds:

$$|\langle u, v \rangle| \leq \|u|W_{\text{bdr}}^+\| \, \|v|W_{\text{bdr}}^-\|. \qquad (16.8)$$

Moreover, for a triple of rigged Hilbert spaces, for example, $W_{\text{bdr}^+}^- \supseteq L_2(\Omega) \supseteq W_{\text{bdr}^+}^+$, there exists an isometric operator J_{bdr^+} taking all of $W_{\text{bdr}^+}^-$ onto $W_{\text{bdr}^+}^+$ and such that for all $u, v \in W_{\text{bdr}^+}^-$

$$(u, v)_{W_{\text{bdr}^+}^-} = \langle J_{\text{bdr}^+}u, v \rangle = \langle u, J_{\text{bdr}^+}v \rangle = (J_{\text{bdr}^+}u, J_{\text{bdr}^+}v)_{W_{\text{bdr}^+}^+}. \qquad (16.9)$$

The operator $A: C_{\text{bdr}}^2(\overline{\Omega}) \to W_{\text{bdr}^+}^-$ can be extended by continuity to all of W_{bdr}^+, since the following estimate for any u in the set $D(A)$ dense in W_{bdr}^+ can be obtained directly from (16.7) and (16.1):

$$\|Au|W_{\text{bdr}^+}^-\| = \sup_{v \neq 0} \frac{|(Au, v)|}{\|v|W_{\text{bdr}^+}^+\|}$$

$$= \sup_{v \neq 0} \frac{\left|\int_\Omega [-(a_{ij}v)_{x_i} u_{x_j} + v a_i u_{x_i} + v a_0 u]\, dx\right|}{\|v|W_{\text{bdr}^+}^+\|}$$

$$\leq C_1 \partial \|u|W_{\text{bdr}}^+\|, \tag{16.10}$$

Similarly,

$$\|A^+ v|W_{\text{bdr}}^-\| \leq C_2 \|v|W_{\text{bdr}^+}^+\| \quad \forall v \in W_{\text{bdr}^+}^+. \tag{16.10'}$$

We use the previous notation A and A^+ for the operators A and A^+ extended in this manner (by continuity) to W_{bdr}^+ and $W_{\text{bdr}^+}^+$. Therefore $D(A) = W_{\text{bdr}}^+$, $R(A) \subseteq W_{\text{bdr}^+}^-$; $D(A^+) = W_{\text{bdr}^+}^+$, $R(A^+) \subseteq W_{\text{bdr}}^-$, and by (16.3) we have

ASSERTION 16.1.

$$\langle Au, v \rangle = \langle u, A^+ v \rangle \quad \forall u \in W_{\text{bdr}}^+, \ v \in W_{\text{bdr}^+}^+.$$

Together with (16.10), an important role in the investigation of solvability of problems (16.5) and (16.6) is played by the a priori estimates of a solution

$$\|u|L_2(\Omega)\| \leq C_3 \|Au|W_{\text{bdr}^+}^-\| \quad \forall u \in W_{\text{bdr}}^+, \tag{16.11}$$

$$\|v|L_2(\Omega)\| \leq C_4 \|A^+ v|W_{\text{bdr}}^-\| \quad \forall v \in W_{\text{bdr}^+}^+. \tag{16.12}$$

LEMMA 16.2 [30], [76]. *There exists a left inverse operator A^{-1} of problem (16.5) for an arbitrary $f \in L_2(\Omega)$ if and only if inequality (16.12) holds. Similarly, there exists a strong solution in $W_{\text{bdr}^+}^+$ of problem (16.6) for an arbitrary $g \in L_2(\Omega)$ if and only if inequality (16.11) holds.*

16.2. With the help of the operator J_{bdr^+} in (16.9) we define an auxiliary operator B by the relation

$$B = J_{\text{bdr}^+} A, \qquad D(B) = D(A) = W_{\text{bdr}}^+. \tag{16.13}$$

Then from (16.9) for all $u, v \in W_{\text{bdr}}^+$

$$\langle Au, Bv \rangle = \langle Au, J_{\text{bdr}^+} Av \rangle = \langle J_{\text{bdr}^+} Au, Av \rangle = \langle Bu, Av \rangle,$$

i.e., the operator A is B-symmetric relative to the bilinear form $\langle \cdot, \cdot \rangle$. Since

$$\langle Au, Bu \rangle = \langle J_{\text{bdr}^+} Au, Au \rangle = \|Au|W_{\text{bdr}^+}^-\|^2 \geq 0 \quad \forall u \in W_{\text{bdr}}^+,$$

$$\langle Au, Bu \rangle = (J_{\text{bdr}^+} Au, J_{\text{bdr}^+} Au)_{W_{\text{bdr}^+}^+} = \|Bu|W_{\text{bdr}^+}^+\|^2,$$

the operator A will be B-positive-definite (Definition 1.3) if, for example, (16.11) holds.

As usual, we define the "energy" space H_{AB} as the completion of $D(A) = W_{bdr}^+$ in the norm

$$\|u\|_{AB} = [u, u]_{AB}^{1/2}, \tag{16.14}$$

$$[u, v]_{AB} = \langle Au, Bv \rangle. \tag{16.15}$$

Hence, in particular,

$$\|u\|_{AB} = \|Au|W_{bdr^+}^-\| \quad \forall u \in W_{bdr}^+. \tag{16.16}$$

If (16.11) is satisfied, then obviously

$$H_{AB} \to L_2(\Omega), \tag{16.17}$$

and by (16.10), (16.16), and the definition of H_{AB}

$$W_{bdr}^+ \to H_{AB}. \tag{16.18}$$

Assuming henceforth that (16.11) is satisfied, we introduce

DEFINITION 16.2 An element $u_0 \in H_{AB}$ is called a *weak solution* of problem (16.1), (16.2) if

$$[u_0, v]_{AB} = \langle f, Bv \rangle \quad \forall v \in W_{bdr}^+. \tag{16.19}$$

COROLLARY 16.1. *Any strong solution of problem* (16.1), (16.2) *is a weak solution of this problem.*

Indeed, since the operator A has been extended to all of W_{bdr}^+ by virtue of (16.10), according to Lemma 16.1, a strong solution is characterized by the equality $Au_0 - f = 0$. Since here $u_0 \in W_{bdr}^+ \subseteq H_{AB}$ and $Au_0 \in W_{bdr^+}^-$, considering (16.15), we have

$$0 = \langle Au_0 - f, Bv \rangle = [u_0, v]_{AB} - (f, Bv) \quad \forall v \in W_{bdr}^+,$$

i.e., u_0 is also a weak solution of problem (16.1), (16.2). ∎

LEMMA 16.3. *Suppose relation* (16.12) *or the following condition is satisfied:*

$$\langle f, v \rangle = 0 \quad \forall v \in N_{bdr^+} = \{v \in W_{bdr^+}^+ : A^+ v = 0\}. \tag{16.20}$$

Then if a weak solution belongs to W_{bdr}^+, it is a strong solution.

PROOF. Suppose u_0 is a weak solution of problem (16.1), (16.2) and $u_0 \in W_{bdr}^+$. Then, since $Au_0 \in W_{bdr^+}^-$, by (16.19) and (16.15)

$$\langle Au_0 - f, Bv \rangle = 0 \quad \forall v \in W_{bdr}^+, \tag{16.21}$$

or, considering (16.13),

$$\langle A^+ J_{\mathrm{bdr}^+}(Au_0 - f), v \rangle = 0.$$

Since v runs through all of W_{bdr}^+, it follows that

$$\|A^+ J_{\mathrm{bdr}^+}(Au_0 - f)\|_{W_{\mathrm{bdr}^+}^-} = 0.$$

Hence

$$J_{\mathrm{bdr}^+}(Au_0 - f) = w, \qquad (16.22)$$

where w is a solution of the homogeneous problem $A^+ w = 0$, $w \in W_{\mathrm{bdr}^+}^+$. Then either by (16.12) it is obvious that $w = 0$, wherefore $J_{\mathrm{bdr}^+}^{-1} w = 0$ and from (16.22) $Au_0 - f = 0$; or if (16.20) is satisfied we have from (16.22)

$$Au_0 - f = J_{\mathrm{bdr}^+}^{-1} w.$$

Taking the inner product of the last equality with $w \in N_{\mathrm{bdr}^+}$, we obtain

$$\langle w, Au_0 \rangle - \langle w, f \rangle = \langle J_{\mathrm{bdr}^+}^{-1} w, w \rangle, \qquad (16.23)$$

and, since $A^+ w = 0$, by Assertion 16.1 $\langle w, Au_0 \rangle = \langle A^+ w, u_0 \rangle = 0$, and with consideration of (16.20) we find from (16.23) that $\langle J_{\mathrm{bdr}^+}^{-1} w, w \rangle = \| w | W_{\mathrm{bdr}^+}^+ \|^2 = 0$, i.e., from (16.22) $Au_0 - f = 0$. ∎

Since

$$\| Bu | W_{\mathrm{bdr}^+}^+ \|^2 = \langle Au, Bu \rangle = \| u \|_{AB}^2 \quad \forall u \in D(B),$$

the operator B can be extended by continuity to all of H_{AB} to some closed operator $B_0 \supseteq B$ (see Theorem 4.3), $D(B_0) = H_{AB}$. This makes it possible to consider the variational problem of minimization in H_{AB} of the functional

$$\Phi(u) = \|u\|_{AB}^2 - 2\langle f, B_0 u \rangle, \qquad u \in H_{AB}, \qquad (16.24)$$

so that for $u \in W_{\mathrm{bdr}}^+$

$$\Phi(u) = \langle Au, Bu \rangle - 2\langle f, Bu \rangle. \qquad (16.24')$$

To prove the existence of an element minimizing the functional (16.24) in H_{AB} for $f \in W_{\mathrm{bdr}^+}^-$ it obviously suffices to prove the inequality

$$|\langle f, B_0 u \rangle| \le C_5 \| f | W_{\mathrm{bdr}^+}^- \| \|u\|_{AB} \quad \forall u \in H_{AB}. \qquad (16.25)$$

Now (16.25) is easily established for an arbitrary (but fixed) $f \in W_{\mathrm{bdr}^+}^-$ and for all u in the set W_{bdr}^+ (dense in H_{AB}) with consideration of (16.9) and (16.16):

$$|\langle f, B_0 u \rangle| = |\langle f, J_{\mathrm{bdr}^+} Au \rangle| = |\langle Au, J_{\mathrm{bdr}^+} f \rangle|$$
$$\le \| Au | W_{\mathrm{bdr}^+}^- \| \, \| J_{\mathrm{bdr}^+} f | W_{\mathrm{bdr}^+}^+ \| = \|u\|_{AB} \| f | W_{\mathrm{bdr}^+}^- \|.$$

From the general scheme of §11.5 we then obtain a special case of Assertion 11.1.

THEOREM 16.1. *Suppose inequality* (16.11) *is satisfied. Then for any* $f \in W_{\mathrm{bdr}^+}^-$ *there exists a unique element* $u_0 \in H_{AB}$ *minimizing the functional* (16.24) *in* H_{AB}. *Moreover, this element* u_0 *coincides with the solution in the weak sense of Definition* 16.2 *of the original boundary value problem, and*

$$\|u_0\|_{AB} \le \|f|W_{\mathrm{bdr}^+}^-\|. \tag{16.26}$$

From this, Corollary 16.1, and Lemmas 16.2 and 16.3 we obtain

COROLLARY 16.2. *If inequalities* (16.11) *and* (16.12) *hold, then for any* $f \in L_2(\Omega)$ *there exists a unique strong solution* $u_0 \in W_{\mathrm{bdr}}^+$ *of problem* (16.5) *which coincides with the unique element minimizing the functional* $\Phi(u)$ *of* (16.24') *in* W_{bdr}^+.

In a number of cases (see [86], [268], and [269]) it is possible to establish estimates stronger than (16.11) and (16.12):

$$\|Au|W_{\mathrm{bdr}^+}^-\| \ge C_6 \|u|W_{\mathrm{bdr}}^+\| \quad \forall u \in W_{\mathrm{bdr}}^+, \tag{16.11'}$$

$$\|A^+v|W_{\mathrm{bdr}^+}^-\| \ge C_7 \|P|v|W_{\mathrm{bdr}^+}^+\| \quad \forall v \in W_{\mathrm{bdr}^+}^+. \tag{16.12'}$$

It is known (see [30] or [76]) that, in analogy to Lemma 16.2, the validity of inequalities (16.11') and (16.12') is a necessary and sufficient condition for the existence of a strong solution in W_{bdr}^+ and $W_{\mathrm{bdr}^+}^+$ of problems (16.5) and (16.6) for any $f \in W_{\mathrm{bdr}^+}^-$, for all $g \in W_{\mathrm{bdr}}^-$. In this case, as follows from (16.10), (16.16), and (16.11), $H_{AB} = W_{\mathrm{bdr}}^+$, and by Lemma 16.3 any weak solution is a strong solution.

COROLLARY 16.3. *If inequalities* (16.11') *and* (16.12') *hold, then for any* $f \in W_{\mathrm{bdr}^+}^-$ *there exists a unique strong solution* $u_0 \in W_{\mathrm{bdr}}^+$ *of problem* (16.1), (16.2) *which coincides with the unique element minimizing the functional* $\Phi(u)$ *of* (16.24') *in* W_{bdr}^+.

It should be borne in mind that it is harder to establish the estimates (16.11') and (16.12') than it is to establish (16.11) and (16.12); it is obvious that (16.11) and (16.12) follow immediately from (16.11') and (16.12'), while from (16.11) and (16.12) estimates of the following form established by many authors (for various special cases of equation (16.11)) are obtained [76]:

$$\|Au\|_{L_2} \ge C_8 \|u|W_2^1\| \quad \forall u \in W_{\mathrm{bdr}}^+ \cap W_2^2,$$

$$\|A^+v\|_{L_2} \ge C_9 \|v|W_2^1\| \quad \forall v \in W_{\mathrm{bdr}^+}^+ \cap W_2^2.$$

REMARK 16.1. V. P. Didenko did not introduce the concept of a weak solution in the space H_{AB} as is customary in the usual variational method (see [106], [107], [209], [225], [250], and [293]. In this subsection we have followed the scheme of §§1–4 and [76].

16.3. We shall show that the functional $\Phi(u)$ of (16.24) differs only by a constant from the functional

$$M(u) = \|Au - f|W_{\text{bdr}^+}^-\|^2 \qquad (16.27)$$

of the method of least squares if $u \in W_{\text{bdr}}^+$ and $Au \in W_{\text{bdr}^+}^-$.

Indeed, by means of (16.9) and (16.13) we find that

$$\begin{aligned}
\Phi(u) &= \|u\|_{AB}^2 - 2\langle f, Bu \rangle \\
&= \langle J_{\text{bdr}^+} Au, Au \rangle - 2\langle f, J_{\text{bdr}^+} Au \rangle \\
&= (Au, Au)_{W_{\text{bdr}^+}^-} - 2(f, Au)_{W_{\text{bdr}^+}^-} \\
&= \|Au - f|W_{\text{bdr}^+}^-\|^2 - \|f|W_{\text{bdr}^+}^-\|^2 = M(u) - C_f^2. \qquad (16.28)
\end{aligned}$$

Therefore, Didenko's variational method can be considered a generalization of MLS to the negative spaces $W_{\text{bdr}^+}^-$.

In connection with the representation (16.28) we note that the condition of unique solvability of the adjoint problem (16.6) is, as we know, a necessary and sufficient condition for conditional solvability of problem (16.5) for $f \in W_{\text{bdr}^+}^-$, i.e., for the existence of a sequence $\{u_n\} \subset W_{\text{bdr}}^+$ such that $Au_n - f \to 0$ $(n \to \infty)$ in $W_{\text{bdr}^+}^-$.

Thus, if $f \in W_{\text{bdr}^+}^-$ and the condition $A^+ w = 0$ implies $w = 0$ in $W_{\text{bdr}^+}^+$, then for the functional $M(u)$ of the method of least squares there exists a sequence $\{u_n\}: M(u_n) \to C_f^2$, $Au_n \to f$ in $W_{\text{bdr}^+}^-$ $(n \to \infty)$, but the sequence $\{u_n\}$ itself may not converge to anything. Inequalities of the type (16.11) and (16.11′) are invoked to established its convergence.

Having in mind concrete applications of Didenko's variational method, we draw the following conclusions.

1. It is necessary to establish the a priori estimate (16.11) in order to obtain by the variational method a weak solution in H_{AB} of problem (16.1), (16.2) which minimizes the functional $\Phi(u)$ of (16.24) for $f \in W_{\text{bdr}^+}^-$.

2. If it is required to obtain a smoother solution u_0 in $W_{\text{bdr}}^+ \subseteq H_{AB}$ for $f \in L_2(\Omega)$, then by Corollary 16.2 of Theorem 16.1 it suffices to further prove the estimate (16.12) (or for $f \in W_{\text{bdr}^+}^-$ the estimates (16.11′) and (16.12′)).

3. The functional $\Phi(u)$ of (16.24) can be written in an explicit form if the operator J_{bdr^+} mapping all of $W_{\text{bdr}^+}^-$ onto $W_{\text{bdr}^+}^+$ is known. This integral operator, as is known [30], is defined by its kernel—the Green function of the problem

$$J_{\text{bdr}^+}^{-1} u \equiv \Delta u - f(x), \qquad x \in \Omega, \qquad (16.29)$$

with boundary conditions corresponding to (16.2). It is clearly impossible to directly construct in explicit form the operator J_{bdr^+}, and hence the functional (16.24), for the majority of domains Ω. The nonconstructive character of the functional (16.23) in the general case is evident also from the right side of (16.28): the norm $\| \cdot |W_{\text{bdr}^+}^-\| $ (see (16.7)) is difficult to compute even for elementary smooth functions.

REMARK 16.2. The estimates (16.11) and (16.12) (and even the stronger estimates (16.11′) and (16.12′)) needed for Didenko's variational method are established in [71]–[76], [268], and [269] for some classes of linear equations with variable coefficients (elliptic, hyperbolic, parabolic, mixed, and other equations) considered in a multidimensional domain $\Omega \subset R_n$. Didenko and his student A. A. Popova in [76] and [269] proposed various ways of numerical realization of the functional (16.24) without writing it in explicit form. A minimizing sequence for the functional (16.24) can be obtained, for example, by the Ritz method, which leads to a system of linear algebraic equations containing in the coefficients bilinear forms of the type $\langle u, J_{\text{bdr}^+} v \rangle$ which, in turn, can be computed by minimizing a functional of the form

$$\Phi_c(u) = (u, u)_{(W_{\text{bdr}}^+} - 2\langle u, v \rangle.$$

It should be noted also that the problem of minimizing the functional (16.24) is equivalent [75] to the problem

$$\min_{\in W_{\text{bdr}}^+} \max_{\varphi \in W_{\text{bdr}^+}^+} \{ 2\langle \varphi, Au - f \rangle - (\varphi, \varphi)_{W_{\text{bdr}^+}^+} \}, \qquad (16.30)$$

and in finding an approximate solution of this problem by the Ritz method a system of linear algebraic equations for the coefficients a_i ($\varphi \approx \sum_1^p a_i \varphi_i$) and b_j ($u \approx \sum_1^m b_j u_j$) is also obtained [76].

However, questions of numerical realization of variational methods of nonsymmetric equations require a separate investigation and are not considered here.

16.4. As is evident from (16.28), Didenko's variational method becomes effective if it is possible to constructively compute the negative norm $\| \cdot |W_{\text{bdr}^+}^-\|$. We remark that passage to a computable equivalent norm (for example, in [30], Chapter I, §3.5) in this case does not produce the desired result: the quantities $\|f(u)|W_1\|$ and $\|f(u)|W_2\|$, where $\| \cdot |W_1\|$ and $\| \cdot |W_2\|$ are equivalent norms, may achieve a minimum for distinct $u_1, u_2 \in W_1 = W_2$.

In §9 the space $\tilde{S}_{\bar{a}}^{-l} W_2(\Omega)$ was introduced whose completion is the space $S^{-l}\overset{\circ}{W}_2(\Omega, \Gamma)$ dual to a space of functions with a dominant mixed derivative and for which the norm $\| \cdot |\tilde{S}_{\bar{a}}^{-l} W_2(\Omega)\|$ can be computed effectively (see Lemma 9.1). We shall clarify these results in the case of a space which we shall use later in this section.

Let $\Omega = \{x, y: 0 < x < a, \ 0 < y < b\}$. We define the space of functions with a dominant mixed derivative $S^{1,1}\overset{\circ}{W}_2(\Omega, \Gamma_3, \Gamma_4)$ for which the following norm exists and is finite:

$$\|v|S^{1,1}W_2(\Omega, \Gamma_3, \Gamma_4)\| = \left[\int_\Omega \left(\frac{\partial^2 v}{\partial y \partial x}\right)^2 d\Omega\right]^{1/2}, \qquad (16.31)$$

$$v(x, y) = 0, \qquad (x, y) \in \Gamma_3 = \{x, y: x = a, \ 0 < y < b\}, \quad (16.32)$$

$$\partial v/\partial x = 0, \qquad (x, y) \in \Gamma_4 = \{x, y: y = b, \ 0 < x < a\}. \quad (16.33)$$

According to results of Nikol'skii [240] (see also §9), the boundary conditions (16.32) and (16.33) are stable in the mean-square sense for functions in $S^{1,1}\overset{\circ}{W}_2(\Omega, \Gamma_3, \Gamma_4)$, and with the help of the Newton-Leibniz equality it is not hard to obtain the estimate

$$\|v|S^{1,1}\overset{\circ}{W}_2(\Omega, \Gamma_3, \Gamma_4)\| \geq \mathrm{const}\,\|v|L_2(\Omega)\|. \qquad (16.34)$$

Similar properties are possessed by the space $S^{1,1}\overset{\circ}{W}_2(\Omega, \Gamma_1, \Gamma_2)$ of functions with a dominant mixed derivative

$$\|u|S^{1,1}W_2(\Omega, \Gamma_1, \Gamma_2)\| = \left[\int_\Omega \left(\frac{\partial^2 u}{\partial x \partial y}\right)^2 d\Omega\right]^{1/2}, \qquad (16.35)$$

$$\partial u/\partial y = 0, \qquad (x, y) \in \Gamma_1 = \{x, y: x = 0, \ 0 < y < b\}, \quad (16.36)$$

$$u(x, y) = 0, \qquad (x, y) \in \Gamma_2 = \{x, y: y = 0, \ 0 < x < a\}. \quad (16.37)$$

On the basis of the positive space $S^{1,1}\overset{\circ}{W}_2(\Omega, \Gamma_3, \Gamma_4)$ of (16.31) the negative space $S^{-1,-1}\overset{\circ}{W}_2(\Omega, \Gamma_4, \Gamma_5)$ is constructed as the closure of $\overset{\circ}{C}^\infty(\Omega)$ in the norm

$$\|u|S^{-1,-1}\overset{\circ}{W}_2(\Omega, \Gamma_3, \Gamma_4)\| = \sup_{v \neq 0} \frac{|(u, v)|}{\|v|S^{1,1}\overset{\circ}{W}_2(\Omega, \Gamma_3, \Gamma_4)\|}. \qquad (16.38)$$

It follows from Lemma 9.1 that the norm (16.38) can actually be effectively computed for certain regular functions by (9.35):

$$\|u|S^{-1,-1}\overset{\circ}{W}_2(\Omega,\Gamma_3,\Gamma_4)\|$$

$$= \left[\int_\Omega \left(\int_0^x d\xi \int_0^y u(\xi,\theta)\,d\theta\right)^2 d\Omega\right]^{1/2}. \quad (16.39)$$

According to the notation of §9, the set of functions for which there exists a finite right side in (16.39) is the space $\widetilde{S}_{0,0}^{-1,-1}W_2(\Omega)$, so that its completion in the norm (16.38) (or in the norm equal to the right side of (16.39)) gives the entire space $S^{-1,-1}\overset{\circ}{W}_2(\Omega,\Gamma_3,\Gamma_4)$.

16.5. With the help of the spaces $\widetilde{S}_a^{-l}W_2(\Omega)$ we construct a constructive modification of Didenko's method.

We consider a general linear hyperbolic equation with variable coefficients in Ω

$$\sum_{i,j=1}^2 a_{ij}(x)u_{x_ix_j} + b_i(x)u_{x_i} + c(x)u(x) = \widetilde{g}(x), \qquad x = (x_1,x_2). \quad (16.40)$$

As is known [176], this equation by a change of the independent variables and the unknown function can be reduced to the form

$$A_0u \equiv \frac{\partial}{\partial\xi}\left(k(\xi,\eta)\frac{\partial u}{\partial\eta}\right) + h(\xi,\eta)u(\xi,\eta) = g(\xi,\eta). \quad (16.41)$$

In §13 we investigated by the variational method a complex boundary value problem for equation (16.41) for $h \equiv 0$ but with a variable function $k(\xi,\eta)$. We now consider the case where $k(\xi,\eta) \equiv$ const in Ω, and h is a given measurable function of the independent variables ξ and η in Ω:

$$Au = \partial^2u/\partial x\partial y + h(x,y)u = g(x,y), \qquad (x,y) \in \Omega, \quad (16.42)$$
$$u(x,y) = 0, \qquad (x,y) \in \Gamma_2, \quad (16.43)$$
$$\partial u/\partial y = 0, \qquad (x,y) \in \Gamma_1. \quad (16.44)$$

For W_{bdr}^+ we take the space $S^{1,1}\overset{\circ}{W}_2(\Omega,\Gamma_1,\Gamma_2)$ of (16.35)–(16.37); then

$$W_{\text{bdr}^+}^+ = S^{1,1}W_2(\Omega,\Gamma_3,\Gamma_4)$$

(see (16.31)–(16.33)),

$$W_{\text{bdr}^+}^- = S^{-1,-1}W_2(\Omega,\Gamma_3,\Gamma_4)$$

of (16.38),

$$w_{\text{bdr}}^- = [S^{1,1}\overset{\circ}{W}_2(\Omega,\Gamma_1,\Gamma_2)]^*,$$

and for "regular" functions $u(x, y) \in \tilde{S}_{0,0}^{-1,-1} W_2(\Omega)$ we have

$$\|u|W_{\mathrm{bdr}^+}^-\| = \|u|S^{-1,-1} \overset{\circ}{W}_2(\Omega, \Gamma_3, \Gamma_4)\|$$

$$= \left[\int_\Omega \left(\int_0^x d\xi \int_0^y u(\xi, \theta) \, d\theta \right)^2 dx \, dy \right]^{1/2}. \quad (16.45)$$

We define the auxiliary operator $B = J_{\mathrm{bdr}^+} A$ on $W_{\mathrm{bdr}^+}^-$, where J_{bdr^+} is the isometric operator mapping $W_{\mathrm{bdr}^+}^-$ onto $W_{\mathrm{bdr}^+}^+$, so that a relation of the type (16.9) is satisfied. It is obvious that here for all $u, v \in W_{\mathrm{bdr}}^+$

$$\langle Au, Bv \rangle = \langle Bu, Av \rangle,$$

$$\langle Au, Bu \rangle = \|Au|W_{\mathrm{bdr}^+}^-\|^2, R(A) \subseteq W_{\mathrm{bdr}^+}^-$$

and the quantity

$$\|u\|_{AB} = \langle Au, Bu \rangle^{1/2} \quad (16.46)$$

defines a norm if an a priori estimate of the type (16.11) holds.

LEMMA 16.4. *Suppose, for some $\varepsilon > 0$, $h(x,y)$ satisfies one of the conditions*

$$h^2(x, y) \le 2 - a^2 b^2 - 2\varepsilon, \quad (x, y) \in \Omega, \quad (16.47)$$

$$\int_0^x d\xi \int_0^y h^2(\xi, \theta) \, d\theta \le \frac{1 - 2\varepsilon}{ab}, \quad (x, y) \in \Omega. \quad (16.48)$$

Then

$$\|Au|W_{\mathrm{bdr}}^-\| \ge \varepsilon \|u|L_2(\Omega)\|. \quad (16.49)$$

PROOF. The basic idea of obtaining a priori estimates of the type (16.11), (16.12), and (16.49) in negative spaces is due to Didenko and consists in the following. If with the help of some integral operator I, $\|Iu|W_{\mathrm{bdr}^+}^+\| \le \|u|L_2(\Omega)\|$, it is possible to prove that

$$\int_\Omega IuAu \, d\Omega \ge \varepsilon \|u|L_2(\Omega)\|^2, \quad (16.50)$$

then, applying the Schwarz inequality (16.8) to the left side in (16.50), we obtain

$$\varepsilon \|u|L_2(\Omega)\|^2 \le \|Au\|_{W_{\mathrm{bdr}^+}^-} \|Iu\|_{W_{\mathrm{bdr}^+}^+}$$

$$\le \|Au\|_{W_{\mathrm{bdr}^+}^+} \|u|L_2(\Omega)\|,$$

whence (16.49) follows.

We set

$$Iu \equiv \int_1^x d\xi \int_b^y u(\xi, \eta) \, d\eta,$$

and $D(I) = D(A)$; then $\||Iu|W^+_{\text{bdr}+}\|| = \||u|L_2(\Omega)\||$ and

$$\int_\Omega IuAu\,d\Omega \equiv \int_\Omega \left(\int_a^x d\xi \int_b^y u(\xi,\eta)\,d\eta \right) \left(\frac{\partial^2 u}{\partial x \partial y} + hu \right) d\Omega$$

$$= \int_\Omega \left[u^2 + \int_a^x d\xi \int_b^y u(\xi,\eta)\,d\eta\, hu \right] d\Omega$$

$$\geq \int_\Omega \left[u^2 - \frac{1}{2}h^2 u^2 - \frac{1}{2} \left(\int_a^x d\xi \int_b^y u(\xi,\eta)\,d\eta \right)^2 \right] d\Omega$$

$$\geq \varepsilon \||u|L_2(\Omega)\||^2,$$

if $1 - \frac{1}{2}a^2 b^2 - \frac{1}{2}h^2 \geq \varepsilon > 0$. Similarly,

$$\int_\Omega IuAu\,d\Omega = \int_\Omega \left\{ u^2 + u \int_0^x d\xi \int_0^y h(\xi,\eta)u(\xi,\eta)\,d\eta \right\} d\Omega$$

$$\geq \int_\Omega \left\{ u^2 - \frac{1}{2}u^2 - \frac{1}{2} \left(\int_0^x d\xi \int_0^y hu\,d\eta \right)^2 \right\} d\Omega$$

$$\geq \varepsilon \||u|L_2(\Omega)\||^2,$$

if (16.48) is satisfied. Defining H_{AB} as the completion of W^+_{bdr} in the norm (16.46), we find by (16.49) that $H_{AB} \to L_2(\Omega)$.

LEMMA 16.5. *If one of the conditions* (16.47) *or* (16.48) *is satisfied, then*

$$\|Au\|_{W^-_{\text{bdr}+}} = \|u\|_{AB} \leq C\||u|W^+_{\text{bdr}}\|| \quad \forall u \in W^+_{\text{bdr}}. \tag{16.51}$$

It obviously suffices to prove this for the set of smooth functions in $C^2_{\text{bdr}}(\overline{\Omega})$; the validity of (16.51) for all W^+_{bdr} can then be obtained by passing to the limit and assuming that A is defined by continuity on all of W^+_{bdr}.

With consideration of (16.45), (16.34) we have for $u \in C^2_{\text{bdr}}(\overline{\Omega})$

$$\|u\|_{AB} = \|Au\|_{W^-_{\text{bdr}+}} = \left\{ \int_\Omega \left[\int_0^x d\xi \int_0^y \left(\frac{\partial^2 u}{\partial \xi \partial \eta} + hu(\xi,\eta) \right) d\eta \right]^2 d\Omega \right\}^{1/2}$$

$$\leq \sqrt{2} \left\{ \int_\Omega \left[u^2 + \left(\int_0^x d\xi \int_0^y h(\xi,\eta)u(\xi,\eta)\,d\eta \right)^2 \right] d\Omega \right\}^{1/2}$$

$$\leq C_9 \||u|L_2(\Omega)\|| \leq C_{10} \||u|W^+_{\text{bdr}}\||.$$

COROLLARY 16.4.

$$W^+_{\text{bdr}} \to H_{AB}. \tag{16.52}$$

Considering now the Didenko functional

$$D(u) = \|u\|^2_{AB} - 2\langle g, Bu \rangle, \tag{16.53}$$

it is not hard to verify in a manner analogous to (16.25) that in this case

$$|\langle g, Bu\rangle| \le \|g\|_{W^-_{\mathrm{bdr}^+}} \|u\|_{AB} \quad \forall u \in W^+_{\mathrm{bdr}}, \ g \in W^-_{\mathrm{bdr}^+}. \tag{16.54}$$

From Theorem 16.1 we thus obtain

ASSERTION 16.2. *If one of the conditions* (16.47) *or* (16.48) *is satisfied, for any* $g \in S^{-1,-1} \overset{\circ}{W}_2(\Omega, \Gamma_1, \Gamma_2)$ *there exists a unique element* $u_0 \in H_{AB}$ *minimizing the functional* $D(u)$ *of* (16.53) *in* H_{AB}. *Moreover, the element* u_0 *coincides with the weak solution* (*in a sense analogous to* (16.19)); *see also* (11.75), *and* $\|u_0\|_{AB} \le \|g|W^-_{\mathrm{bdr}^+}\|$.

If now g, $Au_n \in \tilde{S}^{-1,-1}_{0,0} W_2(\Omega)$, $n = 1, 2, \ldots$, for the sequence $\{u_n\} \subset C^2_{\mathrm{bdr}}(\overline{\Omega})$ minimizing the functional (16.53), then by (16.28) and (16.45)

$$D(u_n) = \|Au_n - g|W^-_{\mathrm{bdr}^+}\|^2 - \|g|W^-_{\mathrm{bdr}^+}\|^2$$
$$= \int_\Omega \left\{ \left[u_n - \int_0^x d\xi \int_0^y h u_n \, d\theta - \int_0^x d\xi \int_0^y g(\xi, \eta) \, d\eta \right]^2 \right.$$
$$\left. - \left(\int_0^x d\xi \int_0^y g(\xi, 0) \, d\theta \right)^2 q \right\} d\Omega \tag{16.55}$$

and the values of the Didenko functional can be constructively computed.

16.6. Thus, the basic condition in Didenko's method involves obtaining a priori estimates of the type (16.11). In the case of general spaces $W_{\mathrm{bdr}} = \overset{\circ}{W}^l_2$ the functional (16.24) is nonconstructive, and obtaining the a priori estimates (16.11) and (16.12) involves considerable technical difficulties (see [71]–[74]).

Invoking as $W^-_{\mathrm{bdr}^+}$ the spaces $S^{-l} \overset{\circ}{W}_2(\Omega, \Gamma)$ dual to the spaces $S^l \overset{\circ}{W}_2(\Omega, \Gamma)$ with a dominant mixed derivative makes it possible to give a constructive representation of the functional (16.24′) and to obtain the a priori estimate in a rather simple way (see Lemma 16.4).

Without repeating the scheme of Didenko's method, we present other examples which demonstrate the utility of the spaces $\tilde{S}^{-l} W_2(\Omega, \Gamma)$ in obtaining a priori estimates of the form (16.11).

EXAMPLE 1. We consider the general linear hyperbolic equation (16.41) in Ω with functions $k(\xi, \eta) \in C^1(\overline{\Omega})$ and $h(\xi, \eta) \in C(\overline{\Omega})$; $D(A_0) = \{u \in C^2(\overline{\Omega})$ satisfying (16.36) and (16.37)$\}$. If we take $W^-_{\mathrm{bdr}^+} = S^{-1,0} \overset{\circ}{W}_2(\Omega, \Gamma_3)$, then by Lemma 9.1 for all $u \in \widetilde{S}_{0,-}^{-1,0} W_2(\Omega)$ we have

$$
\|u|\widetilde{S}_{0,-}^{-1,0} W_2(\Omega)\| \equiv \left[\int_\Omega \left(\int_0^x u(\xi, y)\, d\xi \right)^2 d\Omega \right]^{1/2}
$$

$$
= \sup_{v \neq 0} \frac{|(u, v)|}{\|v|S^{1,0}\overset{\circ}{W}_2(\Omega, \Gamma_3)\|} \equiv \|u|S^{-1,0}\overset{\circ}{W}_2(\Omega, \Gamma_3)\|, \quad (16.56)
$$

and the functionals (16.27) and (16.28) can be constructively computed in any case for $A_0 u \in \widetilde{S}_{0,1}^{-1,0} W_2(\Omega)$.

LEMMA 16.6. *Take the operator A_0 defined in* (16.41), (16.36), *and* (16.37), *and* $k(x, b) \leq 0$ $(0 \leq x \leq a)$. *Then*

$$
\|A_0 u | W_{\mathrm{bdr}^+}^- \| \geq \varepsilon \|u|L_2(\Omega)\| \quad \forall u \in D(A_0), \quad (16.57)
$$

if for some $\varepsilon > 0$ one of the following conditions is satisfied:
a)

$$
\frac{\partial k}{\partial y} - h^2(x, y) \geq a^2 + 2\varepsilon, \quad (x, y) \in \Omega;
$$

b)

$$
\frac{\partial k}{\partial y} - a \sup_{(x,y) \in \Omega} \int_0^x h^2(\xi, y)\, d\xi \geq 1 + 2\varepsilon, \quad (x, y) \in \Omega.
$$

PROOF. We follow Didenko's idea for obtaining estimates of the type (16.49) which was presented in Lemma 16.4. We show that for all $u \in D(A_0)$

$$
\int_\Omega I_a u A_0 u\, d\Omega \geq \varepsilon \|u|L_2(\Omega)\|^2, \quad (16.58)
$$

where $I_a u = \int_a^x u(\xi, y)\, d\xi$. For $u \in D(A_0)$

$$
\int_\Omega I_a u A_0 u\, d\Omega = \int_\Omega \left\{ -uk \frac{\partial u}{\partial y} + \int_a^x u(\xi, y)\, d\xi\, hu \right\} d\Omega
$$

$$
= \int_\Omega \left\{ \frac{1}{2} \frac{\partial k}{\partial y} u^2 + hu \int_a^x u(\xi, y)\, d\xi \right\} d\Omega
$$

$$
- \int_0^a k(x, b) u^2(x, b)\, dx
$$

$$
\geq \int_\Omega \left(\frac{1}{2} \frac{\partial k}{\partial y} u^2 - \frac{1}{2} h^2 u^2 - \frac{1}{2} a^2 u^2 \right) d\Omega \geq \varepsilon \|u|L_2(\Omega)\|^2,
$$

if condition a) of the lemma is satisfied. Similarly,

$$\int_\Omega I_a u A_0 u \, d\Omega = \int_\Omega \left\{ \frac{1}{2} \frac{\partial k}{\partial y} u^2 - \int_0^x h(\xi, y) u(\xi, y) \, d\xi u \right\} d\Omega$$
$$- \int_0^a k(x, b) u^2(x, b) \, dx$$
$$\geq \varepsilon \| u | L_2(\Omega) \|^2,$$

if condition b) is satisfied.

Applying the Schwarz inequality to the left side of (16.58), we find

$$\| I_a u \|_{W^-_{\text{bdr}^+}} \| A_0 u \|_{W^-_{\text{bdr}^+}} = \| u \|_{L_2} \| A_0 u | W^-_{\text{bdr}^+} \| \geq \varepsilon \| u | L_2(\Omega) \|^2,$$

whence the a priori estimate (16.57) follows. ∎

EXAMPLE 2. Didenko's method also has the merit that in the right sides of equations it admits functions of the spaces $\overset{\circ}{W}{}_2^{-l}(\Omega)$, for example, functions with a singularity $(x - a)^{-\alpha} \notin L_1(\Omega)$, or of the type of Dirac delta functions. Therefore, even for elliptic equations with a selfadjoint and positive-definite operator it makes sense to apply this method, for example, in the case of the problem

$$\mathscr{L}_e u \equiv (k_1(x) u_x)_x + (k_2(y) u_y)_y = g(x, y), \qquad (x, y) \in \Omega, \quad (16.59)$$
$$\partial u / \partial x = 0, \qquad (x, y) \in \Gamma_3 \cup \Gamma_4, \qquad (16.60)$$
$$u = 0, \qquad (x, y) \in \Gamma_1 \cup \Gamma_2, \qquad (16.61)$$
$$k_1(x), k_2(y) \in C^1(\overline{\Omega}), \qquad k_1, k_2 \geq \kappa > 0,$$
$$\kappa = \text{const}, \qquad (x, y) \in \Omega; \qquad (16.62)$$
$$g(x, y) \in W^-_{\text{bdr}^+} = S^{-2,0} \overset{\circ}{W}_2(\Omega, \Gamma),$$

$$\| g | \widetilde{S}^{-2,0}_{(a,0)} W_2(\Omega) \| = \left[\int_\Omega \left(\int_a^x \frac{d\theta}{k_1(\theta)} \int_0^\xi g(\xi, y) \, d\xi \right)^2 d\Omega \right]^{1/2} < \infty. \qquad (16.63)$$

We remark that the function $\widetilde{g}(x, y) = (x - a)^{-\alpha} \in \widetilde{S}^{-2,0}_{(a,0),-} W_2(\Omega)$ for $\alpha < 2$, while $\widetilde{g}(x, y) \in L_1(\Omega)$ only for $\alpha < 1$.

For this problem we obtain the a priori estimate

$$\| \mathscr{L}_e u | W^-_{\text{bdr}^+} \| = \| \mathscr{L}_e u | S^{-2,0} \overset{\circ}{W}_2(\Omega, \Gamma) \| \geq \varepsilon \| u | L_2(\Omega) \| \qquad (16.64)$$

for all $u \in D(\mathscr{L}_e) \equiv C^2_{\text{bdr}}(\overline{\Omega})]\{ u \in C^2(\overline{\Omega})$ satisfying (16.60) and (16.61)$\}$.

Integrating by parts on the set $D(\mathscr{L}_e)$, we obtain

$$\int_\Omega I_e u \mathscr{L}_e u \, d\Omega \equiv \int_\Omega \left[\int_a^x \frac{d\theta}{k_1(\theta)} \int_0^\theta u(\xi, y) \, d\xi \right] |(k_1 u_x)_x + (k_2 u_y)_y| \, d\Omega$$

$$= \int_\Omega \left[u^2 + \frac{k_2}{k_1} \left(\int_0^x \frac{\partial u}{\partial y}(\xi, y) \, d\xi \right)^2 \right] d\Omega$$

$$\geq \|u|L_2(\Omega)\|^2. \tag{16.65}$$

Estimating the left side of (16.65) by the Schwarz inequality (16.8), we find that

$$\|I_e u|S^{2,0}\overset{\circ}{W}_2(\Omega, \Gamma)\| \cdot \|\mathscr{L}_e u|S^{-2,0}\overset{\circ}{W}_2(\Omega, \Gamma)\|$$

$$= \|\mathscr{L}_e u|S^{-2,0}\overset{\circ}{W}_2(\Omega, \Gamma)\| \cdot \|u|L_2(\Omega)\| \geq \|u|L_2(\Omega)\|^2,$$

whence (16.64) follows.

Commentary to Chapter III

To §11. 1. For equations with nonpotential operators we have analyzed the mathematical aspects of formulating variational principles without considering their physical content, in contrast to V. L. Berdichevskiĭ's monograph [28] where no special attention is given to "nonphysical" variational principles ([27], Chapter III, §11). In his book a systematic exposition is given of variational principles of continuum mechanics based on the statement that "it is possible to describe all *reversible physical* phenomena by variational principles—assertions that in actually realizable processes certain functionals have a stationary value" ([28], foreword; emphasis added). From the results of Chapter 1 it follows that if a physical phenomenon is described by a mathematical model which is a linear equation $Au = f$ with a *mathematically invertible* operator A, then it is always possible to construct for the given model a variational principle analogous in variational properties to the Dirichlet principle for the Laplace equation. In this approach for irreversible physical phenomena as well (for example, heat conduction) it is possible to construct variational principles which, however, in the majority of cases do not have a real physical meaning. The following remarks, quoted by Berdichevskiĭ in [28], on the nonlocal-in-time character of "physical" variational principles for dynamical problems are interesting in this connection.

"L'énoncé même du principe de moindre action a quelque chose de choquant pour l'esprit. Pour se rendre d'un point à un autre, une molécule matérielle, soustraite à l'action de toute force, mais assujettie à se mouvoir sur une surface, prendra la ligne géodésique, c'est-à-dire le chemin le plus court.

"Cette molécule semble connaître le point où l'on veut la mener, prévoir le temps qu'elle mettra à l'atteindre en suivant tel et tel chemin, et choisir ensuite le chemin le plus convenable. L'énoncé nous la présente pour ainsi dire comme un être animé et libre. Il est clair qu'il vaudrait mieux le remplacer par un énoncé moins choquant, et où, comme diraient les philosophes, les causes finales ne sembleraient pas se substituer aux causes efficientes." (Poincaré, [265])

"Nicht allein mache es [das Hamilton'sche Prinzip] die gegenwärtige Bewegung abhängig von Folgen, welche erst in der Zukunft hervortreten können und mute dadurch der leblosen Natur Absichten zu, sondern, was schlimmer sei, es mute der Natur sinnlose Absichten zu. Denn das Integral, dessen Minimum das Hamilton'sche Prinzip fordert, habe keine einfache physikalische Bedeutung; es sei aber für die Natur ein unverständliches Ziel, einen mathematischen Ausdruck zum Minimum zu machen oder seine Variation zum Verschwinden zu bringen." (Hertz, [136]).

These consequences of nonlocalness in time of variational principles for dynamical problems are apparently connected with the purely mathematical analysis of models of physical phenomena.

A generalization of the variational approach with physical content to irreversible processes was proposed by Sedov [290]. A certain basic variational equation, which represents a writing of the laws of thermodynamics in variational form, is taken as the foundation (for more details see [28]).

2. We do not consider methods of constructing functionals by means of increasing the dimension of the space of solutions of the original problem; these approaches are based on the construction for the equation $\mathscr{L}u = f$ with the help of the adjoint operator $\mathscr{L}^*$ of a functional of the form $I(u, v) = (u, \mathscr{L}^*v - g) - (v, f)$, where u is the unknown solution of the original problem and v is a new unknown auxiliary function (see [28], Chapter III, §11 and [102]).

3. Although for rather general nonparabolic linear PDE (11.18) the question of the existence of an equivalent functional in the class of Euler functionals can be answered affirmatively according to the scheme of Copson [54], this approach is actually not always constructive: to determine the unknown functions of the desired functional (11.19) it is necessary to obtain in *explicit* form a solution of the system (11.21)–(11.23) of linear PDE of first order with variable coefficients.

4. The question of the existence of a classical solution of the inverse problem of the calculus of variations in the formulation of Balatoni [18]

remains unstudied. For the equation $\mathscr{L}u - G = 0$ of (6.8) there have been obtained for it necessary and sufficient conditions for the existence of a functional (11.3) such that

$$\delta\Phi(u) = \int_\Omega \mu(x, y, u(x, y), u_x, u_y)(\mathscr{L}u - G)\delta u\, dx\, dy$$

with some function μ, $\mu \neq 0$ in Ω. In what cases is it possible to find here a functional (11.3) which is bounded below? The answer to this question is unknown. We note that in the case when $\mu(x, y, u, u_x, u_y)$ does not reduce to some function $\mu(x, y)$ Copson's approach [54]), from the right side of the relation presented here it is not hard to conclude that even for linear operators $\mathscr{L}$ and G in (6.8) the corresponding functional $\Phi(u)$ may fail to be quadratic.

To §12. The functionals constructed by the author for parabolic operators were discussed by Finlayson (see [103]) in a comparison with functionals of other variational principles for parabolic operators. We note, however, that all other variational principles considered by Finlayson do not satisfy all of our conditions A–C (see the Introduction).

To §13. The possibility of constructing a functional (13.27) for the wave equation which does not contain derivatives of the unknown function is a special case of a fact noted earlier: if for a boundary value problem $Au = f$ an auxiliary symmetrizing operator B is chosen in the form $B = (A^{-1})^*C$ (1.7), where C is an arbitrary integral operator symmetric and positive in the usual sense, then the corresponding functional

$$D[u] = (Au, Bu) - 2(f, Bu) = (u, Cu) - 2(f, (A^{-1})^*Cu)$$

may fail to contain derivatives of the unknown function u. In the case of the wave equation (13.7) it is possible to construct an operator B of the form (1.6), proceeding from the following known [57], Vol. 2) representation of a general solution of (13.7):

$$u(\xi, \eta) = \int_{\xi_0}^{\xi} \frac{d\theta}{k(\theta, \eta)} \int_{\eta_0}^{\eta} f(\theta, \tau)\, d\tau + w(\xi) + v(\eta).$$

Such functionals, which do not contain derivatives of the unknown function, admit effective numerical realization, but the author is unaware of other constructive approaches to an operator in the form (1.7) for substantial boundary value problems for equations of mathematical physics.

To §14. 1. We emphasize that the functionals (14.16) and (14.17) constructed in this section do not belong in general either to the class of

Euler functionals (6.2) or to the class (6.18), but are functionals of the
form

$$F[u] = \int_\Omega F(x, u(x), u'_{x_1},, \ldots, u'_{x_n}, u * g, \nabla u * g) \, d\Omega,$$

depending on the convolution of the unknown function and the given function. Closely related functionals depending on convolution were considered in [282] and [283] in a manner unrelated to variational methods for
boundary value problems.

2. In contrast to all other techniques presented in Chapter 3 for constructing classical solutions of the inverse problem of the calculus of variations a way of generalizing the approach of §14 to equations with variable
coefficients is not evident, although various estimates of fundamental solutions for such equations are known.

3. The method presented in §14 of constructing a functional for a
general linear PDE with constant coefficients is constructive if it is possible
to construct at least one fundamental solution of the adjoint equation; for
many equations of mathematical physics such fundamental solutions are
known (see [57], [89], [119], [157], [281], [297], [318], and [327]). A
constructive approach to fundamental solutions for other rather general
PDE can be carried out by methods of [41], [120], [179], [219], and [281].

We mention the method of [178] for approximate solution of linear
PDE when at least one fundamental solution of the original differential
equation is known.

4. In connection with the boundary value problem considered in §14
for a general PDE (14.1) of hypoelliptic type with conditions (14.2) on
the entire boundary $\partial\Omega$ of the domain Ω, we remark that back in 1961
Dezin and Mikhaĭlov [65] pointed out the following: "... one natural
approach to the study of boundary value problems for general hypoelliptic
operators is the study of boundary value problems for parabolic equations
with conditions given on the entire boundary".

To §12-16.   1. To obtain smoother solutions from an analysis of the
identity

$$(Au_0 - f, Bv) = 0 \text{ for all } v \in D(A, B)$$

defining the generalized solution u_0, denseness of the set $R_A(B) = \{Bv : v \in D(A, B)\}$ in $L_2(\Omega) \times L_2(\partial\Omega)$ is important. Since in many cases the operator B is closely related to the operator A of the original boundary value
problem (see (14.10), (15.21), and (16.13)), proofs (which are not compli-
cated but are technically cumbersome) that $\overline{R_A(B)} = L_2(\Omega) \times L_2(\partial\Omega)$ can
often be based on the condition of solvability of the adjoint problem for

smooth functions—the right sides of the equations and boundary conditions. This technique—establishing solvability of the original problem by means of the solvability of the adjoint problem—is well known.

2. In many examples of §§12–16 it is possible to formulate stronger assertions regarding the existence and uniqueness of generalized solutions of the boundary value problems studied for functions $f(x)$—the right sides of the equations—in negative spaces $S^{-\bar{l}}W_2(\Omega)$ (the completions of the spaces $\tilde{S}_{\bar{a}}^{-\bar{l}}\overset{\circ}{W}_2(\Omega)$ considered), since by Lemma 9.1.

$$\||u|\tilde{S}_{\bar{a}}^{-\bar{l}}W_2(\Omega) = \||u|S^{-\bar{l}}\overset{\circ}{W}_2(\Omega)\|| \quad \forall u \in \tilde{S}_{\bar{a}}^{-\bar{l}}W_2(\Omega).$$

In other words, the linear functional (f, Bu) in the quadratic functional

$$D[u] = \|u\|_{AB}^2 - 2(f, Bu)$$

can be extended by continuity not only in u to all of H_{AB} but also in f to all of $S_{\bar{a}}^{-\bar{e}}\overset{\circ}{W}_2(\Omega)$, since in the cases of §§12–16 an estimate of the type

$$|(f, Bu)| \le \text{const} \|u\|_{AB} \||f|S^{-\bar{l}}\overset{\circ}{W}_2(\Omega)\|| \tag{$*$}$$

holds. Extension of the functional (f, Bu) in u to all of H_{AB} is necessary for consideration of the problem of minimizing the functional $D[u]$ in H_{AB}; extension of the functional (f, Bu) in f is not necessary, and we gave preference to a more constructive notation for the functional. We note that extension by continuity of the functional (f, Bu) in f and u by virtue of $(*)$ to $S_{\bar{a}}^{-\bar{l}}\overset{\circ}{W}_2(\Omega) \times H_{AB}$ admits in the right sides of the equations investigated generalized functions f of the type of the Dirac delta function in those cases where $S_{\bar{a}}^{-\bar{l}}\overset{\circ}{W}_2(\Omega) \to C(\overline{\Omega})$.

CHAPTER IV

Variational Principles For Nonlinear Equations

In this chapter we consider solution of the inverse problem of the calculus of variations for a nonlinear operator equation

$$N(u) = 0.(^{33}) \qquad (17.0)$$

We discuss questions of existence and of the construction for the given equation (17.0) of a functional $\Phi(u)$ with the following properties:

$\Phi(u)$ is bounded below (or above) on some subset of $D(N)$;   (17.1)

(17.0) is the Euler equation for $\Phi(u)$ or   (17.2)

the set of critical points of $\Phi(u)$ coincides with   (17.2′)
the set of solutions of the original problem (17.0);

$\Phi(u)$ contains derivatives of the unknown function u (17.3)
of lower order than equation (17.0).

The last condition is obviously hard to verify for such a degree of generality of the nonlinear operator N of (17.0). We shall take this condition into account in the sense that the functionals $\Phi(u)$ constructed in the case of a linear (generally speaking, nonsymmetric, i.e., nonpotential) operator N must also satisfy condition (17.3).

In §17 we present a result of Tonti [324] regarding the construction for the nonlinear equation (17.0) of a functional $\Phi(u)$ with property (17.2′); for a broad class of integro-differential equations we indicate sufficient conditions for the existence of a solution of this inverse problem of the

---

(^{33})For the time being it is more convenient to write the equation in the form (17.0), understanding by the operator N the operator $N(u) \equiv N(x, u(x)) \equiv N(u) - f(x)$, i.e., equation (17.0) is inhomogeneous unless otherwise mentioned. Therefore, $N(0) \not\equiv 0$ in general.

calculus of variations and also carry out the construction of functionals with properties (17.1) and (17.3).

In §18 a classical variational method developed mainly for equations with potential operators [331] is extended to equation (17.0). The principle of such a generalization was proposed independently by Nashed [235] and Lyashko [202]. We generalize these results to broader classes of nonlinear equations and consider questions of the existence of functionals with properties (17.1), (17.2'), and (17.3). The connection between generalized (weak, strong) solutions (17.0) and the corresponding variational problem is established by means of a construction of Petryshyn for extension of a nonlinear operator.

Although the exposition of the material is mainly independent, for a better understanding of it the reader should be familiar with Chapters 1–3 of Vaĭnberg's monograph [331], which is already classical in this area.

§17. On the inverse problem of the calculus of variations
for nonlinear equations

17.1. In considering the inverse problem of the calculus of variations, it makes sense first of all to verify the potential criterion from which for a concrete equation (17.0) one can draw a conclusion regarding the existence or nonexistence of a functional with property (17.2).

For completeness of the exposition and unification of notation we first present some known facts that will be needed later.

Suppose an operator N acting from a real Banach space U to the dual Banach space $V = U^*$ is defined on a linear manifold $D(N)$ dense in U. If for some $u \in D(N)$ and for any $h \in D(N)$ there exists the limit

$$\lim_{t \to 0} \frac{N(u + th) - N(u)}{t} = DN(u, h), \tag{17.4}$$

which is a linear expression in h, $DN(u, h) = N'_u h$, then the linear operator N'_u is called the *Gâteaux derivative* at the point u (the "weak derivative" in the terminology of [331]) of the operator N. The limit in (17.4) is understood in the sense of convergence in the norm of the Banach space V; various conditions for the existence of the Gâteaux derivative for nonlinear operators are presented in [331]. It is obvious that if $N = A$ is a linear operator, then $A' = A$. For the quantity $N'_u h \equiv DN(u, h)$ the (Gâteaux differential) in analogy with the classical calculus of variations the notation $\delta N(u) = N'_u \delta u = DN(u, \delta u)$ with $h = \delta u$ is used [236].

We suppose that on $U \times V$ there is defined a real, symmetric nondegenerate, bilinear functional $\langle \cdot, \cdot \rangle : U \times V \to R$, so that the spaces U and V are considered as dual spaces relative to this form. A operator $F : U \to U^*$ is called a *potential operator* on some set $\omega \subset u$ if there exists a functional $f(u)$ defined on ω such that

$$\lim_{t \to 0} \frac{f(u + th) - f(u)}{t} = \langle F(u), h \rangle \quad \forall h \in U.$$

Writing $f'_u = F(u) = \operatorname{grad} f(u)$, we call the functional f the *potential* of the operator F, while the operator F is called the *gradient* of the functional f.

THEOREM (VOLTERRA [342], KERNER [159], VAĬNBERG [331]). *Suppose there is given an operator $F: U \to U^*$ having a Gâteaux derivative in a simply connected domain $D \subset U$ such that the form $\langle F'_u h_1, h_2 \rangle$ is continuous in u, $u \in D$. The operator F is a potential operator in D if and only if the following symmetry condition is satisfied:*

$$\langle F'_u h_1, h_2 \rangle = \langle h_1, F'_u h_2 \rangle \quad \forall u \in D, \ \forall h_1, h_2 \in U. \tag{17.5}$$

Moreover, $F(u) = \operatorname{grad} f(u)$, where

$$f(u) = \int_0^1 \langle F(tu), u \rangle \, dt + \text{const.}(^{34}) \tag{17.6}$$

Thus, (17.5) is a necessary and sufficient condition for the existence for a given operator equation $F(u) = 0$ of a functional (17.6) possessing property (17.2).

In the case where F is the operator of a boundary value problem for a differential equation the operators F and F' are understood as vectorial operators (the operator of the equation plus the operators of the boundary conditions) or are defined on some set of functions satisfying the boundary conditions, and criterion (17.5) must be satisfied for such operators on this set of functions. At the same the concept of a formal potential operator turns out to be useful for verification of condition (17.5). An operator N defined on a set of functions $u(x)$, $x \in \Omega \subset R_n$, is called a *formal potential operator* if

$$\langle N'_u \varphi, \psi \rangle = \langle \varphi, N'_u \psi \rangle \quad \forall u, \varphi, \psi \in \overset{\circ}{C}{}^\infty(\Omega). \tag{17.7}$$

In the case where the functions in $D(N'_u)$ satisfy some homogeneous local boundary conditions it is obvious that validity of the simpler identity (17.7) is necessary in order that condition (17.5) be satisfied. Proceeding from this, many authors have established various criteria for formal potentiality of concrete nonlinear differential operators, for example, in the following manner [321].

(^{34})Regarding some historical aspects of this result, see [324] and [331].

17.2. Suppose N is the operator of a nonlinear partial differential equation of second order:

$$N(u) \equiv f(u(x), u_k, u_{hk}) = 0, \qquad x \in \Omega \subset R_n, \qquad (17.8)$$

$$\text{where } u_k = \frac{\partial u}{\partial x_k}, \quad u_{hk} = u_{kh} = \frac{\partial^2 u}{\partial x_k \, \partial x_h}, \quad h, k = 1, \ldots, h;$$

$$D(N) = C^2(\Omega); \qquad a_s \nabla_s \varphi := \sum_{s=1}^{n} a_s \frac{\partial \varphi}{\partial x_s};$$

and f is a smooth function in all its variables. Considering that

$$N_u' \varphi \equiv \left[\frac{\partial}{\partial \varepsilon} f(u + \varepsilon \varphi) \right]_{\varepsilon=0} = \frac{\partial f}{\partial u} \varphi + \frac{\partial f}{\partial u_k} \varphi_k + \frac{\partial f}{\partial u_{hk}} \varphi_{hk},$$

we rewrite (17.7) in the form

$$\int_\Omega \psi \left[\frac{\partial f}{\partial u} + \frac{\partial f}{\partial u_k} \nabla_k + \frac{\partial f}{\partial u_{hk}} \nabla_h \nabla_k \right] \varphi \, d\Omega$$

$$= \int_\Omega \varphi \left[\frac{\partial f}{\partial u} - \nabla_k \frac{\partial f}{\partial u_k} + \nabla_h \nabla_k \frac{\partial f}{\partial u_{hk}} \right.$$

$$\left. + \left(-\frac{\partial f}{\partial u_k} + 2\nabla \frac{\partial f}{\partial u_{hk}} \right) \nabla_k + \frac{\partial f}{\partial u_{hk}} \nabla_h \nabla_k \right] \psi \, d\Omega.$$

$$(17.9)$$

Since $\varphi, \psi \in \overset{\circ}{C}{}^2(\Omega)$ are arbitrary, from this we obtain the system of three equations

$$\frac{\partial f}{\partial u} = \frac{\partial f}{\partial u} - \nabla_k \frac{\partial f}{\partial u_k} + \nabla_k \nabla_h \frac{\partial f}{\partial u_{hk}},$$

$$\frac{\partial f}{\partial u_k} = -\frac{\partial f}{\partial u_k} + 2\nabla \frac{\partial f}{\partial u_{hk}}, \qquad k = 1, \ldots, n, \qquad (17.10)$$

$$\frac{\partial f}{\partial u_{hk}} = \frac{\partial f}{\partial u_{hk}}, \qquad k, h = 1, \ldots, n,$$

which must be satisfied identically for all $u \in \overset{\circ}{C}{}^2(\Omega)$. From (17.10) it is not hard to conclude that for this the single relation

$$\frac{\partial f}{\partial u_k} - \nabla_h \frac{\partial f}{\partial u_{hk}} = 0 \quad \forall u \in \overset{\circ}{C}{}^2(\Omega), \ k = 1, \ldots, n, \qquad (17.11)$$

is sufficient, which condition is a criterion for formal potentiality of the operator N of (17.8).

Criteria for formal potentiality for equations more general than (17.8) can be obtained in a similar way (see Remark 17.1; tables of necessary and sufficient conditions for formal potentiality of some ordinary differential equations, partial differential equations, and systems of such equations are presented in [168] and [321]). However, it is not hard to verify that even for an ordinary differential equation of any odd order

$$f(u(x), u'(x), \ldots, u^{(2k+1)}(x)) = 0, \qquad u \in \overset{\circ}{C}{}^{2k+1}(a, b) \qquad (17.12)$$

condition (17.7) is never satisfied.

At the same time it should be kept in mind that, for example, for a "very good" elliptic equation

$$Au = \Delta^2 u + u = f(x), \qquad x \in \Omega, \qquad (17.13)$$

symmetric on $\overset{\circ}{C}{}^2(\Omega)$, i.e., satisfying the condition of formal potentiality (17.7), the criterion of potentiality (17.5) is not satisfied on the set of functions satisfying the well-defined boundary conditions

$$u|_{\partial\Omega} = 0, \qquad \left.\frac{\partial(\Delta u)}{\partial n}\right|_{\partial\Omega} = 0.(^{35}) \qquad (17.14)$$

REMARK 17.1. The importance of obtaining conditions of potentiality (formal potentiality) of differential operators has attracted the efforts of many mathematicians (pure and applied) and physicists to this problem over the course of almost a hundred years.

For an ordinary differential equation of second order condition (17.11) was obtained by Helmholtz [133]. In 1896 Hirsch [138] established a criterion of formal potentiality for a broad class of nonlinear ordinary differential equations and partial differential equations of order $m \geq 1$.

For a general nonlinear multidimensional PDE of second order condition (17.11) was obtained by Rapoport [273], [274], and it was then independently rediscovered by Tonti [321] and Berdichevskiĭ [26]. V. I. Zaplatniĭ [351] generalized this result to nonlinear PDE of fourth order. For a nonlinear PDE of arbitrary finite order $p \geq 1$ Berdichevskiĭ ([28], pp. 70–72) presented formulas obtained in a manner analogous to (17.11) from (17.5) and expressing a criterion of formal potentiality on $\overset{\circ}{C}{}^p(\Omega)$. Some other approaches to the investigation of potentiality of nonlinear PDE are known (see, for example, [8], [16], [121], [132], [141], [177], [286], and [335]. An investigation of formal potentiality of broad classes of integro-differential equations was recently carried out in the dissertation

---

(³⁵)This example is presented in [194], Chapter 2, §9, where symmetrization of the operator A of (17.13) and (17.14) is carried out with the help of an operator B, $Bv = -\Delta v$, $D(B) = D(A)$.

of V. M. Savchin (*Ostrogradsky's method and inverse problems of dynamics*, Peoples Friendship University, Moscow, 1984).

17.3. We shall consider the inverse problem of the calculus of variations in a broad sense: it is required to find for a nonlinear equation (17.0) a functional $\Phi(u)$ whose set of extremals coincides with the set of solutions of this equation. It is natural, especially in the case where $N = A$ is a linear operator

$$Au = f, \tag{17.15}$$

to use the functional of the method of least squares

$$\Phi_A = \|Au - f\|^2 \to \min. \tag{17.16}$$

However, in a number of cases of differential operators A the condition $f \in D(A^*)$ in the justification of this method may require additional smoothness of the function $f \in R(A)$, caused only by the special nature of this method, and also may require that f satisfy certain additional boundary conditions in considering an initial boundary value problem. It therefore makes sense to first apply to both sides of (17.15) an auxiliary operator $K: R(A) \to D(A^*)$,

$$KAu = Kf, \tag{17.17}$$

and then obtain the equation

$$A^*KAu = A^*Kf. \tag{17.18}$$

If $A^* K$ are invertible linear operators defined on $R(K)$ and $R(A)$ respectively, then any solution of (17.15) is a solution of (17.18), and conversely. If the linear operator K is chosen to be symmetric, then A^*KA will also be symmetric on $D(A)$, and by virtue of satisfying the criterion of potentiality (17.5) the functional (17.6) for $N = A^*KA$ gives a solution of this inverse problem of the calculus of variations.

The idea is due to E. Tonti, and he established it in the case of a general nonlinear equation (17.0) in [324].

THEOREM 17.1. *Suppose the nonlinear operator N of the equation*

$$N(u) = \theta_v \text{(}^{36}\text{)} \tag{17.19}$$

acting from a Banach space U to a Banach space $V = U^*$ satisfies the following conditions

1) $D(N) \subseteq U$, $R(N) \subseteq V$, *and on $U \times V$ there is defined a real, symmetric, nondegenerate, bilinear functional* $\langle \cdot, \cdot \rangle: U \times V \to R$.

---

[36] Here and for the rest of this subsection θ_u (θ_v) denotes the zero element in U (in V).

2) $D(N)$ is a convex set in U.

3) $D(N)$ is dense in U.

4) $R(N) \ni \theta_v$.

5) The Gâteaux derivative N'_u exists for all $u \in D(N)$.

6) The operator $(N'_u)^*$ is invertible for all $u \in D(N)$.

Suppose also that K is any operator satisfying the following conditions:

7) $D(K) \supseteq R(N)$.

8) $R(K) \subseteq D[(N'_u)^*]$ for all $u \in D(N)$.

9) K is linear.

10) K is invertible on $R(K)$.

11) K is symmetric.

Then the operator $\overline{N}$ defined by

$$\overline{N}(u) = (N'_u)^* K N(u), \tag{17.20}$$

possesses the following properties:

a) It is defined on $D(N)$.

b) An element $u_0 \in D(N)$ is a solution of (17.19) if and only if $\overline{N}(u_0) = \theta_v$.

c) $\overline{N}$ is a potential operator; the set of solutions in $D(N)$ of (17.19) coincides with the set of critical points of the functional

$$\overline{F}(u) = \tfrac{1}{2}\langle N(u), KN(u)\rangle, \qquad D(\overline{F}) = D(N) \tag{17.21}$$

and the gradient of the functional $\overline{F}$ is the operator $\overline{N}$.

PROOF. a) It follows directly from conditions 7) and 8) of the theorem for all $u \in D(N)$ there exists $\overline{N}(u)$ of (17.20); i.e., $D(\overline{N}) = D(N)$.

b) If $N(u_0) = \theta_v$, $u_0 \in D(N)$, then by conditions 7) and 8) and the linearity of K and $(N'_u)^*$ we obtain $\overline{N}(u_0) = \theta_v$.

Conversely, if $\overline{N}(u_0) = \theta_v$, from the invertibility of the operators $(N'_u)^*$ and K it follows that $N(u_0) = \theta_v$.

c) Using condition 11), we obtain

$$\begin{aligned}
\delta\overline{F}(u) &= \langle \delta N(u), KN(u)\rangle = \langle N'_u \delta u, KN(u)\rangle \\
&= \langle (N'_u)^* KN(u), \delta u\rangle = \langle \overline{N}(u), \delta u\rangle. \tag{17.22}
\end{aligned}$$

Since $D(\overline{F}) = D(N)$, it follows that $D(\overline{F})$ by condition 2) is simply connected set, and hence $\overline{F}(u)$ is a potential. From (17.22) we conclude also, because of the denseness of the set $D(N) = D(\overline{N})$ in U, that

$$\delta\overline{F}(u) = 0 \Leftrightarrow \overline{N}(u) = 0 \qquad (u \in D(N)). \quad \blacksquare$$

Thus, for a given nonlinear equation (17.19) considered under conditions 1)–6) the construction of a functional with property (17.2′) reduces to the construction of an operator K satisfying conditions 7)–11) of Theorem 7.1.

We shall consider the question of the existence of such an operator K for a general nonlinear equation (17.19) with an operator N acting in $L_2(\Omega) = U = V$, so that the bilinear functional indicated in Tonti's theorem is simply the inner product in $L_2(\Omega)$. Suppose

$$R(N) \subseteq C(\overline{\Omega}); \qquad \exists l \geq 0: \qquad D(N) \supseteq \overset{\circ}{C}{}^l(\Omega) \qquad (17.23)$$

and the operator N also satisfies conditions 2), 5) and 6) of Theorem 17.1.

The class of problems (17.19), (17.23) includes many boundary value problems for a PDE of finite order with sufficiently smooth coefficients $(R(N) \subseteq C(\overline{\Omega})$ and local boundary conditions (or nonlocal conditions of the form

$$\left. \sum_{|\alpha| \leq l} \sum_{(k)} a_\alpha D^\alpha \right|_{\Gamma_k} = 0,$$

$\Gamma_k \subseteq \Gamma = \partial\Omega$).

ASSERTION 17.1. *For the operator N of problem* (17.19), (17.23) *there exists a positive definite operator K satisfying the conditions of Theorem* 17.1.

To construct a positive-definite operator K satisfying conditions 9) and 10) of the theorem and such that (see conditions 7) and 8))

$$D(K) \supseteq C(\overline{\Omega}), \qquad R(K) \subseteq \overset{\circ}{C}{}^l(\Omega), \qquad (17.24)$$

we consider the Dirichlet problem for the polyharmonic equation

$$\begin{cases} (-1)^{l+1}\Delta^{l+1}u(x) = \varphi(x), & x \in \Omega, \\ u = \frac{\partial u}{\partial n} = \cdots = \frac{\partial^l u}{\partial n^l} = 0, & x \in \partial\Omega. \end{cases} \qquad (17.25)$$

As is known [314], for all $\varphi(x) \in C(\overline{\Omega})$ there exists $u(x) \in C^{2(l+1)}(\Omega) \cap \overset{\circ}{C}{}^l(\Omega)$—the solution of problem (17.25)—which can be represented in the form

$$u(x) - \int_\Omega G(x,\xi)\varphi(\xi)\,d\xi - L^{-1}\varphi. \qquad (17.26)$$

where $G(x,\xi)$ is the Green function of problem (17.25) whose operator $\mathcal{L}$ is symmetric and positive-definite on $D(\mathcal{L}) = C^{2(l+1)}(\Omega) \cap \overset{\circ}{C}{}^l(\Omega)$. If we now define the operator K by

$$Kv - \mathcal{L}^{-1}v - \int_\Omega G(x,\xi)v(\xi)\,d\xi, \qquad D(K) = C(\overline{\Omega}), \qquad (17.27)$$

then the linear operator K will be symmetric and positive-definite because of known (see [57], vol. II and [314]) analogous properties of the operator $\mathscr{L}$ of problem (17.25), i.e., K satisfies conditions 9)–11). Condition 7) follows from (17.27) and the first relation in (17.23).

If $\psi \in R(K)$, then by construction the function ψ is a solution of problem (17.25) in $D(\mathscr{L})$, i.e., considering (17.23), $\psi \in \overset{\circ}{C}{}^{l}(\Omega) \subseteq D[(N'_u)^*]$ for all $u \in D(N)$, which proves condition 8): $R(K) \subseteq D[(N'_u)^*]$ for all $u \in D(N)$. ∎

COROLLARY 17.1. *If the integral operator K in the conditions of Theorem 17.1 is not only symmetric but also positive-definite (see Assertion 17.1 and the examples in [324]), while the operator $N: D(N) \subseteq H \to H$, so that the bilinear functional $\langle \cdot, \cdot \rangle$ used in Theorem 17.1 is the inner product in H, then it is possible to define a "negative" space H_K as the completion of $D(N)$ in the norm*

$$\|u|H_K\| \equiv (\langle u, u \rangle_K)^{1/2} = \langle u, Ku \rangle^{1/2}. \tag{17.28}$$

The functional (17.21) is then the functional of the method of least squares in the space H_K:

$$\overline{F}(u) = \tfrac{1}{2}\langle N(u), N(u) \rangle_K = \tfrac{1}{2}\|N(u)|H_K\|^2. \tag{17.29}$$

It is obvious here that the functional $\overline{F}(u)$ of (17.29) satisfies not only condition (17.2′) but also (17.1).

Condition (17.3) for a linear differential operator can also be satisfied, for example, if $H = L_2(\Omega)$, since in this case the space H_K generated by the Green function of problem (17.25) is equivalent to some space $\overset{\circ}{W}{}_2^{-l}(\Omega)$ [30]. We shall verify condition (17.3) in the case of the operator N of the problem

$$-\Delta u = f(x), \quad x \in \Omega \subset R_n, \quad u|_{\partial\Omega} = 0, \tag{17.30}$$

for which the Dirichlet functional, possessing properties (17.1)–(17.3), is known. Assuming that $R(N) \subseteq C(\overline{\Omega})$, we see that the operator $N = N'$ of problem (17.30) satisfies (17.23) for $l = 0$. Therefore, the operator K of (17.27) defined by the Green function of problem (17.25) for $l = 0$ coincides with the operator $N^{-1}, K = N^{-1} = (-\Delta)^{-1}$, of problem (17.30) and by Assertion 17.1 satisfies conditions 7)–11) of Theorem 17.1. In this case the functional (17.29) has the form

$$\overline{F}(u) = \frac{1}{2}\| - \Delta u - f|H_K\|^2$$
$$= \frac{1}{2}\int_\Omega (-\Delta u - f)(-\Delta)^{-1}(-\Delta u - f)\,d\Omega$$
$$= \frac{1}{2}\int_\Omega \{|\nabla u|^2 - 2fu\}\,dx + C_f,$$

i.e., differs only by the constant $C_f = (f, Kf)$ from the familiar Dirichlet functional for problem (17.30).

REMARK 17.2. Corollary 17.1 demonstrates the closeness of the approaches of Tonti and Didenko (see (16.28)) in the case of a linear operator N. Didenko's symmetrizing operator is constructed (see [75] and [76]) in explicit form with the help of the Green function of a boundary value problem for the Laplace equation; in [324] for a nonlinear ordinary differential equation of second order auxiliary operators K were also constructed by means of the Green function for the operator d^2/dt^2; we generalized the idea of these constructions in the proof of Assertion 17.1. Although in a number of cases [324] it is easy to construct operators K in this manner for nonlinear ordinary differential equations, for boundary value problems for partial differential equations in rather general domains $\Omega \subset R_n$ this approach to the construction of operators K is complicated.

17.4. On the justification of MLS for nonlinear, nonpotential, nonmonotone operators*. The quasiclassical solution of the inverse problem of the calculus of variations obtained in the preceding subsection is the functional of MLS in a certain negative Hilbert space H_K, but the constructions of that subsection did not provide a justification of the method of least squares itself; this is done as follows.*

Suppose the operator N of the equation

$$N(u) = f \tag{17.31}$$

acts in some Hilbert space H (possibly a negative space) with inner product $(\cdot, \cdot)$, and

$$\begin{cases} D(N) \text{ is a convex dense set in a Hilbert space } H_1, H_1 \subseteq H; \\ R(N) \text{ is dense in } H; \\ N(0) = 0; \\ \text{the system } \varphi_1, \varphi_n \ldots \text{ forms a basis in } H_1. \end{cases} \tag{17.32}$$

---

*Subsection 17.4 has been added in translation.

The method of least squares is to take as an approximate solution of (17.31) a linear combination

$$u_n = \sum_{k=1}^{n} a_k \cdot \varphi_k, \qquad (17.33)$$

of the given elements $\varphi_1, \ldots, \varphi_n$ whose unknown coefficients $a_1, \ldots, a_n$ are determined from the condition

$$\|N(u_n) - f\|_H^2 \to \min_{a_1, \ldots, a_n} \qquad (17.34)$$

THEOREM 17.2 (LAGENBACH). ([37]) *Suppose condition* (17.32) *is satisfied, and*([38])

$$(N(u_1) - N(u_2), u_1 - u_2) \geq C_1 \|u_1 - u_2\|_{H_1}^2 \quad \forall u_1, u_2 \in D(N), \qquad (17.35)$$

$$\|u\|_H \leq C_2 \|u\|_{H_1} \quad \forall u \in D(N). \qquad (17.36)$$

Then the sequence of elements u_n *of* (17.33) *and* (17.34) *converges in* H_1.

The monotonicity condition (17.35) in the linear case is the condition of positive-definiteness, and therefore Langenbach's theorem did not permit justification of MLS for many linear and nonlinear PDE. We shall prove

THEOREM 17.3. *Suppose conditions* (17.32) *are satisfied, and there exists a linear operator* B *acting in* H, $D(B) \supseteq D(N)$, *such that*

$$(N(u_1) - N(u_2), B(u_1 - u_2)) \geq C_3 \|u_1 - u_2\|_{H_1}^1 \quad \forall u_1 u_2 \in D(N), \qquad (17.37)$$

$$\|Bu\|_H \leq C_4 \|u\|_{H_1} \quad \forall u \in D(N). \qquad (17.38)$$

Then the sequence of elements u_n *of* (17.33) *and* (17.34) *converges in* H_1.

PROOF. From the definition of u_n and (17.32) it follows that there exists at least one sequence $\{u_k\}$ for which

$$\|N(u_n) - N(u_m)\|_H \to 0 \qquad (m, n \to \infty). \qquad (17.39)$$

With the help of (17.37) and (17.38) we find that

$$\|u_n - u_m \cdot \|_{H_1}^2 \leq \frac{1}{C_3} \|N(u_n) - N(u_m)\| \cdot \|B(u_m - u_n)\|_H$$

$$\leq \frac{C_4}{C_3} \cdot N(u_n) - N(u_m)\| \cdot \|u_m - u_n\|_{H_1},$$

([37])See [185].

([38])The positive constants $C_1, C_2, \ldots$ in this subsection do not depend on u, v, u_i, or v_i $(i = 1, 2, \ldots)$.

that is

$$\|u_n - u_m\|_{H_1} \leq \frac{C_4}{C_3} \cdot \|N(u_n) - N(u_m)\|_H. \tag{17.40}$$

From this by (17.39) it follows that the sequence $\{u_n\}$ is fundamental in the Hilbert space H_1, as required.

Use of the auxiliary operator B makes it possible to generalize, obtaining a number of other useful properties for the MLS.

ASSERTION 17.2. a) *If on* $D(N)$ *there exists a linear Gâteaux differential continuous on any line*

$$N'(u) \cdot h \equiv \left\{ \frac{d}{d\alpha} N(u + \alpha \cdot h) \right\}_{\alpha=0}, \tag{17.41}$$

then (17.37) *follows from the condition*

$$(N'(u) \cdot h, Bh) \geq C_3 \cdot \|h\|^2_{H_1} \quad \forall u_1 h \in D(N). \tag{17.42}$$

b) *Suppose* (17.37) *and* (17.38) *are satisfied. Then on* $R(N)$ *the operator* N *has an inverse operator satisfying a Lipschitz condition.*

PROOF. a) For all $u_1, u_2 \in D(N)$ we have

$$N(u_1) - N(u_2) = \int_0^1 N'(t \cdot u_1 + (1 - t)u_2) \cdot (u_1 - u_2) \, dt. \tag{17.43}$$

Taking the inner product in H of (17.43) with $B(u_1 - u_2)$ and interchanging integration and the inner product, by (17.42) we obtain

$$(N(u_1) - N(u_2), B(u_1 - u_2)) = \int_0^1 (N'(t \cdot u_1 + (1 - t)u_2) \cdot (u_1 - u_2),$$
$$B(u_1 - u_2)) \, dt \geq C_3 \cdot \|u_1 - u_2\|^2_{H_1}.$$

b) Let v_1 and v_2 be arbitrary elements in $R(N)$. By (17.37) to them there correspond unique elements u_1 and u_2 in $D(N)$ such that $v_1 = Nu_1$ and $v_2 = Nu_2$. From (17.40) we have

$$\|u_1 - u_2\|_{H_1} \leq \frac{C_4}{C_3} \cdot \|v_1 - v_2\|_H \quad \forall u_1, u_2 \in D(N), \tag{17.44}$$

which gives an estimate of the Lipschitz constant. ∎

Thus the operator N^{-1}, having according to (17.32) a domain $R(N)$ dense in H, admits a continuous extension $\tilde{N}^{-1}$ to all of H. More precisely, we can introduce

DEFINITION 17.1. An element $u_0 \in H_1$ is called a *generalized solution* of equation (12.13) if there exists a sequence $\{u_m\}, u_m \in D(N), m = 1, 2, \ldots,$

such that

$$\|N(u_n) - f\|_H \to 0 \qquad (m \to \infty), \tag{17.45}$$

$$\|u_m - u_0\|_{H_1} \to 0 \qquad (m \to \infty), \tag{17.46}$$

From Theorem 17.3 and Assertion 17.2b) we obtain

COROLLARY 17.2. *Suppose conditions* (17.32), (17.37), *and* (17.38) *are satisfied for a nonlinear operator* N. *Then for any* $f \in H$ *there exists a unique generalized solution* $u_0 \in H_1$ *of equation* (17.31).

The approximation of u_n to the exact solution u_0 can be determined for MLS from the system of nonlinear algebraic equations

$$\frac{\partial}{\partial a_k}\|N(u_k) - f\|_H^2 = 0, \qquad k = 1,\ldots,n, \tag{17.47}$$

which can be written in the form

$$(N'(u_n)\varphi_k, N(u_n) - f) = 0 \qquad k = 1,\ldots,n. \tag{17.48}$$

For completeness we shall establish the solvability of this nonlinear system under the generalized conditions (17.37) and (17.38).

LEMMA 17.1. *Suppose the operator* N *has a continuous Gâteuax derivative,* $N(0) = 0$, *conditions* (17.37) *and* (17.38) *are satisfied, and the system of coordinate functions* $\varphi_1, \varphi_2, \ldots$ *are orthonormal in* H_1. *Then the system* (17.48) *of algebraic equations of MLS is solvable for any* $n \in \mathbf{N}$.

PROOF. Considering (17.38) and (17.37), we find that for all $u \in D(N)$

$$C_4 \cdot \|N(u_n)\|_H \cdot \|u_n\|_{H_1} \geq (N(u_n), Bu_n) \geq C_3 \cdot \|u_n\|_{H_2}^2,$$

where

$$\|N(u_n)\| \geq \frac{C_3}{c_4} \cdot \|u_n\|_{H_1}. \tag{17.49}$$

The coordinate functions $\varphi_1, \varphi_2, \ldots$ are orthonormal in H_1, and hence for the u_n of (17.33) we have $\|u_n\|_{H_1} = (\sum_1^n a_k^2)^{1/2}$. By the conditions on N the function $F(a_1,\ldots,a_n) = \|(N(u_n) - f\|_H^2$ is a continuously differentiable function of the coefficients $a_1, \ldots, a_n$, and

$$\|N(u_n)\| \geq \frac{C_3}{C_4} \cdot \|u_n\|_{H_1} = \frac{C_3}{C_4} \cdot \left(\sum_{K=1}^n a_k^2\right)^{1/2} \geq 2 \cdot \|f\|_H$$

on a sphere of sufficiently large radius R_f. From this we obtain

$$\|N(u_n) - f\|_H^2 \geq \|N(u_n)\|_H \cdot \{\|N(u_n)\|_H - 2\|f\|_H\} + \|f\|_H^2 \geq \|f\|_H^2,$$

that is, the continuous function $F(a_1, \ldots, a_n)$ is bounded below in the ball $\sum_1^n a_k^2 \leq R_f^2$ and therefore has a minimum in this ball. The system (17.48) is another form of writing the necessary condition (17.47) for a minimum $F(a_1, \ldots, a_n)$; hence, this system is solvable, as required. $\blacksquare$

From the results of this subsection we obtain the corresponding results of Langenbach indicated above if we set $B = I$. However, in this case (for $B = I$) even if $H = L_2(\Omega)$ Langenbach's results did not justify MLS, for example, for a number of ordinary differential operators of odd order and for parabolic operators. As an application of our generalizations of the results, we justify MLS[39] for the nonlinear parabolic problem

$$N(u) \equiv \frac{\partial u}{\partial t} - \sum_{i=1}^{n-1} \frac{\partial^2 u}{\partial x_i^2} - \frac{\partial^2}{\partial x_n^2} k(x, t, u) = g(x, t), \qquad (x, t) \in \Omega, \quad (17.50)$$

where the bounded domain

$$\Omega = \{(x, t) \equiv (x_1, \ldots, x_n, t): x_i \in (a_i, b_i), \ i = 1, \ldots, n, \ t \in (0, T)\},$$

$$u(x, t)|_{t=0} = 0, \tag{17.51}$$

$$u(x, t)|_{x_i = B_i} = 0 \qquad (i = 1, \ldots, n), \tag{17.52}$$

$$\left. \frac{\partial u}{\partial n} \right|_{x_i = a_i} = 0 \qquad (i = 1, \ldots, n), \tag{17.53}$$

$$\begin{cases} g(x, t) \in L_2(\Omega); \quad k(\xi, \eta, \zeta) \in C^2(R_3); \quad \exists \alpha \equiv \text{const}: \partial k/\partial u \geq \alpha^2 > 0 \\ \forall u \in D(N); \quad \partial k/\partial u \in C(\overline{\Omega}) \quad \forall u \in D(N); \end{cases}$$
$$\tag{17.54}$$

here $D(N) = \{u(x, t) \in C_{x,t}^{2,1}(\overline{\Omega})$ satisfying (17.51)–(17.53)$\}$.

We define the auxiliary operator B by the relation

$$Bh(x, t) = -\int_{B_n}^{x_n} d\xi \int_{a_n}^{\xi} h(x_1, x_2,, \ldots, x_{n-1}, \theta, t) \, d\theta, \qquad D(B) = D(N). \tag{17.55}$$

ASSERTION 17.3. *Suppose* (17.54) *is satisfied, and the Hilbert space* H_1 *is the completion of* $D(N)$ *in the metric*

$$\|u\|_{H_1} = [u, u]^{1/2}, \tag{17.56}$$

$$[u, v] = \int_\Omega \left\{ \alpha^2 \cdot u \cdot v + \sum_{i=1}^{n-1} \int_{a_n}^{x_n} \frac{\partial u}{\partial x_i}(x_1, \ldots, x_{n-1}, \xi, t) \, d\xi \right.$$
$$\left. \cdot \int_{a_n}^{x_n} \frac{\partial v}{\partial x_i}(x_1, \ldots, x_{n-1}, \xi, t) \, d\xi \right\} d\Omega. \tag{17.57}$$

---

[39] For which if (17.32) is satisfied it suffices by Theorem 17.3. to establish (17.37) and (17.38).

Then relations (17.37) and (17.38) are satisfied for the operator N of (17.50)—(17.53) and the operator B of (17.55).

PROOF. By integration by parts on $D(N)$ we obtain, by (17.54),

$$
(N'_u h, Bh) \equiv \int_\Omega \left\{ \frac{\partial h}{\partial t} - \sum_{i=1}^{n-1} \frac{\partial^2 h}{\partial x_i^2} - \frac{\partial^2}{\partial x_n^2}\left(\frac{\partial k}{\partial u}\cdot h\right) \right\}
$$

$$
\cdot \left\{ -\int_{B_n}^{x_n} d\xi \int_{a_n}^{\xi} h(x_1, \ldots, x_{n-1}, \theta, t)\, d\theta \right\} d\Omega
$$

$$
= \int_\Omega \left\{ \int_{a_n}^{x_n} h(x_1, \ldots, x_{n-1}, \theta, t)\, d\theta \right.
$$

$$
\cdot \int_{a_n}^{x_n} \frac{\partial h}{\partial t}(x_1, \ldots, x_{n-1}, \theta, t)\, d\theta
$$

$$
\left. + \frac{\partial k}{\partial u}\cdot h^2 + \sum_{i=1}^{n-1} \left(\int_{a_n}^{x_n} \frac{\partial h}{\partial x_i}(x_1, \ldots, x_{n-1}, \theta, t)\, d\theta \right)^2 \right\} d\Omega
$$

$$
\leq \|h|H_1\|^2.
$$

Thus, (17.37) is satisfied.

That (17.38) is satisfied can be established by direct computation: for all $h \in D(N)$

$$
\|Bh\|_H^2 \equiv \int_\Omega \left(-\int_{b_n}^{x_n} d\xi \int_{a_n}^{\xi} h(x_1, \ldots, x_{n-1}, \theta, t)\, d\theta \right)^2 d\Omega
$$

$$
\leq (b_n - a_n)^2 \cdot \int_\Omega h^2(x, t)\, dx \cdot dt \leq \frac{(b_n - a_n)^2}{\alpha^2}\cdot \|h|H_1\|^2. \quad \blacksquare
$$

The theoretical results of this subsection make it possible to formulate the following corollary for the concrete problem (17.50)–(17.53).

COROLLARY 17.3. *Suppose condition (17.32) is satisfied for the operator N of boundary value problem (17.50)—(17.53).*[40] *Then the following assertions are true.*

a) *For any $g \in L_2(\Omega)$ there exist a unique element $u_0 \in H_1$ and a sequence $\{u_n\}$, $u_n \in D(N)$, $n = 1, 2, \ldots$, such that, as $n \to \infty$,*

$$
\|u_n - u_0|H_1\| \to 0, \qquad \|N(u_n) - g|L_2(\Omega)\| \to 0.
$$

b) *The sequence $\{u_n\}$, $u_n = \sum_1^n a_k \varphi_k$, $n = 1, 2, \ldots$, can be found from the condition of a minimum of MLS,*

$$
F(a_1, \ldots, a_n) \equiv \|N(u_n) - g|L_2(\Omega)\|^2 \to \min_{a_1, \ldots, a_n},
$$

---

[40] Since $D(N)$ is dense in $H = L_2(\Omega)$, this condition (17.32) on $k(\cdots)$ (in addition to the orthogonality of the basis $\varphi_1, \varphi_2, \ldots$) is sufficient for $R(N)$ to be dense in $L_2(\Omega)$.

which leads to the system of algebraic equations

$$\int_\Omega N'(u_n) \cdot \varphi_k(x) \cdot \{N(u_n) \to g\} \, dx = 0 \qquad (k = 1, \ldots, n); \qquad (17.58)$$

c) *The system of nonlinear algebraic equations* (17.58) *is solvable for all* $g \in L_2(\Omega)$ *for any* $n = 1, 2 \ldots$.

§18. A variational method of solving nonlinear equations with nonpotential operators

18.1. We consider the equation

$$N(x) = f \qquad (18.1)$$

with a nonlinear operator N acting in some real Hilbert space H with inner product $(\cdot, \cdot)$ and norm $\|x\|_H = \|x\| = (x, x)^{1/2}$; f is a given function in H.

We introduce the following conditions.[41]

α) The operator N is defined on a linear manifold of functions $D(N)$ dense in H; $N(0) = 0$; and for all $x \in D(N)$ the continuous (in x) Gâteaux derivative N'_x exists, so that $H'(t) = (N'_{x+ty}\varphi, x\psi) \in C_t[0, 1]$ for all $x, y, \varphi, \psi \in D(N)$.

β) There exists a closable linear operator B, $D(B) \supset D(N)$, $\overline{R_N(B)} = H$, such that

$$(N'_x u, Bv) = (Bu, N'_x v) \quad \forall x, u, v \in D(N).$$
$$\gamma)(N'(0)u, Bu) \geq C_0\|u\|^2 \ \forall u \in D(N);$$
$$\gamma')(N'_x u, Bu) \geq C(N'(0)u, Bu) \ \forall x, u \in D(N);$$
$$\gamma'')|(f, Bu)| \leq C_f(N'(0)u, Bu)^{1/2} \ \forall u \in D(N), \ f \in H,$$

where C, C_0, and C_f are positive constants not depending on x or u.

As in Chapter 1, we shall henceforth assume that if the operator $N = N'(0)$ is B-symmetric and B-positive [i.e., γ) and β) are satisfied for $x = 0$], then it must also be weakly closable (§1). If the B-symmetric and B-positive operator $A = N'(0)$ also satisfies condition γ'') (according to Lemma 4.2, it is then B-positive-definite), then the condition of weak closability of A is not needed.

---

[41]In this subsection we follow the scheme of Lyashko [202], but in the beginning we do not have to require that the operator N satisfy the entire collection of these conditions; the assertions proved below are obtained under various combinations of conditions α)–γ'').

LEMMA 18.1. *Suppose conditions* α), γ), *and* γ') *are satisfied for the operator* N. *Then, if there exists a solution* $u_0 \in D(N)$ *of equation* (18.1), *it is unique.*

Indeed, if x_1 and x_2 are two solutions (in $D(N)$) of (18.1), then by conditions α), γ'), and γ)

$$
\begin{aligned}
0 &= (Nx_1 - Nx_2, B(x_1 - x_2)) \\
&= \int_0^1 (N'(x_2 + t(x_1 - x_2))(x_1 - x_2), B(x_1 - x_2))\, dt \\
&\geq C(N'(0)(x_1 - x_2), B(x_1 - x_2)) \geq CC_0\|x_1 - x_2\|^2,
\end{aligned}
$$

whence it follows that $x_1 = x_2$. ∎

THEOREM 18.1. 1. *If conditions* α)—γ') *are satisfied and an element* $x_0 \in D(N)$ *is a solution of equation* (18.1), *then* x_0 *realizes in* $D(N)$ *a minimum of the functional*

$$
\Phi(x) = \int_0^1 (N(tx), Bx)\, dt - (f, Bx). \tag{18.2}
$$

2. *If conditions* α) *and* β) *are satisfied, then an element* $x_0 \in D(N)$ *minimizing the functional* $\Phi(x)$ *of* (18.2) *in* $D(N)$ *is a solution of equation* (18.1).

PROOF. 1. Suppose $x_0 \in D(N)$ is a solution of (18.1). For any $x, h \in D(N)$

$$
\begin{aligned}
\Phi(x + h) &- \Phi(x) \\
&= \int_0^1 [(N(t(x + h)), B(x + h)) - (N(tx), Bx)]\, dt - (f, Bh) \\
&= \int_0^1 (N(t(x + h)), B(x + h))\, dt \\
&\quad + \int_0^1 (N(t(x + h)) - N(tx), Bx)\, dt - (f, Bh). \tag{18.3}
\end{aligned}
$$

We transform the second integral on the right with the help of condition β):

$$\int_0^1 (N(t(x+h)) - N(tx), Bx)\, dt$$

$$= \int_0^1 dt \int_0^t \frac{d}{ds}(N(tx+sh), Bx)\, ds$$

$$= \int_0^1 ds \int_s^1 (N_t'(tx+sh)h, Bx)\, dt$$

$$= \int_0^1 ds \int_0^1 (N'(tx+sh)x, Bh)\, dt$$

$$= \int_0^1 (N(x+sh) - (N(sx+sh), Bh)\, ds$$

$$\equiv \int_0^1 (N(x+th) - N(tx+th), Bh)\, dt.$$

From the last two relations we obtain

$$\Phi(x+h) - \Phi(x) = \int_0^1 (N(x+th), Bh)\, dt - (f, Bh)$$

and, since $N(x_0) = f$, with the help of condition γ) and γ') we obtain

$$\Phi(x_0 + h) = \Phi(x_0) + \int_0^1 (N(x_0 + th) - N(x_0), Bh)\, dt$$

$$= \Phi(x_0) + \int_0^1 \frac{dt}{t} \int_0^1 (N'(x_0 + sth)th, B(th))\, ds \geq \Phi(x_0).$$

Hence, since $h \in D(N)$ is arbitrary, the element x_0 minimizes the functional $\Phi(x)$ in $D(N)$.

2. Suppose, conversely, than an element $x_0 \in D(N)$ miminizes the functional $\Phi(x)$ of (18.2) in $D(N)$; then from the necessary condition for a minimum of $\Phi(x)$, by integrating by parts and using the symmetry condition β), we find that

$$0 = \left[\frac{d}{d\alpha} \Phi(x_0 + \alpha\eta) \right]_{\alpha=0} = (N(x_0) - f, B\eta) \quad \forall \eta \in D(N).$$

From the density in H of the set $R_N(B)$ we conclude that $N(x_0) = f$. ∎

It is not hard to verify

ASSERTION 18.1. *Under conditions* $\alpha)$-$\gamma)$ *the relations*

$$[u, v]_{NB} = (N'(0)u, Bv), \tag{18.4}$$

$$\|u\|_{NB} = [u, u]_{NB}^{1/2} \tag{18.5}$$

define an inner product and norm on $D(N)$.

Hence, in analogy to Chapter 1 we can introduce the Friedrichs space H_{NB} as the completion of the set $D(N)$ in the norm (18.5).

COROLLARY 18.1. *Because of condition* γ) *convergence of some sequence* $\{u_n\}$ *in* H_{NB} *implies its convergence in* H, *and hence* $H_{NB} \subseteq H$ (*see Corollary* 3.1).

COROLLARY 18.2. *Condition* γ'') *is equivalent to the inequality*
$$\tilde{\gamma}'') \quad \|u\|_{NB} \geq \tilde{C}\|Bu\| \quad \forall u \in D(N)$$
with a constant $\tilde{C} > 0$ *not depending on* u.

Indeed, $\tilde{\gamma}''$) is satisfied, then for all $u \in D(N)$

$$|(f, Bu)| \leq \|f\|_H \|Bu\|_H \leq \|f\| \frac{1}{\tilde{C}} (N'(0)u, Bu)^{1/2}.$$

Conversely, if γ'') is true for all $f \in H$, then

$$\left| \left(f, \frac{Bu}{\|n\|_{NB}} \right) \right| \leq C_f \quad \forall u \in D(N),$$

where the constant $C_f > 0$ does not depend on u. Then by a familiar theorem of functional analysis the set

$$\left\{ \frac{Bu}{\|u\|_{NB}} : u \in D(N) \right\}$$

is bounded in H, i.e., there exists a constant $C_1 > 0$, not depending on u, such that $\|Bu\| \leq C_1\|u\|_{NB}$ for all $u \in D(N)$. □

Theorem 18.1 established the equivalence of the variational problem of minimizing the functional (18.2) and problem (18.1) under the condition of the existence in $D(N)$ of at least one element minimizing (18.2) in $D(N)$. Directly in terms of the operators introduced it is also possible to establish conditions of existence and uniqueness of a solution of the following variational problem.

DEFINITION 18.1. An element $u_0 \in H_{NB}$ is called a *solution of the variational problem of minimizing the functional* $\Phi(x)$ *in* H_{NB} if there exists a sequence $\{u_n\}$ of elements in $D(N)$ such that

$$\|u_n - u_0\| \to 0 \quad (n \to \infty),$$
$$\lim_{n \to \infty} \Phi(u_n) = d \equiv \inf_{D(N)} \Phi(u).$$

It follows from Theorem 18.1 that if u_0 is a solution of the variational problem and $u_0 \in D(N)$, then u_0 is also a solution of (18.1). Therefore, in general any solution $u_0 \in H_{NB}$ of the variational problem is sometimes called [95] a generalized solution of equation (18.1).$(^{42})$

THEOREM 18.2. *Under conditions* α)—γ''), *for any* $f \in H$ *there exists a unique solution in* H_{NB} *of the variational problem of minimizing the functional* (18.2).

PROOF. We first show that the function $\Phi(x)$ is bounded below:

$$\Phi(x) = \int_0^1 (N(tx), Btx)\frac{dt}{t} - (f, Bx)$$

$$= \int_0^1 \frac{dt}{t} \int_0^1 (N'(stx)tx, Btx)\, dx - (f, Bx)$$

$$\leq \frac{C}{2}\|x\|_{NB}^2 - |(f, Bx)| \geq -C_f^2\frac{1}{2C} > -\infty,$$

where C and C_f are the positive constants of γ') and γ'').
   For arbitrary $u, v \in D(N)$ we define

$$\rho(u, v) \equiv \frac{1}{2}\Phi(u) + \frac{1}{2}\Phi(v) - \Phi\left(\frac{u+v}{2}\right)$$

$$= \frac{1}{2}\left[\Phi(u) - \Phi\left(\frac{u+v}{2}\right)\right] + \frac{1}{2}\left[\Phi(v) - \Phi\left(\frac{u+v}{2}\right)\right]$$

and transform it, setting $h = v - u$:

$$\rho(u, v) = \frac{1}{2}\left[\Phi(u) - \Phi\left(u + \frac{h}{2}\right)\right] + \frac{1}{2}\left[\Phi(u+h) - \Phi\left(u + \frac{h}{2}\right)\right]$$

$$= \frac{1}{2}\int_0^1 \left(N\left(u + \frac{h}{2} + s\frac{h}{2}\right) - N\left(u + s\frac{h}{2}\right), B\frac{h}{2}\right) ds$$

$$= \frac{1}{2}\int_0^1 \int_0^1 ds \int_0^1 \left(N'\left(u + (s+t)\frac{h}{2}\right)\frac{h}{2}, B\frac{h}{2}\right) dt$$

$$\leq \frac{C}{8}\|h\|_{NB}^2 = \frac{C}{8}\|u - v\|_{NB}^2. \tag{18.6}$$

Since $\Phi(x)$ is bounded below on $D(N)$, there exists a sequence $\{x_n\}, x_n \in D(N), n = 1, 2, \ldots$, for which

$$\lim_{n\to\infty} \Phi(x_n) = d = \inf_{D(N)} \Phi(x),$$

i.e., for every $\varepsilon > 0$ there exists $n_0 = n_0(\varepsilon) > 0$ such that for all $m, n \geq n_0$

$$\Phi(x_n) < d + \varepsilon, \qquad \Phi(x_m) < d + \varepsilon.$$

$(^{42})$The concepts of weak and strong solutions of (18.1) are introduced in §18.3, and their connection with a solution of the variational problem is established.

For such n and m, from (18.6) we obtain

$$\|x_n - x_m\|_{NB}^2 \leq \frac{8}{C}\varepsilon, \tag{18.7}$$

and therefore the sequence $\{x_n\}$ converges in H_{NB} to an element $\bar{x} \in H_{NB}$.

From condition γ) we then find that the sequence $\{x_n\}$ also converges in H, and hence $\bar{x}$ is a solution of the variational problem.

If we assume the existence in H_{NB} of two solutions $\bar{x}$ and $\bar{\bar{x}}$ of the variational problem minimizing $\Phi(x)$, then, choosing two minimizing sequences $\{x_n\}$ and $\{x_n'\}$ in $D(N)$, which converge in H_{NB} to $\bar{x}$ and $\bar{\bar{x}}$ respectively, we find from (18.7) (where it is necessary to set $x_m = x_m'$ for all m)

$$\|x_m' - x_n\|_{NB} \to 0 \qquad (m, n \to \infty),$$

i.e., the sequences $\{x_n\}$ and $\{x_n'\}$ converge in H_{NB} (and hence in H) to a single element $\bar{x} = \bar{\bar{x}}$. $\square$

Existence and uniqueness of a solution of the variational problem has been established under the condition of the existence of an auxiliary operator B with properties β)–γ''). If N' is a symmetric positive-definite operator, then it suffices to set $B = I$ (see [181], [184], and [331]). For an invertible linear operator N the existence of an auxiliary operator B is established in Theorem 1.1. In the more general case of nonlinear operator N we have

THEOREM 18.3. *If the operator N has on $D(N)$, $\overline{D(N)} = H$, a Gâteaux derivative N', $R(N') = H$, and if $(N')^{-1}$ exists on $R(N')$, then the linear operator*

$$B = [(N')^{-1}]^* C \tag{18.8}$$

exists where C is an arbitrary, linear, symmetric, and positive-definite operator on $D(C) \supseteq D(N')$, and the operator N is B-symmetric, B-positive, and weakly closable (relative to B).

PROOF. From the existence of $(N')^{-1}$ and the density in H of the sets $D(N')$ and $R(N')$ there follows the existence of $(N')^*$, $[(N')^*]^{-1}$, $[(N')^{-1}]^*$, and $[(N')^*]^{-1} = [(N')^{-1}]^*$.

We select an arbitrary operator C satisfying the conditions of the theorem; it suffices, for example, to set $C = \kappa \cdot I$, $\kappa = \text{const} > 0$, $D(C) = D(N)$.

We consider the product of operators $C(N')^{-1}$, which is meaningful, since $D(C) \supseteq D(N') = R[(N')^{-1}]$. From the density of the set $R(N') = D[(N')^{-1}]$ in H it then follows that $D[C(N')^{-1}] = H$, and hence the operator

$$B = [C(N')^{-1}]^* = [(N')^{-1}]^* C \tag{18.9}$$

exists. Here

$$(N'u, Bv) = (N'_u, [(N')^{-1}]^*Cv) = (u, Cv)$$

and the operator N' is B-symmetric, B-positive, and weakly closable, since by construction the operator C is symmetric and positive-definite. ∎

Remark 18.1. Because of the density in H of the sets $D(N)$ and $R_N(B)$, condition γ' holds if in H the inequality

$$N'_x u \geq CN'(0)u \quad \forall x, u \in D(N), \tag{18.10}$$

holds in the sense that $(N'_x u, \varphi) \geq C(N'(0)u, \varphi)$ for all $\varphi \in D(N)$; here x and u are arbitrary but fixed elements in $D(N)$.

The assumption of uniform (in x) boundedness of the operator $(N'_x)^{-1}$,

$$\|(N'_x)^{-1}u\| \leq \alpha_1 \|u\|_{NB} \quad \forall x \in D(N), \ u \in R(N'),$$

where the constant $\alpha_1 > 0$ does not depend on x or u, is sufficient for the validity of condition γ''). Indeed, since the arbitrary operator C in the proof of Theorem 18.1 can be chosen to satisfy

$$\|Cu\| \leq \alpha_2 \|u\|, \qquad \alpha_2 \equiv \text{const} > 0,$$

it follows that

$$\begin{aligned}|(f, Bu)| &\leq \|f\|_H \|C(N')^{-1}u\|_H \\ &\leq \alpha_2 \|f\|_H \|(N')^{-1}u\|_H \leq \alpha_1\alpha_2 \|f\|_H \|u\|_{NB},\end{aligned}$$

which proves γ'').

18.2. In the preceding subsection a generalized solution of the nonlinear equation (18.1) was understood to be an element $x_0 \in H_{NB}$ which is a solution of the variational problem of minimizing the functional (18.2) in H_{NB}. This approach was used in [182], [184], and [202], where the domain $D(\tilde{N})$ of some extension $\tilde{N} \supset N$ of the operator N was thus defined a priori as the set of solutions of the variational problem of minimizing the functional $\Phi(x)$ of (18.2).

In the case of a linear operator $N \equiv A$ a generalized solution was defined in Chapter 1 on the basis of an extension of the operator, and its was then proved that such a generalized solution coincides with the element minimizing the corresponding functional.

In §18.3 we shall present an analogous investigation (by means of the solvable extension of nonlinear operators developed by Petryshyn in [258] and [52]) of the generalized solution constructed in §18.1 for equation (18.1), and we now present the required facts, following [52] and [258].

Suppose N is a nonlinear operator, A is some linear, B-symmetric, and B-positive-definite operator, $D(A) = D(N)$, and B is a closable operator: $D(B) \subseteq H \to R(B) \subseteq H$, $D(B) \supseteq D(A)$, $\overline{D(A)} = H$, $\overline{R_A(B)} = H$, and also

$$(Au, Bu) \geq \alpha^2 \|u\|^2 \quad \forall u \in D(A), \tag{18.11}$$

$$(Au, Bu) \geq \beta^2 \|Bu\|^2 \quad \forall u \in D(A). \tag{18.12}$$

For the space H_{AB}—the completion of $D(A)$ in the norm

$$\|u\|_{AB} = [u, u]_{AB}^{1/2}, \tag{18.13}$$

$$[u, v]_{AB} = (Au, Bv) \tag{18.14}$$

—it was shown in Chapter 1 that the relations

$$\|u\|_{AB} \geq \alpha \|u\|_H, \tag{18.15}$$

$$\|u\|_{AB} \geq \beta \|B_0 u\|_H \tag{18.16}$$

remain valid for all $u \in H_{AB}$ if the operator B is extended by continuity to all of H_{AB}, i.e., to a bounded (by (18.11) and (18.16)) linear operator B_0 mapping all of H_{AB} into H, $B \subseteq B_0 \subseteq \overline{B}$, where $\overline{B}$ is the closure of B in H. A closed B_0-symmetric and B_0-positive-definite extension of the operator A, $A_0 \supseteq A$, such that A_0 has a bounded inverse A_0^{-1} defined on all of H was also constructed in §4.

DEFINITION 18.2. An operator N is called H_{AB}-*semicontinuous* if the conditions $\{u_n\} \subset D(N)$, $u \in D(N)$, and $u_n \to u$ strongly in H_{AB} imply that $Nu_n \rightharpoonup Nu$ weakly in H.

DEFINITION 18.3. An operator N is called H_{AB}-*semiclosed* if the conditions $\{u_n\} \subset D(N)$, $u_n \to u$ strongly in H_{AB}, and $Nu_h \rightharpoonup g$ weakly in H imply that $u \in D(N)$ and $Nu = g$.

For $A = B = I$ these definitions coincide with the definitions of semicontinuous and semiclosed operators introduced and investigated in [45], [46], [152], [228], and [352].

We have the following important result, due to Petryshyn [258].

THEOREM 18.4. *Suppose A is a B-symmetric and B-positive-definite operator and N is a nonlinear mapping in H, $D(A) = D(N)$. If for some constant $\eta > 0$*

$$(N(u) - N(v), B(u - v)) \geq \eta \|u - v\|_{AB}^2 \quad \forall u, v \in D(N) \tag{18.17}$$

and for any sequence $\{u_n\} \subset D(N)$ converging in H_{AB}

$$(Nu_n - Nu_m, B_0 h) \to 0 \quad \forall h \in H_{AB} \quad (m, n \to \infty), \tag{18.18}$$

then in H_{AB} the operator N admits a unique H_{AB}-semiclosed extension N_0 such that $N_0 \supseteq N$, N_0 maps $D(N_0)$ onto all of H in a one-to-one manner, and structurally $N_0 = A_0 W_0$, where W_0 is an extension of the operator $A_0^{-1} N$ in H_{AB}.

We present Petryshyn's detailed proof of this theorem, since precisely this proof provides a nontrivial construction of a solvable extension of a nonlinear operator N which in the case of a selfadjoint linear operator $N \equiv A$ coincides (for $B = I$) with the Friedrichs solvable extension of an operator.

Suppose A_0 is the B_0-positive-definite extension of the linear operator A indicated above and constructed in §4. We set $W = A_0^{-1} N$; then W is an operator acting in H_{AB}, $D(W) = D(N) \subseteq H_{AB}$, and $R(W) \subseteq H_{AB}$.

From (18.18), (18.14), and the definition of W we find that

$$[W u_n - W u_m, h]_{AB} = (N u_n - N u_m, B_0 h) \to 0 \qquad (m, n \to \infty) \quad (18.19)$$

for any sequence $\{u_n\} \subset D(N)$ converging strongly in H_{AB}, i.e., W takes any sequence strongly convergent in H_{AB} of elements of $D(N)$ into the sequence $\{W u_n\}$ converging weakly in H_{AB}.

We now construct an extension $\hat{W}: H_{AB} \to H_{AB}$ of W in the following manner. If $u \in D(W)$, the $\hat{W} u = W u$; if $u \in \overline{D(W)} = H_{AB}$, then in $D(W)$ there exists $\{u_n\}$, $u_n \to u$ strongly in H_{AB}, and hence $\{W u_n\}$ converges weakly in H_{AB}. Since H_{AB} is a complete space, there exists a unique element

$$u_0 \in H_{AB}: \quad u_0 = \widetilde{\lim_{n \to \infty}} W u_n,$$

where $\widetilde{\lim}$ denotes the limit in the sense of weak convergence in H_{AB}. It is easy to see that the element u_0 does not depend on the choice of sequence $\{u_n\} \subset D(W)$, and hence for any $u \in H_{AB}$ it is possible to set

$$\hat{W} u = \widetilde{\lim_{n \to \infty}} W u_n.$$

Thus, by construction the operator $\hat{W}$ is semicontinuous and maps all of H_{AB} into H_{AB}. Moreover,

$$\|\hat{W} u - \hat{W} v, u - v]_{AB} \geq \eta \|u - v\|_{AB}^2, \quad \forall u, v \in H_{AB}. \quad (18.20)$$

For the proof of the last relation we note that if u and v are arbitrary elements of H_{AB} and $\{u_n\}$ and $\{v_n\}$ are sequences in $D(W)$ such that $\|u_n - u\|_{AB} \to 0$ and $\|v_n - v\|_{AB} \to 0$ as $n \to \infty$, then from the semicontinuity of $\hat{W}$ in H_{AB} it follows that $(W u_n - W v_n) \to \hat{W} u - \hat{W} v$ weakly in H_{AB}. Since by (18.14) and (18.18)

$$|[\hat{W} u_n - \hat{W} v_n, u_n - v_n]| \geq \eta \|u_n - v_n\|_{AB},$$

on passing to the limit in the last relation, we obtain (18.20).

For a semicontinuous mapping $\hat{W}: H_{AB} \to H_{AB}$ satisfying (18.20) it follows from a theorem of Felix Browder [45] that $\hat{W}$ maps H_{AB} onto all of H_{AB}, and the bounded operator $\hat{W}^{-1}$ defined on $H_{AB} = R(\hat{W})$ exists.

It is then possible to consider an operator W_0 acting from H_{AB} to H_{AB} such that $W \subseteq W_0 \subseteq \hat{W}$ and $R(W_0) = D(A_0) \subset H_{AB}$.

If we now define

$$N_0 = A_0 W_0, \qquad D(N_0) = D(W_n), \qquad (18.21)$$

then $N_0 \supseteq N$, and the operator N_0 maps $D(N_0)$ onto all of H in a one-to-one manner. Indeed, $W_0 u = W u = A_0^{-1} N u$ for $u \in D(N)$, and hence $N_0 u = A_0 W_0 u = N u$, i.e., $N_0 \supset N$. Since $R(W_0) = D(A_0)$ and A_0 maps $D(A_0)$ onto all of H, the operator $N_0 = A_0 W_0$ maps $D(W_0) = D(N_0)$ onto all of H. Moreover, if $N_0 u = g$ and $N_0 v = g$, then from (18.20) and the definition of N_0 we get

$$0 = |(N_0 u - N_0 v, B_0(u - v))|$$
$$= |[W_0 u - W_0 v, u - v]_{AB}| \geq \eta \|u - v\|_{AB}^2,$$

i.e., $u = v$, and hence the mapping N_0 is one-to-one.

We shall show that N_0 is H_{AB}-semiclosed. If $\{u_n\} \subset D(N): u_n \to u_0 \in H_{AB}$ strongly in H_{AB} and $N_0 u_n \rightharpoonup f$ weakly in H, then by the semicontinuity of $\hat{W}$ in H_0 we have $\hat{W} u_n \rightharpoonup \hat{W} u_0$ weakly in H_{AB}, and from (18.21) and the continuity of A_0^{-1} in H (Theorem 4.3)

$$A_0^{-1} N_0 u_n = W_0 u_n \rightharpoonup A_0^{-1} f, \qquad (18.22)$$

i.e.,

$$|\hat{W} u_n, h]_{AB} \to |\hat{W} u_0, h]_{AB} \quad \forall h \in H_{AB}, \qquad (18.23)$$

while $(N_0 u_n, z) \to (f, z)$ for all $z \in H$, and, in particular, for $z = Bh$, $h \in D(N)$. Since $\hat{W} = W_0 = W$ on $D(N)$, it follows from (18.21)—(18.23) that $[\hat{W} u_0, h]_{AB} = [A_0^{-1} f, h]_{AB}$ for all $h \in D(N)$.

Now $D(N)$ is dense in H_{AB}, and hence $\hat{W} u_0 = A_0^{-1} f$ and $W u_0 \in D(A_0)$, i.e, $u_0 \in D(W_0) = D(N_0)$ and $N_0 u_0 = A_0 W_0 u = f$, which proves that the operator N is H_{AB}-semiclosed.

To prove uniqueness([43]) of the extension constructed we note first that $(N_0 u, Bv) = [\hat{W} u, v]$ is continuous with respect to $u \in D(N_0)$ for any v in H_{AB}. Suppose there exists another extension N_1 of the operator N, $D(N_1) \subseteq H_{AB}$, such that N_1 is a one-to-one mapping of $D(N_1)$ onto H and $(N_1 u, Bv)$ is continuous with respect to $u \in D(N_1)$ for any $v \in H_{AB}$.

([43])This proof was kindly communicated to us by Professor W. V. Petryshyn as a supplement to his paper [258].

We shall show first that $N_0 \supseteq N_1$. Indeed, suppose $w \in D(N_1)$, and let $f = N_1 w$. Let $\{u_n\} \subset D(N)$ be such that $u_n \to w$ in H_{AB}. Then

$$(N_1 u_n, Bv) = (Nu_n, Bv) = [Wu_n, v] = [\hat{W} u_n, v]$$

for all n and any v in H_{AB}. Passing to the limit as $n \to \infty$, we obtain

$$[\hat{W} w, v] = (N_1 w, Bv) = (f, Bv) = [A_0^{-1} f, v].$$

Therefore, $\hat{W} w = A_0^{-1} f$ and $\hat{W} w \in D(A_0)$, i.e., $w \in D(W_0)$, $\hat{W} w = W_0 w$, and $A_0 W_0 w = N_0 w = f$. Hence, $N_0 \supseteq N_1$.

Suppose now that there exists $u_0 \in D(N_0)$, but $u_0 \notin D(N_1)$; we set $f_0 = N_0 u_0$. Then there exists $u_1 \in D(N_1)$ such that $f_0 = N_1 u_1$, and, moreover, $N_0 u = f_0$, since $N_0 \supseteq N_1$. Therefore,

$$0 = (N_0 u_1 - N_0 u_0, B_0(u_1 - u_0)) \geq \eta \|u_1 - u_0\|_{AB}^2,$$

i.e., $u_1 = u_0$ in H_{AB} and in H. Hence, $N_1 = N_0$; in other words, the extension N_0 is unique. ∎

As noted in [258], sometimes in applications (for example, for the non-linear equations of elastoplasticity; see [181]–[183], [114], and [52]) it is possible to establish conditions stronger than (18.18): for any $M > 0$ there exists $C(M) > 0$ such that

$$|(N_w' u, Bh)| \leq C(M)\|u\|_{AB}\|h\|_{AB} \quad \forall u, h, w \in D(N), \|w\|_{AB} \leq M \quad (18.24)$$

or the closely related but even stronger condition: there exists $\theta > 0$ such that for all $u, v, h \in D(N)$

$$|(Nu - Nv, Bh)| \leq \theta \|u - v\|_{AB}\|h\|_{AB}. \qquad (18.25)$$

In application to the nonlinear operator considered in §18.1 it is then possible to show that the following result holds.

THEOREM 18.4′. *Suppose the nonlinear operator N satisfies conditions α), β), and γ') of (18.24) with a B-positive-definite operator $A = N'(0)$. Then the assertion of Theorem 18.4 holds. If in place of condition (18.24) condition (18.25) is satisfied, then the weak extension $\hat{W}$ constructed in Theorem 18.4 can be replaced by an extension $\tilde{W}$ based on the Lipschitz operator $W = A_0^{-1} N$, $\hat{W} = \tilde{W} \supseteq W$, and for all $u, v \in H_{AB}$*

$$\|\tilde{W} u - \tilde{W} v\|_{AB} \leq \theta \|u - v\|_{AB}, \qquad (18.26)$$

$$[\tilde{W} u - \tilde{W} v, u - v]_{AB} \geq C \|u - v\|_{AB}^2. \qquad (18.27)$$

PROOF. From the existence (see α)) of a continuous Gâteaux derivative N' and condition γ') we have for all $u, v \in D(N)$

$$(Nu - Nv, B(u - v)) = \int_0^1 (N'_{v+t(u-v)}(u - v), B(u - v))\, dt$$

$$\geq C(N'(0)(u - v), B(u - v)) = C\|u - v\|_{AB}^2$$

$$(18.28)$$

and hence condition (18.17) of Theorem 18.4 is satisfied. Moreover, for all $u, v, h \in D(N)$ such that $\|u\|_{AB} \leq M$ and $\|v\|_{AB} \leq M$, by (18.24), where $w = v + t(u - v)$ and $\|w\| \leq 2M$, for $t \in (0, 1)$ we obtain

$$|(Nu - Nv, Bh)| = \left| \int_0^1 (N'_{v+t(u-v)}(u - v), B(u - v))\, dt \right|$$

$$\leq C(2M)\|u - v\|_{AB}\|h\|_{AB}.$$

$$(18.29)$$

The validity of the second condition of Theorem 18.24 follows from this, and therefore all assertions and constructions in its proof remain valid. Under condition (18.25) the operator $W = A_0^{-1}N$ also satisfies a Lipschitz condition: there exists a constant $\theta > 0$ such that

$$\|Wu - Wv\|_{AB} \leq 0\|u - v\|_{AB} \quad \forall u, v \in D(N). \qquad (18.30)$$

Indeed, setting $h = Wu - Wv$, by (18.14) and (18.25) we obtain

$$\|h\|_{AB}^2 = [Wu - Wv, h]_{AB} = (Nu - Nv, Bh) \leq \theta\|u - v\|_{AB}\|h\|_{AB},$$

whence the Lipschitz condition (18.30) follows.

Therefore, there exists a unique (strong, in the norm of H_{AB}) Lipschitz extension $\tilde{W} = \hat{W}$ of W such that $D(\tilde{W}) = H_{AB}$ and $\tilde{W}u = Wu$ for all $u \in D(N)$. Inequalities (18.26) and (18.27) for all $u, v, \in H_{AB}$ can be established with consideration of the definition of $\tilde{W}$ be passing to the limit from (18.28) and (18.30) in the norm of the space H_{AB}. ∎

We present also some corollaries from [258] and [52] which establish the theorem for special cases of the nonlinear operator N and sometimes simplify the description of $D(N_0)$.

COROLLARY 18.3 [258]. *If A is a B-symmetric and B-positive-definite operator, $N = A + S$, where $D(S) \supseteq D(A)$, (18.17) is satisfied, and for any sequence $\{u_n\} \subset D(N)$ converging in H_{AB}.*

$$(Su_n - Su_m, B_0h) \to 0 \quad \forall h \in H_{AB}, \qquad (m, n \to \infty), \qquad (18.31)$$

then the assertions of Theorem 18.4 hold with $N_0 = A_0(I + R_0)$, where R_0 is some extension in H_{AB} of the operator $R = A_0^{-1}S$.

Indeed, since conditions (18.17) and (18.31) ensure the validity of Theorem 18.4, the operator N has a solvable extension $N_0 = A_0 W_0$, where $W_0 \supset W = A_0^{-1} N$ is a restriction of $\hat{W}$ such that $R(W_0) = D(A_0)$. We obtain

$$W = A_0^{-1}(A + S) = A^{-1}A + A_0^{-1}S = I + R \qquad (18.32)$$

for $D(R) = D(w) = D(A) \subseteq H_{AB}$, and by (18.31) and (18.32) the operator $R = A_0^{-1}S$ has a semicontinuous extension $\hat{R} = \hat{W} - I$ from $R_0 = W_0 - I$. ∎

COROLLARY 18.4 [52]. *If the operator A is B-symmetric and B-positive-definite and the nonlinear operator $N = A + S$ is such that $D(N) = D(A) \subset H_{AB}$ and there exist $\eta_2 > 0$ and $\theta_1 > 0$ such that for all $u, v \in D(B)$*

$$|(Nu - Nv, B(u - v))| \geq \eta_2 \|u - v\|_{AB}^2, \qquad (18.17_2)$$

$$\|Su - Sv\|_H \leq \theta_1 \|u - v\|_{AB}, \qquad (18.18')$$

then $D(N_0) = D(A_0)$ and

$$N_0 = A_0 + S_0, \qquad (18.33)$$

where S_0 is some extension of S in H_{AB}.

Since (18.31) follows from (18.18'), by Corollary 18.3 we have $N_0 = A_0(I + P_0)$, where P_0 is a restriction of the operator $P = A_0^{-1}S$ (an extension $\tilde{S} \supset S$ in H_{AB} exists by (18.18')). Then for $\tilde{W} = I + A_0^{-1}\tilde{S}$ we have $\tilde{W}u \in D(A_0)$ if and only if $u \in D(A_0)$, since $A_0^{-1}\tilde{S}u \in D(A_0)$ for all $u \in D(A_0)$. Therefore, $D(W_0) = D(A_0)$, and hence for $N_0 = A_0 W_0$ we have $D(N_0) = D(A_0)$ and $N_0 = A_0 + S_0$ if we set $S_0 = A_0 P_0$.

18.3. We return to the investigation of the solvability by a variational method of the nonlinear equation (18.1), using the results of §18.2. Following [52], we introduce several definitions.

DEFINITION 18.4. An element $\bar{u}$ is called a *classical solution* of equation (18.1) if $\bar{u} \in D(N)$ and $N\bar{u} - f = 0$ in H.

DEFINITION 18.5. If the nonlinear operator N satisfies the conditions of Theorem 18.4, then an element $u_0 \in D(N_0) \subseteq H_{AB}$ satisfying

$$N_0 u - f = 0 \qquad (18.34)$$

in the sense of H is called a *strong solution* of equation (18.1).

Here N_0 is the extension of the nonlinear operator N constructed in §18.2.

Suppose now that the operator N is such that the form $p(u, v) = (Nu, B_0 v)$ can be extended by continuity to all of $H_{AB} \times H_{AB}$. The following definition is then meaningful.

DEFINITION 18.6. An element $\tilde{u} \in H_{AB}$ is called a *weak solution* of equation (18.1) if

$$p(\tilde{u}, v) = (f, B_0 v) \quad \forall v \in D(N). \tag{18.35}$$

If the operator N satisfies the conditions of Theorem 18.4, then the indicated extension of the form $p(u, v)$ can be realized by means of the operator $\hat{W} \supset W = A_0^{-1} N$ constructed in Theorem 18.4 if we set

$$p(u, v) = [\hat{W}u, v]_{AB}, \quad (f, B_0 v) = [h, v] \quad \forall u, v \in H_{AB},$$

where $h = A_0^{-1} f$. It is therefore possible to give an equivalent definition

DEFINITION 18.6'. An element $\tilde{u} \in H_{AB}$ is a *weak solution* of equation (18.1) if and only if $\tilde{u}$ satisfies in H_{AB}

$$\hat{W}u - h = 0. \tag{18.36}$$

It is obvious that any classical solution of (18.1) is a strong solution, and any strong solution is a weak solution; in general the converse assertions are false.

LEMMA 18.2. *Suppose the nonlinear operator N satisfies the conditions of Theorem 18.4'. Then an element $\tilde{u} \in H_{AB}$ is a weak solution of equation (18.1) if and only if $\tilde{u}$ is a critical point of the functional*

$$\Phi(u) = \int_0^1 |W(tu) - h, u]_{AB}\, dt, \tag{18.37}$$

where $\hat{W}$ is the extension in H_{AB} of the operator $W = A_0^{-1} N$ ($D(W) = D(N)$) constructed in Theorem 18.4.

PROOF. With the help of (18.14) we note first that on $D(N)$ the functional $\hat{\Phi}(u)$ coincides with the functional $\Phi(u)$ of (18.1),

$$\Phi(u) = \Phi(u) = \int_0^1 (N(tu), Bu)\, dt - (f, Bu), \quad u \in D(N),$$

and, using conditions α) and β) and the definition of W, it is not hard to justify the existence of the integral in (18.37).[44]

We compute the first variation of the functional $\hat{\Phi}(u)$ on some element $\tilde{u} \in H_{AB}$. By the construction in Theorem 18.4 we have $D(\hat{W}) = R(\hat{W}) = H_{AB}$, and for all $\tilde{u} \in H_{AB}$ there exists $\{u_n\} \subset D(N)$ such that $u_n \to \tilde{u}$

---

[44] The constructions carried out in this proof are easily justified if in Theorem 18.4' condition (18.25) is satisfied, since then $W = \hat{W}$, where $\hat{W}$ is the strong (in the norm of H_{AB}) Lipschitz extension of W.

strongly in H_{AB} and $\tilde{W} u_n \to \hat{W} \tilde{u}$ weakly in H_{AB}. Then for all $\eta \in D(N)$

$$
\begin{aligned}
\delta\Phi(\tilde{u}) &= \left[\frac{d}{d\alpha} \Phi(\tilde{u} + \alpha\eta) \right]_{\alpha=0} \\
&= \left\{ \frac{d}{d\alpha} \lim_{n\to\infty} \int_0^1 |\hat{W}(t(u_n + \alpha\eta)) - h, u_n + \alpha\eta|_{AB}\, dt \right\}_{\alpha=0} \\
&= \left\{ \frac{d}{d\alpha} \lim_{n\to\infty} \left[\int_0^1 (N(t(u_n + \alpha\eta)), B(u_n + \alpha\eta))\, dt \right. \right. \\
&\qquad\qquad\qquad\qquad\qquad\qquad\left.\left. - (f, B(u_n + \alpha\eta)) \right] \right\}_{\alpha=0} \\
&= \lim_{n\to\infty} \int_0^1 [(N'_{tu_n} t\eta, Bu_n) + (N(tu_n), B\eta)]\, dt - (f, B\eta) \\
&= \lim_{n\to\infty} \int_0^1 [(N'_{tu_n} tu_n, B\eta) + (N(tu_n), \beta\eta)]\, dt - (f, B\eta) \\
&= \lim_{n\to\infty} (N(u_n) - f, B\eta) = \lim_{n\to x} [W(u_n) - h, \eta]_{AB} \\
&= [\hat{W}(\bar{u}) - h, \eta]_{AB},
\end{aligned}
$$

where the second and eighth equalities follow from the definition of $\hat{W}$; the third and seventh follow from (18.14); the fourth is justified by means of condition α), the theorem on the differentiability of a sequence of differentiable functions convergent at at least one point, and the theorem on the differentiability of an integral depending on a parameter; the fifth equality follows from the symmetry condition β; the sixth is obtained by integration by parts.

Thus,

$$
\delta\hat{\Phi}(\tilde{u}) = [\hat{W}(\tilde{u}) - h, \eta]_{AB} \quad \forall \eta \in D(N) \tag{18.39}
$$

and hence $\delta\hat{\Phi}(\tilde{u}) = 0$ if and only if $\tilde{u}$ satisfies equation (18.36) defining a weak solution of (18.1). ∎

COROLLARY 18.5. *If the operator N satisfies the conditions of Theorem 18.4′, then for any $f \in H$ there exists a unique weak solution in H_{AB} of equation (18.1), and hence there exists a unique critical point of the functional (18.37) in H_{AB}.*

This assertion can be obtained directly from Lemma 18.2, since the existence and uniqueness in H_{AB} of a solution of (18.36) (i.e., a weak solution of (18.1)) follow from the facts established in Theorem 18.4: the equalities $D(\hat{W}) = R(\hat{W}) = H_{AB}$ and the existence of a bounded $\tilde{W}^{-1}$ defined on all of H_{AB}, $h = A_0^{-1}f \in H_{AB}$. ∎

THEOREM 18.5. *Suppose the operator N satisfies conditions α), β), and γ') of (18.18) with a B-positive-definite operator $A = N'(0)$. Then for any $f \in H$ there exists a unique strong solution of equation (18.1) in H_{AB}; it coincides with the unique element minimizing the functional $\hat{\Phi}(u)$ of (18.37) and depends continuously on the given function f.*

PROOF. The conditions are obviously sufficient for the validity of Theorem 18.4. The existence and uniqueness in H_{AB} of a strong solution u_0 of (18.1) for any $f \in H$ then follows from Theorem 18.4: the operator N_0 maps $D(N_0) \subseteq H_{AB}$ in a one-to-one manner onto all of H, and hence for $f \in H$ there exists a unique solution of (18.34) in $D(N_0) \subseteq H_{AB}$. Uniqueness of the strong solution in H_{AB} can be obtained from uniqueness of a weak solution, since any strong solution (18.1) is a weak solution. From this and Corollary 18.5 we conclude that the functional (18.37) has a unique critical point u_0 in H_{AB} and that $u_0 \in D(N_0)$.

On the other hand, it follows([45]) from Theorem 18.2 that there exists a unique element $\bar{u}$ in H_{AB} minimizing the functional $\hat{\Phi}(u)$ of (18.37) in H_{AB}. If $\bar{u} \neq u_0$, then by Lemma 18.2 the element $\bar{u}$ is a weak solution of (18.1) distinct from u_0, which contradicts the uniqueness of a weak solution.

We obtain the last assertion of the theorem from the boundedness on H_{AB} of the operator $\hat{W}^{-1}$ of (18.36), i.e, from the existence of $C_1, C_2 > 0$ such that

$$\|u_0\|_{AB}^2 = \|W^{-1}h\|_{AB}^2 \leq C_1\|h\|_{AB}^2 = C_1(A_0A_0^{-1}f, B_0A_0^{-1}f)$$
$$\leq C_1\|f\|_H\|B_0A_0^{-1}f\|_H \leq \frac{C_1}{\beta}\|f\|_H\|A_0^{-1}f\|_{AB} \leq C_2\|f\|_H^2,$$

$$(18.40)$$

where we have used: in the second equality the replacement $h = A_0^{-1}f$ and a property of the inner product in H_{AB} (see (18.14) and also §§3 and 4); in the third inequality the B-positive-definiteness of the operator A_0 (see (1.5) and (18.16)); and in the last equality the property of boundedness of the operator A_0^{-1} on H (Theorem 4.3).

From (18.40) we obtain

$$\|u_0\|_{AB} \leq C_2\|f\|_H \qquad (18.41)$$

([45])It is not hard to see that conditions γ) and γ'') used in Theorem 18.2 are a consequence of γ'), the B-positivity of the operator $A = N'(0)$, and Corollary 18.2.

i.e., the continuous dependence on $f \in H$ of a weak and hence a strong solution of (18.1) and of the element minimizing the functional (18.37) in H_{AB}. ∎

Remark 18.3. Since under the conditions of Theorem 18.5 for all $f \in H$ there exists a unique element u_0 in H_{AB} minimizing the functional $\hat{\Phi}(u)$ of (18.37) and $u_0 \in D(N_0) = D(W_0)$ for the operator W_0 constructed in Theorem 18.4,

$$\hat{W} \supset W_0 \supset W, \quad W_0 = A_0^{-1} N_0, \quad D(W_0) = D(N_0), \tag{18.42}$$

it follows that the problem of minimizing the functional $\hat{\Phi}(u)$ of (18.37) in H_{AB} is equivalent to the problem of minimizing the functional

$$\Phi_0(u) = \int_0^1 [W_0(tu) - h, u]_{AB}\, dt$$

$$= \int_0^1 (N_0(tu) - f, B_0 u)\, dt \tag{18.43}$$

on the set $D(N_0) \subset H_{AB}$.

If, for example, the nonlinear operator $N = A + S$ satisfies the conditions of Corollary 18.4, then $D(N_0) = D(A_0)$—the domain of the B_0-positive-definite extension of the linear operator A (see Theorem 4.3). Generally speaking, $D(A_0) \subseteq H_{AB}$, but in a number of cases $D(A_0) = H_{AB}$ (see Theorem 4.4, Corollary 4.3, and also [257]). Examples of a constructive construction of spaces H_{AB} are presented in §§12–16.

From Theorem 18.5 and Remark 18.3 we obtain

COROLLARY 18.6. *Under the conditions of Theorem* 18.5 *the set* $D(N_0)$ *coincides with the set of elements which are solutions of the variational problem of minimizing the functional* $\Phi_0(u)$ *of* (18.43) *in* H_{AB}.

Therefore, by what was said at the beginning of §18.2 our extension N_0 of the operator N coincides with the extension of N in [202] and [182], where the domain $D(\tilde{N})$ f the extension $\tilde{N} \supset N$ was defined a priori as the set of elements which are solutions of the variational problem of minimizing the functional $\Phi(u)$ of (18.2) in H_{AB}. In the case of a linear operator $N = A$ Corollary 18.6 is contained in Theorem 4.3.

Commentary to Chapter IV

From the results of §§17.3, 18.1, and 18.3 it is evident that the practical construction of well-posed variational principles([46]) for a given nonlinear

---

([46])I.e., functionals for which there exists a unique element u_0 minimizing this functional in some Hilbert space $\tilde{H}$; u_0 coincides with the unique generalized solution in $\tilde{H}$ of the original boundary value problem and depends continuously on the initial data.

initial boundary value problem is an involved procedure. However, if variational methods of investigating equations are intended as approximate—projection—methods of solving equations, then the approaches presented in Chapter 3 to symmetrization of operators make it possible to constructively extend many of these methods to broad classes of nonlinear operator equations of the form

$$Au + \lambda Ku = f, \qquad \lambda \in Z. \tag{1}$$

Here A is a linear operator for which an auxiliary operator B with properties (1.1)–$(1.5)(^{47})$ is constructed, while K is a linear or nonlinear, bounded or unbounded operator satisfying certain conditions (frequently depending on the method applied).

Martynyuk [209]–[214], [217], Lyashko [196]–[201] Petryshyn [250], [255], and Dzhishkariani [82], [83] extended the Galerkin method and the method of moments to broad classes of linear and nonlinear equations (1) with a B-positive-definite operator A. Many approximate methods of solving eigenvalue problems (with linear, unbounded, and symmetric operators A and K) (1) in the case of a B-positive-definite operator A have been investigated by Petryshyn [250]–[253], [255], Bradley and Sleeman [42] and Andrushkiw [11]. For a nonlinear eigenvalue problem (1) with a B-positive-definite operator A, Petryshyn [258], [260]–[263] and Martynyuk [216], [217] developed iterative and projection-iterative methods in which an approximate solution u_m of the nonlinear equation (1) can easily be computed in terms of the approximation $m - 1$ or can be constructed at each iteration as a solution of a system of algebraic equations.

Thus, for constructive application of approximate methods to equation (1) it is important to be able to construct in explicit form an operator B relative to which the linear operator A is B-positive-definite. Solution of this problem for many PDE is possible by means of the results presented in §§12–16, and for the equation $Au + N(u) = f$ the construction of an operator B can sometimes be simplified, since the choice of the linear operator A is relatively arbitrary—linear terms "inconvenient" for symmetrization can be assigned to the operator N. After construction of a symmetrizing operator B effective numerical realization can be accomplished by a projection-grid method Marchuk and Agoshkov [207]).

To §17. In this section some approaches to the solution of inverse problems of the calculus of variations are discussed, and therefore the exposition of the material is incomplete if it is considered from the standpoint

---

$(^{47})$Here, since it is not a question of a variational principle for equation (1), frequently the condition of B-positive-definiteness of the linear operator A is sufficient.

of a variational method for investigating nonlinear equations. For completeness of the exposition in §17.3 it would be necessary, as in §18, to construct an extension of the operator N for which for all $f \in H$ there exists a unique element $u_0 \in H_1$ in some Hilbert space $H_1 \supseteq D(N_0)$ such that it minimizes the functional $F_0(u) = \|N_0(u)\|_{H_K}$ of (17.31) in H_1. Regarding the application of the method of least squares to nonlinear equations, see Langenbach [185].

To §18. 1. The scheme presented in §18.1 for extending the variational method to nonlinear equations with nonpotential operators was proposed, as already noted, independently by Nashed [235] and Lyashko [202]. In [235] the basic problem was not to construct a functional for a given equation, but to develop a method of steepest descent for nonlinear equations. Therefore, in [235] many additional conditions not due to the essence of the variational method were required of the nonlinear operator N and the auxiliary operator B.

The results §18.1 (except for Theorem 18.3) were obtained by Lyashko under the stronger conditions

$$(N'_x u, Bu) \geq C(N'(0)u, Bu) \geq \beta^2 \|Bu\|^2 \geq \alpha^2 \|u\|^2 \quad \forall x, u \in D(N). \quad (2)$$

Validity of conditions γ), γ'), and γ'') (§18.1) is obvious from (2); the converse is not true. For example, boundedness of the operators B^{-1} and $A^{-1} = (N'_x)^{-1}$ for all $x \in D(N)$ follows from (2), which is not necessary under our conditions γ)—γ'').

2. The results of Lyashko [202] were a direct generalization of some results of Langenbach [181], [182], and [184], so that for $B = I$ (i.e., for a potential operator N) the corresponding results of [181], [182], and [184] follow from §18.1. If $N = A$ is a linear operator, then the constructions of §18 coincide with the results of Chapter 1 for a B-symmetric and B-positive-definite operator A.

3. An extension of a nonlinear potential operator N needed for completeness of the variational method is carried out in [182]; the possibility of an analogous extension for a nonpotential operator N was noted in [202]. We have used Petryshyn's construction for extension of a nonlinear operator which generalizes the corresponding constructions of §§3 and 4 and, in particular, is a generalization of a Friedrichs solvable extension of selfadjoint operators.

Other constructions of extending nonlinear operators are known which are also based on results of Browder [45], [46] and Zarantonello [352]. For example, for a closed linear operator $\mathscr{L}$ and a nonlinear monotone operator Φ in [346] an extension of the operator $\mathscr{L}^*\Phi\mathscr{L}$ is constructed which generalizes the corresponding extension of von Neumann [237] of the operator $\mathscr{L}^*\mathscr{L}$.

4. Necessary and sufficient conditions for the existence of a positive functional $\varphi(x) > 0$ and a functional $f(x)$ such that grad $f(x) = \varphi(x)F(x)$ are obtained in Minty's dissertation [227], in particular, for a given nonpotential operator F. Apparently, these results remain a so-far-unused possibility for constructively treating variational principles for a given nonlinear boundary value problem.

On the Nonexistence of Semibounded Solutions of Inverse Problems of the Calculus of Variations

V. M. FILIPPOV AND V. M. SAVCHIN

Suppose there is given a boundary value problem

$$\mathscr{L}u \equiv \sum_{i,j=1}^{n} p^{ij}u_{ij} + \sum_{i=1}^{n} q^i u_i + r \cdot u = f(x), \qquad x \in \Omega, \qquad (1)$$

$$u(x) = 0, \qquad x \in \partial\Omega. \qquad (2)$$

Here Ω is a bounded domain in R^n with piecewise smooth boundary $\partial\Omega$; the $p^{ij} = p^{ji}$ and q^i $(i,j = 1,\ldots,n)$ are constant coefficients in $\overline{\Omega}$; $r(x) \in C(\overline{\Omega})$ and $f(x) \in C(\overline{\Omega})$ are given functions; $u = u(x) \in M = C^2(\Omega) \cap C^1(\overline{\Omega}) \cap \overset{\circ}{C}(\overline{\Omega})$, $x = (x_1,\ldots,x_n)$; $u_i = \partial u/\partial x_i \equiv D_i u$, and $u_{ij} = D_j D_i u$ $(i,j = 1,\ldots,n)$.

We introduce the class of Euler-Lagrange functionals

$$\mathscr{I}_{[u]} = \int_{\Omega} L(x,u,u_1,\ldots,u_n)\,dx. \qquad (3)$$

The following formulation of the inverse problem of the calculus of variations (IPCV) for equation (1) is known (see [54] and [18]).

FORMULATION 1. Find a function $\mu(x,u,u') \equiv \mu(x,u,u_1,\ldots,u_n)$ which is continuously differentiable in R^{2n+1}, with $\mu(x,u,u') \neq 0$ $(x \in \overline{\Omega})$ for all $u \in M$, and in the class of functionals (3) find a functional $F_\mu[u]$ such that for all $u \in M$

$$\delta F_\mu[u] = \int_{\Omega} \mu(x,u,u') \cdot (Lu - f)\delta u\,dx. \qquad (4)$$

If $\partial\mu/\partial u \not\equiv 0$ or $\partial\mu/\partial u_i \not\equiv 0$ for some $i = 1,\ldots,n$, then because of the nonlinearity of the integrand in (4) it is obvious that the corresponding functional $F[u]$ is nonquadratic. Since for a linear equation (1) it is expedient to restrict attention to functionals (3) quadratic in u and $u_1,\ldots,u_n$, we henceforth consider the following IPCV (see [54]).

FORMULATION 2. Find a function $\mu(x) \in C^1(\overline{\Omega})$, $\mu(x) \neq 0$ $(x \in \overline{\Omega})$, and a functional $F_\mu[u]$ in the class (3) quadratic in u and $u_1,\ldots,u_n$ such that for all $u \in M$

$$\delta F_\mu[u] = \int_\Omega \mu(x) \cdot (\mathscr{L}u - f) \cdot \delta u \, dx. \tag{5}$$

The usual IPCV follows from this for $\mu(x) \equiv 1$.

Results of Copson [54] were presented in §11.

LEMMA 1. *For any equation* (1) *nonparabolic in* Ω *the general solution of the IPCV in Formulation 2 is given by*([47])

$$V[u] = \int_\Omega \exp\left(\sum_1^n k_i x_i\right) \cdot \left\{ \sum_{i,j=1}^n p^{ij} D_i u D_j u \right.$$

$$\left. - \left[r(x) - \sum_{i=1}^n b^i k_i \right] \cdot u^2 + 2f \cdot u \right\} \, dx, \tag{6}$$

where

$$k_j = \det P_j / \det P, \qquad j = 1,\ldots,n, \tag{7}$$

and the matrix P_j *is obtained from the matrix* $P = (p^{\alpha\beta})$ *by replacing the* j*th column by the column* $(q^1,\ldots,q^n)$*; the constants* $b^1,\ldots,b^n$ *do not depend on* x *or* u.

COROLLARY 1. *For any nonparabolic equation* (1) *with constant coefficients* p^{ij} $(i,j = 1,\ldots,n)$ *and* $q^i = 0$ $(i = 1,\ldots,n)$, *and* $r(x), f(x) \in C(\overline{\Omega})$, *solutions of the IPCV in Formulation 2 are given by*

$$\tilde{V}[u] = \int_\Omega \left[\sum_{i,j=1}^n p^{ij} D_i u D_j u - r(x) \cdot u^2 + 2f \cdot u \right] dx.$$

The functional $\tilde{V}[u]$ is obtained from (6), since for $q^i = 0$ $(i = 1,\ldots,n)$ from (7) we have $k_i = 0$ $(i = 1,\ldots,n)$.

([47])Sets of functions are henceforth given up to a factor and term not depending on x or u.

COROLLARY 2. *An element $u_0 \in M$ is a solution of the Dirichlet problem for the nonparabolic equation* (1) *if and only if u_0 is a critical point of the functional $V[u]$* (6), (7).

Thus, for any nonparabolic operator $\mathscr{L}$ of problem (1), (2) the IPCV in Formulation 2 always has a solution—this should be borne in mind with regard to assertions of "nonvariational" nonselfadjoint elliptic or hyperbolic equations.

We now consider in more detail the question of semiboundedness (that is, boundedness above or below) of the functionals (6) constructed for nonparabolic equations (1).

It is known ([57], vol. II, Chapter 3, §3) that any PDE (1) with constant coefficients (we further assume that $r(x) \equiv \text{const}, x \in \overline{\Omega}$) by a change of the independent variables

$$x_i = \sum_{j=1}^n t_{ij} y_j, \qquad i = 1, \ldots, n \qquad (x = Ty) \tag{8}$$

can be transformed to the form

$$\mathscr{L}_1 u \equiv \sum_{i=1}^n a^i \frac{\partial^2 u}{\partial y_i^2} + \sum_{i=1}^n b_1^i \frac{\partial u}{\partial y_i} + r_1 \cdot u = g_1(y). \tag{9}$$

Here $\det(t_{ij})_{i,j=1}^n \neq 0$, the t_{ij} $(i, j = 1, \ldots, n)$ are constant real coefficients in $\overline{\Omega}$, the constants a^i $(i = 1, \ldots, n)$ assume the values $0, 1$, or -1, r_j is a constant in $\overline{\Omega}$, and $g_1(y) = f(Ty)$.

For a nonparabolic equation (1) in Ω it is possible ([57], vol. II) by a change of the unknown function

$$u(y) = v(y) \cdot \exp(-\tfrac{1}{2} \cdot z), \qquad z = \sum_{i=1}^n b_1^i y_i / a^i \tag{10}$$

to transform equation (9) into

$$\mathscr{L}_2 v \equiv \sum_{i=1}^n a^i \partial^2 v / \partial y_i^2 + \lambda \cdot v = \tilde{g}, \tag{11}$$

where

$$\lambda = r_1 - \frac{1}{4} \cdot \sum_{i=1}^n \frac{(b_1^i)^2}{a^i}, \qquad \tilde{g} = g_1 \cdot \exp(\tfrac{1}{2} z). \tag{12}$$

Thus, any equation (1) elliptic in Ω can be transformed into([48])

$$\Delta u + \lambda u = f(\xi), \qquad \lambda \equiv \text{const}, \ \xi \in \tilde{\Omega}, \tag{13}$$

([48])Appearance of the variables ξ_i $(i = 1, \ldots, n)$ is connected with a possible renumbering of the variables in passing from (11) to (13) or (14).

while any ultrahyperbolic (including properly hyperbolic) equation (1) can be transformed into

$$\sum_{i=1}^{m} \frac{\partial^2 u}{\partial \xi_i^2} - \sum_{i=m+1}^{n} \frac{\partial^2 u}{\partial \xi_i^2} + \lambda \cdot u(\xi) = f(\xi), \qquad \xi \in \tilde{\Omega}, \qquad (14)$$

where $1 \leq m < n$, $n \geq 2$, and $\lambda = \text{const}$ in $\tilde{\Omega}$.

For equations (13) and (14) functionals are known which are solutions of the IPCV. A natural question arises: if for equation (13) or (14) a solution of the IPCV has been constructed, will the functional obtained by transformations inverse to (8) and (10) be a solution of the IPCV for the original equation (1) of the corresponding type? It is convenient to first consider the question of transformation of functionals of this type in a more general case.

Suppose we are given a functional

$$\mathscr{F}[u] = \int_{\Omega} F(x_k, u^j, D_k u^j) \, dx, \qquad k = 1, \ldots, n; \; j = 1, \ldots, m, \qquad (15)$$

where

$$F(x_k, u^j, D_k u^j) \equiv F(x_1, \ldots, x_n, u^1, \ldots, u^m, D_1 u^1,$$
$$\ldots, D_n u^1, \ldots, D_1 u^m \ldots, D_n u^m)$$

is a function continuously differentiable in all its arguments in R^{n+m+mn}, and $u^j(x) \in M$ $(j = 1, \ldots, m)$.

We introduce a nonsingular transformation $y = y(x)$,

$$\begin{cases} y_k = y_k(x_i), & i, k = 1, \ldots, n, \; y_k \in C^1(\overline{\Omega}), \\ \det(\partial y_k / \partial x_i)_{i,k=1}^{n} \neq 0, & x \in \overline{\Omega}. \end{cases} \qquad (16)$$

We denote the inverse transformation by $x = x(y)$.

The variational derivative $\delta F / \delta u^j$ for the functional $\mathscr{F}[u]$ is the left side of the jth Euler equation for the functional (15). We have

LEMMA 2. *For any* $u^{j'} \in M$ $(j' = 1, \ldots, m)$

$$\left(\frac{\partial F}{\partial u^{j'}} \right)_{x=x(y)} = \det \left(\frac{\partial y_r}{\partial x_s} \right)_{r,s=1}^{n}$$
$$\cdot \frac{\delta}{\delta \tilde{u}^{j'}} \left[\tilde{F}(y_k, \tilde{u}^j(y), D_k \tilde{u}^j(y)) \cdot \det \left(\frac{\partial x_r}{\partial y_s} \right)_{r,s=1}^{n} \right],$$
$$j' = 1, \ldots, m, \qquad (17)$$

where

$$\tilde{F}(y_k, \tilde{u}^j(y), D_k \tilde{u}^j(y)) = F\left[x_i(y_k), u^j(x_i(y_k)), \sum_{k=1}^{n} \frac{\partial \tilde{u}^j}{\partial y_k} \cdot \frac{\partial y_k}{\partial x_i}\right],$$

$$u^j(x(y)) = \tilde{u}^j(y) \qquad (j = 1, \ldots, m).$$

PROOF. For any $u^j \in M$ $(j = 1, \ldots, m)$ we have

$$\frac{\partial u^j}{\partial x_i} = \sum_{k=1}^{n} \frac{\partial \tilde{u}^j}{\partial y_k} \cdot \frac{\partial y_k}{\partial x_i},$$

$$\int_{\Omega} (x_k, u^j(x), D_k u^j(x)) \, dx$$

$$= \int_{\tilde{\Omega}} \tilde{F}(y_k, \tilde{u}^j(y), D_k \tilde{u}^j(y)) / \det\left(\frac{\partial x_r}{\partial y_s}\right)^n_{r,s=1} / dy, \tag{18}$$

where $\tilde{\Omega}$ is the domain obtained from Ω by means of the mapping (16). From this we obtain

$$\int_{\Omega} \frac{\delta F}{\delta u^{j'}} \cdot \eta^{j'} \, dx = \left[\frac{\partial}{\partial \alpha^{j'}} \int_{\Omega} F(x_k, u^j + \alpha^j \eta^j, D_k u^j + \alpha^j \cdot D_k \eta^j) \, dx\right]_{\alpha=0}$$

$$= \left[\frac{\partial}{\partial \alpha^{j'}} \int_{\tilde{\Omega}} \tilde{F}(y_k, \tilde{u}^j(y) + \alpha^j \cdot \tilde{\eta}^j(y), D_k \tilde{u}^j(y) + \alpha^j \cdot D_k \tilde{\eta}^j(y))\right.$$

$$\left. \times \left|\det\left(\frac{\partial x_r}{\partial y_s}\right)^n_{r,s=1}\right| \, dy\right]_{\alpha=0}$$

$$= \int_{\tilde{\Omega}} \frac{\delta}{\delta \tilde{u}^{j'}} \left[\tilde{F}(y_k, \tilde{u}^j(y), D_k \tilde{u}^j(y)) \cdot \left|\det\left(\frac{\partial x_r}{\partial y_s}\right)^r_{r,s=1}\right|\right]$$

$$\cdot \eta^j(x(y)) \cdot dy, \qquad j' = 1, \ldots, m. \tag{19}$$

Here $\alpha = (\alpha^1, \ldots, \alpha^m)$ is an arbitrary m-dimensional vector of real numbers, and $\eta^{j'} \in M$ $(j' = 1, \ldots, m)$.

Using (19), we obtain

$$\int_{\Omega} \frac{\delta F}{\delta u^{j'}} \cdot \eta^{j'} \, dx = \int_{\tilde{\Omega}} \left(\frac{\delta F}{\delta u^{j'}}\right)_{x=x(y)} \cdot \eta^{j'}(x(y)) \cdot \left|\det\left(\frac{\partial x_r}{\partial y_s}\right)^n_{r,s=1}\right| dy$$

$$= \int_{\tilde{\Omega}} \frac{\delta}{\delta \tilde{u}^{j'}} \left[\tilde{F}(y_k, \tilde{u}^j, D_k \tilde{u}^j) \cdot \left|\det\left(\frac{\partial x_r}{\partial y_s}\right)^n_{r,s=1}\right|\right] \eta^{j'}(x(y)) \, dy,$$

$$j' = 1, \ldots, m.$$

The required equality (17) follows from this, since $\eta^{j'} \in M$ $(j' = 1, \ldots, m)$ is arbitrary. ∎

Suppose now that equation (1) is transformed into (9) by means of the transformation (8) (see (16)).

LEMMA 3. *A function $\tilde{E}[v]$ is a solution of the IPCV for the nonparabolic equation* (11), (12) *if and only if the functional*

$$\mathscr{F}[u(x)] = \tilde{E}[u \cdot \exp(\tfrac{1}{2} \cdot z)]|_{y=y(x)} \tag{20}$$

is a solution of the IPCV for equation (1).

PROOF. By Lemma 1 we find that the set of solutions of the IPCV for (11) and (12) is given by

$$\tilde{E}[v] = \int_{\hat{\Omega}} \left[\sum_{i=1}^{n} a^i \left(\frac{\partial v}{\partial y_i} \right)^2 - \lambda \cdot v^2 + 2 \cdot \tilde{g} \cdot v \right] dy. \tag{21}$$

From (10) we have

$$v(y) = u(y) \cdot \exp(\tfrac{1}{2} \cdot z), \tag{22}$$

$$\frac{\partial v}{\partial y_i} = \left(\frac{\partial u}{\partial y_i} + \frac{1}{2} \frac{b_1^i}{a^i} u \right) \cdot \exp(\tfrac{1}{2} z), \qquad i = 1, \ldots, n. \tag{23}$$

Substituting (22), (23) into (21), we obtain

$$\mathscr{F}_1[u] = \tilde{E}[u \cdot \exp(\tfrac{1}{2} \cdot z)]$$
$$= \int_{\hat{\Omega}} \left[\sum_{i=1}^{n} a^i \cdot \left(\frac{\partial u}{\partial y_i} + \frac{1}{2} \cdot \frac{b_1^i}{a^i} u \right)^2 - \lambda \cdot u^2 + 2g_1 \cdot u \right] \cdot \exp z \, dy.$$

Hence

$$\delta \mathscr{F}_1[u] = 2 \cdot \int_{\hat{\Omega}} \exp z \cdot \left[\sum_{i=1}^{n} a^i \cdot \left(\frac{\partial u}{\partial y_i} + \frac{1}{2} \cdot \frac{b_1^i}{a^i} u \right) \right.$$
$$\left. \times \left(\delta u_{y_i} + \frac{1}{2} \cdot \frac{b_1^i}{a^i} \cdot \delta u \right) - \lambda u \cdot \delta u + g_1 \cdot \delta u \right] dy$$

$$= 2 \cdot \int_{\hat{\Omega}} \exp z \cdot \left\{ \sum_{i=1}^{n} a^i \frac{\partial}{\partial y_i} \left[\left(\frac{\partial u}{\partial y_i} + \frac{1}{2} \cdot \frac{b_1^i}{a^i} u \right) \delta u \right] \right. \tag{24}$$
$$- \sum_{i=1}^{n} a^i \left(\frac{\partial^2 u}{\partial y_i^2} + \frac{1}{2} \frac{b_1^i}{a^i} \cdot \frac{\partial u}{\partial y_i} \right) \delta u - \lambda \cdot u \cdot \delta u$$
$$\left. + \frac{1}{2} \cdot \sum_{i=1}^{n} b_1^i \left(\frac{\partial u}{\partial y_i} + \frac{1}{2} \cdot \frac{b_1^i}{a^i} u \right) \cdot \delta u + g_1 \cdot \delta u \right\} dy.$$

Noting that for all $u \in M$

$$\int_{\tilde{\Omega}} \exp z \cdot \sum_{i=1}^{n} a^i \frac{\partial}{\partial y_i} \left[\left(\frac{\partial u}{\partial y_i} + \frac{1}{2} \cdot \frac{b_1^i}{a^i} \cdot u \right) \delta u \right] dy$$

$$= \int_{\tilde{\Omega}} \left\{ \sum_{i=1}^{n} a^i \frac{\partial}{\partial y_i} \left[\exp z \cdot \left(\frac{\partial u}{\partial y_i} + \frac{1}{2} \cdot \frac{b_1^i}{a^i} \cdot u \right) \delta u \right] \right.$$

$$\left. - \sum_{i=1}^{n} b_1^i \left(\frac{\partial u}{\partial y_i} + \frac{1}{2} \cdot \frac{b_1^i}{a^i} u \right) \delta u \cdot \exp z \right\} dy,$$

from (24) we obtain, for all $u \in M$,

$$\delta \mathcal{F}_1[u] = -2 \cdot \int_{\tilde{\Omega}} \exp z \cdot \left\{ \sum_{i=1}^{n} b_1^i \cdot \left(\frac{\partial u}{\partial y_i} + \frac{1}{2} \frac{b_1^i}{a^i} \cdot u \right) + \right.$$

$$+ \sum_{i=1}^{n} a^i \left(\frac{\partial^2 u}{\partial y_i^2} + \frac{1}{2} \cdot \frac{b_1^i}{a^i} \cdot \frac{\partial u}{\partial y_i} \right)$$

$$- \frac{1}{2} \cdot \sum_{i=1}^{n} b_1^i \cdot \left(\frac{\partial u}{\partial y_i} + \frac{1}{2} \cdot \frac{b_1^i}{a^i} u \right)$$

$$\left. + r_1 \cdot u - \frac{1}{4} \cdot \sum_{i=1}^{n} \frac{(b_1^i)^2}{a^i} \cdot u - g_1 \right\} \delta u \, dy$$

$$= -2 \cdot \int_{\tilde{\Omega}} \exp z \cdot (\mathcal{L}_1 u - g_1) \cdot \delta u \cdot dy.$$

The lemma follows from this via (17) and (20). ∎

From the lemmas proved above we obtain

COROLLARY 2. *Suppose problem* (11), (12), (2) *is obtained from* (1) *and* (2) *by means of transformations* (8) *and* (10). *Then the general solution of the IPCV in Formulation 2 for problem* (1), (2) *is the functional* $\mathcal{F}[u]$ *obtained from the functional* $\tilde{E}[v]$ *of* (21) *by the transformations* (22) *and* (16).

In theoretical investigations of differential equations by a variational method, and also in justifying direct methods of the calculus of variations, in applications it is important that the corresponding functional should have not a stationary point but an extremal point (a maximum or minimum). For the elliptic equation

$$-\Delta u + \lambda u = g(x), \qquad \lambda \equiv \text{const} < -\lambda_1^2, \ x \in \Omega \subset R^n, \tag{25}$$

or for the wave equation

$$u_{tt} - \Delta u = g(x, t), \qquad (x, t) \in Q \subset R^{n+1} \tag{26}$$

functionals which are solutions of the IPCV are well known: respectively,

$$V_1[u] = \int_\Omega [|\nabla u|^2 + \lambda \cdot u^2 - 2 \cdot g \cdot u] \, dx, \tag{27}$$

$$V_2[u] = \int_Q \{-u_t^2 + |\nabla u|^2 - 2 \cdot g \cdot u\} \, dx \, dt. \tag{28}$$

It can be shown, however (see below), that these functionals are not bounded on M above and below, and the following question arises: for equations (25) and (26) and for the more general nonparabolic equation (1) do there exist semibounded functionals which are solutions of the IPCV in Formulation 2?

THEOREM 1. *If*

$$\sum_{i,j=1}^n p^{ij} \xi_i \xi_j \geq \mu |\xi|^2 > 0,$$

then for an equation (1) *of elliptic type in* Ω *for*

$$r > r_0(\Omega) > 0 \tag{29}$$

in the class of Euler-Lagrange functionals (3) *there are no solutions of the IPCV which are semibounded on* M.

PROOF. *Step* 1. According to Corollary 1 and Lemma 3, the general solution of the IPCV in Formulation 2 for equation (9) is given by

$$\mathcal{J}_1[u] = \int_{\hat\Omega} \left\{ \sum_{i=1}^n a_i \cdot \left(\frac{\partial u}{\partial y_i} + \frac{1}{2} \cdot \frac{b_1^i}{a^i} \cdot u \right)^2 \right.$$
$$\left. - \left(r_1 - \frac{1}{4} \cdot \sum_{i=1}^n \frac{(b_1^i)^2}{a^i} \right) \cdot u^2 + 2 \cdot g_1 \cdot u \right\} \exp z \, dy. \tag{30}$$

Here (see (20))

$$\mathcal{J}_1[u] = \tilde{E}[u \cdot \exp(\tfrac{1}{2} \cdot z)], \tag{31}$$

where $\tilde{E}[v]$ is the functional (21) constructed for equation (11). Therefore, from (31) it is obvious by (10) that for unboundedness above and below on M of the functional $\mathcal{J}_1[u]$ of (31) it is necessary and sufficient that the functional $\tilde{E}[v]$ be unbounded above and below on M.

Step 2. We establish the unboundedness on M of the functional

$$E[v] = \int_{\hat\Omega} \cdot \left\{ \sum_{i=1}^n \left(\frac{\partial v}{\partial y_i} \right)^2 - \lambda \cdot v^2 + 2 \cdot \tilde{g} \cdot v \right\} dy \tag{32}$$

of the form (21) (for an elliptic equation, as stipulated in (9), $a^i = 1$, $i = 1, \ldots, n$). It is obvious that for all $v \in M$

$$E[v] - \int_{\tilde{\Omega}} \tilde{g}^2 \, dy \leq \int_{\tilde{\Omega}} \left\{ \left(\frac{\partial v}{\partial y_i} \right)^2 + (1 - \lambda) \cdot v^2 \right\} dy \equiv E_2[v], \quad (33)$$

$$E[v] + \int_{\tilde{\Omega}} \tilde{g} \, dy \geq \int_{\tilde{\Omega}} \left\{ \sum_{i=1}^{n} \left(\frac{\partial v}{\partial y_i} \right)^2 - (1 + \lambda) \cdot v^2 \right\} dy \equiv E_1[v]. \quad (34)$$

Since with a piecewise smooth boundary $\partial\Omega$ the domain $\Omega \subset R^n$ contains some n-dimensional parallelepiped, because of the nondegeneracy of the transformation (16) taking Ω into $\tilde{\Omega}$, the domain $\tilde{\Omega}$ also contains some parallelepiped

$$Q = (a, b) \equiv (a_1, b_1) \times \cdots \times (a_n, b_n), \qquad |Q| = \prod_{i=1}^{n} (b_i - a_i) > 0.$$

We set

$$v_m(y) = \begin{cases} \displaystyle\prod_{i=1}^{n} \sin[m(l_i y_i + c_i)], & y \in Q, \\ 0, & y \in \tilde{\Omega} \backslash Q, \end{cases} \quad (35)$$

where $m = 1, 2, \ldots,$ and

$$l_i = 2\pi/(b_i - a_i), \quad c_i = -a_i \cdot l_i, \quad i = 1, \ldots, n. \quad (36)$$

It is obvious that $v_m(y) \in \overset{\circ}{W}{}^1_2(\tilde{\Omega})$ and

$$\|v_m|\overset{\circ}{W}{}^1_2(\tilde{\Omega})\| = \left\{ \int_{\tilde{\Omega}} \sum_{i=1}^{n} \left(\frac{\partial v_m}{\partial y_i} \right)^2 dy \right\}^{1/2} < \infty, \quad m = 1, 2, 3, \ldots,$$

and by direct computations we find

$$E_2[v_m] = \left(\frac{1}{2} \right)^n \cdot |Q| \cdot \left[1 - \lambda + m^2 \cdot \sum_{j=1}^{n} l_j^2 \right]. \quad (37)$$

Setting $m = 1$, from this we conclude that for any

$$\lambda > 1 + \sum_{j=1}^{n} l_j^2 \quad (38)$$

we have $E_2[v_1] < 0$. Then for any ρ, not depending on y or v, for the quadratic functional $E_2[v]$ of (33) we have $E_2[\rho v_1] = \rho^2 E_2[v_1] \to -\infty$ $(\rho \to \infty)$. Therefore, the functional $E_2[v]$ and by (33) also $E[v]$ are unbounded below on $\overset{\circ}{W}{}^1_2(\tilde{\Omega})$.

Similarly, for the functional $E_1[v]$ of (34) by direct computations we find that

$$E_1[v_m] = \left(\frac{1}{2}\right)^n \cdot |Q| \cdot \left\{ -1 - \lambda + m^2 \sum_{j=1}^{n} l_j^2 \right\}.$$

Obviously for any $\lambda \in R^1$ we can find $N = N(\lambda)$ such that $E_1[v_N] > 0$. Therefore, $E_1[\rho v_N] = \rho^2 E_1[v_N] \to +\infty$ as $\rho \to \infty$. From this and (34) we conclude also that the functional $E[v]$ is not bounded above on $\overset{\circ}{W}{}_2^1(\Omega)$.

It is now not hard to obtain the unboundedness above and below on the set M of the functional $E[v]$ of (32) and (38) from the fact that the functional $E[v]$ is continuous on $\overset{\circ}{W}{}_1^2(\Omega)$, the set M is dense in $\overset{\circ}{W}{}_2^1(\Omega)$, and $E[v]$ of (32) and (38) is unbounded on $\overset{\circ}{W}{}_2^1(\Omega)$.

Since the functional $E[v]$ of (32) by Corollary 1 is the general solution of the IPCV for the equation

$$\Delta v + \lambda \cdot v = \tilde{g}(y), \quad y \in \tilde{\Omega}, \quad v|_{\partial \tilde{\Omega}} = 0, \tag{39}$$

it is useful to distinguish the result of this second step separately.

COROLLARY 4. *For problem* (39), (38) *in the class of Euler-Lagrange functionals there are no solutions of the IPCV which are semibounded on* M.

Step 3. Since the functional $\mathcal{F}_1[u]$ of (30) is the general solution of the IPCV for equation (9), from the results of Step 2 and the arguments of Step 1 we also obtain

COROLLARY 5. *For the elliptic equation* (9) $(a^i = 1, i = 1, \ldots, n)$, *where*

$$\lambda \equiv r_1 - \frac{1}{4} \cdot \sum_{i=1}^{n} (b_1^i)^2 > 1 + \sum_{j=1}^{n} l_j^2. \tag{40}$$

in the class of Euler-Lagrange functionals there do not exist solutions of the IPCV which are semibounded on M.

Thus, the functional $\mathcal{F}_1[u]$ of (30) is unbounded on M above and below under condition (40). The functional $V[u]$, which is the general solution of the IPCV for equation (1), is obtained, according to Lemma 2, by a nondegenerate change of the independent variables $y = y(x)$ in the functional $\mathcal{F}_1[u(y)]$, i.e., a change of the independent variable occurs under the integral which does not change the value of the integral (see (18)). Therefore, the functional $V[u(x)] = \mathcal{F}_1[u(y)]|_{y=y(x)}$, which is the general solution of the IPCV for equation (1), is also unbounded on M above and below for $r > r_0(\Omega)$. The theorem is proved. ∎

THEOREM 2. *For any ultrahyperbolic (including properly hyperbolic) equation* (1) *in the class of Euler-Lagrange functionals there are no semibounded solutions of the IPCV.*

From the proof of Theorem 1 it is not hard to see that we need only establish unboundedness on M of the functional

$$G[v] = \int_{\tilde{\Omega}} \left\{ \sum_{i=1}^{m} v_{y_i}^2 - \sum_{k=m+1}^{n} v_{y_k}^2 - \lambda \cdot v^2 + 2 \cdot \tilde{g} \cdot v \right\} dy \qquad (41)$$

which is the general solution (see Corollary 1) of the IPCV for the ultra-hyperbolic equation (14).

For all $v \in M$ we have

$$G[v] - \int_{\tilde{\Omega}} \tilde{g}^2 \, dy \le \int_{\tilde{\Omega}} \left\{ \sum_{i=1}^{m} v_{y_i}^2 - \sum_{k=m+1}^{n} v_{y_k}^2 + (1-\lambda) \cdot v^2 \right\} dy \qquad (42)$$

$$\equiv G_1[v];$$

$$G[v] + \int_{\tilde{\Omega}} \tilde{g}^2 \, dy \ge \int_{\tilde{\Omega}} \left\{ \sum_{i=1}^{m} v_{y_i}^2 - \sum_{k=m+1}^{n} v_{y_k}^2 - (1+\lambda) \cdot v^2 \right\} dy \qquad (43)$$

$$\equiv G_2[v].$$

Since the domain Ω contains some parallelepiped and the transformation (16) mapping Ω to $\tilde{\Omega}$ is nondegenerate, the domain $\tilde{\Omega}$ also contains some parallelepiped

$$Q_1 = (a_1, b_1) \times \cdots \times (a_m, b_m) \times (\overline{a}_{m+1}, \overline{b}_{m+1}) \times \cdots \times (\overline{a}_n, \overline{b}_n),$$

$$|Q_1| = \prod_{i=1}^{m} (b_i - a_i) \prod_{i=m+1}^{n} (\overline{b}_i - \overline{a}_i) > 0.$$

For $t, s = 1, 2, \ldots$ we set

$$v_{ts}(y) = \begin{cases} \displaystyle\prod_{i=1}^{m} \sin[t(l_i y_i + c_i)] \prod_{i=m+1}^{n} \sin[s(l_i y_i + c_i)], & y \in Q_1, \\ 0, & y \in \tilde{\Omega} \backslash Q_1. \end{cases} \qquad (44)$$

Here the l_i and c_i for $i = 1, \ldots, m$ are defined in (36), while

$$l_i = \frac{2\pi}{\overline{b}_i - \overline{a}_i}, \qquad c_i = -\overline{a}_i \cdot l_i, \qquad i = m+1, \ldots, n. \qquad (45)$$

Obviously $v_{ts}(y) \in \overset{\circ}{W}{}^1_2(\tilde{\Omega})$ for all $t, s = 1, 2, \ldots$, and by direct computations we find that

$$G_1[v_{ts}] = \left(\frac{1}{2}\right)^n \cdot |Q_1| \cdot \left\{ (t^2 - s^2) \cdot \sum_{j=1}^n l_j + 1 - \lambda \right\}, \qquad (46)$$

$$G_2[v_{ts}] = \left(\frac{1}{2}\right)^n \cdot |Q_1| \cdot \left\{ (t^2 - s^2) \cdot \sum_{j=1}^n l_j^2 - 1 - \lambda \right\}. \qquad (47)$$

From this it is easy to see that the functional $G_1[v_{ts}]$, $(G_2[v_{ts}])$ for any $\lambda \in R^1$ is unbounded below (respectively, above), and from (42) (from (43)) we find that the functional $G[v]$, which is the general solution of the IPCV for equation (14), is unbounded below (respectively, above) on $\overset{\circ}{W}{}^1_2(\Omega)$.

Further, repeating the arguments of Steps 1 and 3 of the proof of Theorem 1, we can show that the functional $\tilde{G}[u]$ obtained from $G[v]$ of (41) by the changes inverse to (10) and (8) (see Lemmas 2 and 3), which is the general solution of the IPCV for the ultrahyperbolic equation (1), is unbounded on M both above and below for any constant coefficients p^{ij}, b_1^i $(i, j = 1, \ldots, n)$, and r. ∎

For a parabolic equation there does not exist a solution of the IPCV in the class of Euler-Lagrange functionals (3) [54]. For any nonparabolic equation (1) in Ω such solutions exist (Lemma 1). However, for any ultrahyperbolic (including hyperbolic) equation (1) and also for an elliptic equation (1) for $r \geq r_0 > 0$ (assuming that $\sum_{i,j=1}^r p^{ij}\xi_i\xi_j > \mu|\xi|^2 > 0$) in the class of Euler-Lagrange functionals there are no functionals bounded above or below which are solutions of the IPCV in Formulation 2. It is therefore necessary to invoke other classes of functionals distinct from the class of Euler-Lagrange functionals (3) to develop a direct variational method of investigating even these basic equations of mathematical physics and to apply minimization methods to the functionals of the corresponding variational problem.

Bibliography

1. György Adler, *Sulla caratterizzabilità dell'equazione del calore dal punto di vista del calcolo delle variazioni*, Magyar Tud. Akad. Mat. Kutató Int. Közl. **2** (1957), 153–157.

2. L. Ainola, *Variational principles and general formulas for a mixed problem for the wave equation*, Eesti NSV Teod. Akad. Toimetised Füüs.-Mat. **18** (1969), 48–56. (Russian)

3. Yu. R. Akopyan and L. A. Oganesyan, *A variational-difference method for the solution of two-dimensional linear parabolic equations*, Zh. Vychisl. Mat. i Mat. Fiz. **17** (1977), 109–118; English transl. in USSR Comput. Math. and Math. Phys. **17** (1977).

4. S. J. Aldersley, *Higher Euler operators and some of their applications*, J. Mathematical Phys. **20** (1979), 522–531.

5. M. A. Aleksidze et al., *On the automation of variational methods for solving boundary value problems*, Trudy Vychisl. Tsentra Akad. Nauk Gruzin. SSR **10** (1971), no. 4, 16–25. (Russian)

6. T. I. Amanov, *Representation and imbedding theorems for the function spaces $S_{p,\theta}^{(r)}B(R^n)$ and $S_{p^*,\theta}^{(r)}B(0 \leq x_j \leq 2\pi;\ j = 1,\ldots,n)$*, Trudy Mat. Inst. Steklov. **77** (1965), 3–34; English transl. in Proc. Steklov Inst. Math. **77** (1975).

7. I. N. Anan'ev, Yu. I. Nyashin, and A. N. Skorokhodov, *A variational method for calculating a time-dependent temperature field*, Inzh.-Fiz. Zh. **26** (1974), 470–476; English transl. in J. Engrg. Phys. **26** (1974).

8. I. M. Anderson and T. Duchamp, *On the existence of global variational Principles*, Amer. J. Math. **102** (1980), 781–868.

9. R. S. Anderssen, *Variational methods and parabolic differential equations*, Ph. D. thesis, University of Adelaide, Adelaide, 1967.

10. T. A. Andreeva and L. D. Gordinskiĭ, *On a boundary value problem of filtration theory*, Boundary Value Problems of the Theory of Heat

Conduction (Yu. A. Mitropol'skiĭ and A. A. Berezovskiĭ, editors), Izdanie Inst. Mat. Akad. Nauk Ukrain. SSR, Kiev, 1975, pp. 5–16. (Russian)

11. R. I. Andrushkiev, *On the approximate solution of K-positive eigenvalue problems* $Tu - \lambda Su = 0$, J. Math. Anal. Appl. **50** (1975), 511–529.

12. A. M. Arthurs, *Dual extremum principles and error bounds for a class of boundary value problems*, J. Math. Anal. Appl. **41** (1973), 781–795.

13. ____, *Complementary variational principles for linear equations*, J. Math. Anal. Appl. **49** (1975), 237–239.

14. ____, *Complementary variational principles*, 2nd ed., Clarendon Press, Oxford, 1980.

15. T. M. Atanacković and Dj. S. Djukić, *An extremum variational principle for a class of boundary value problems*, J. Math. Anal. Appl. **93** (1983), 344–362.

16. R. W. Atherton and G. M. Homsy, *On the existence and formulation of variational principles for nonlinear differential equations*, Studies in Appl. Math. **54** (1975), 31–60.

17. Marc Authier, *Espaces du type Sobolev associés à une seule dérivée partielle, théorème de traces, applicatoin à certains problèmes aux limites*, C. R. Acad. Sci. Paris Sér. A-B **262** (1966), A1158–A1161.

18. F. Belatoni, *Über die Charakterisierbarkeit partieller Differentialgleichungen zweiter Ordnung mit Hilfe der Variationsrechnung*, Magyar Tud. Akad. Mat. Kutató Int. Közl. **5** (1960), 229–233.

19. M. F. Barnsley, *Padé approximant bounds for the difference of two series of Stieltjes*, J. Mathematical Phys. **17** (1976), 559–565.

20. M. F. Barnsley and George A. Baker, Jr., *Bivariational bounds in a complex Hilbert space, and correction terms for Padé approximants*, J. Mathematical Phys. **17** (1976), 1019–1027.

21. M. F. Barnsley and P. D. Robinson, *Dual variational principles and Padé-type approximants*, J. Inst. Math. Appl. **14** (1974), 229–249.

22. ____, *Bivariational bounds*, Proc. Roy. Soc. (London) Ser. A **338** (1974), 527–533.

23. Martin Becker, *The principles and applications of variational methods*, MIT Press, Cambridge, Mass., 1964.

24. Gian Cesare Belli, *Variational formulation for scalar wave and diffusion equations*, Ist. Lombardo Accad. Sci. Lett. Rend. A **106** (1972), 158–166.

25. G. Benthien and M. E. Gurtin, *A principle of minimum transformed energy in linear elastodynamics*, Trans. ASME Ser. E: J. Appl. Mech. **37** (1970), 1147–1149.

26. V. I. Berdichevskiĭ, *A variational equation of continuum mechanics*, Problems of the Mechanics of a Solid Deformable Body (V. V. Novozhilov Sixtieth Birthday Vol.; L. I. Sedov and Yu. N. Robotnov, editors), "Sudostroenie", Leningrad, 1970, pp. 55–66. (Russian)

27. ____, *On a variational principle*, Dokl. Akad. Nauk SSSR **215** (1974), 1329–1332; English transl. in Soviet Phys. Dokl. **19** (1974/75).

28. ____, *Variational principles of continuum mechanics*, "Nauka", Moscow, 1983. (Russian)

29. Yu. M. Berezanskiĭ, *On boundary value problems for general partial differential operators*, Dokl. Akad. Nauk SSSR **122** (1958), 959–962. (Russian)

30. ____, *Expansion in eigenfunctions on selfadjoint operators*, "Naukova Dumka", Kiev, 1965; English transl., Amer. Math. Soc., Providence, R. I., 1968.

31. P. W. Berg, *Calculus of variations*, Handbook of Engineering Mechanics (W. Flügge, editor), McGraw-Hill, 1962, Chapter 16.

32. O. V. Besov and A. D. Dzhabrailov, *Interpolation theorems for some spaces of differentiable functions*, Trudy Mat. Inst. Steklov. **105** (1969), 15–20; English transl. in Proc. Steklov Inst. Math. **105** (1969).

33. O. V. Besov, V. P. Il'in, and S. M. Nikol'skiĭ, *Integral representations of functions and imbedding theorems*, "Nauka", Moscow, 1975; English transl., Vols. 1, 2, Wiley, 1979.

34. Maurice A. Biot, *Variational principles in heat transfer*, Clarendon Press, Oxford, 1970.

35. M. Sh. Birman, *On minimal functionals for elliptic differential equations of second order*, Dokl. Akad. Nauk SSSR **93** (1953), 953–956. (Russian)

36. ____, *On Trefftz's variational method for the equation $\Delta^2 u = f$*, Dokl. Akad. Nauk SSSR **101** (1955), 201–204. (Russian)

37. ____, *Variational methods analogous to Trefftz's for solving boundary value problems*, Vestnik Leningrad. Univ. **1956**, no. 13 (Ser. Mat. Mekh. Astr. vyp. 3), 69–89. (Russian)

38. L. Bittner, *Abschätzungen bei Variationsmethoden mit Hilfe von Dualitätssätzen*. I, Numer. Math. **11** (1968), 129–143.

39. Philippe Blanchard and Erwin Brüning, *Direkte Methoden der Variationsrechnung. Ein Lehrbuch*, Springer-Verlag, 1982.

40. Oskar Bolza, *Lectures in the calculus of variations*, Univ. of Chicago Press, Chicago, Ill., 1904; reprints, Chelsea and Dover, New York, 1960; German transl., Teubner, Leipzig, 1909.

41. V. A. Borovikov, *Fundamental solutions of linear partial differential equations with constant coefficients*, Trudy Moskov. Mat. Obshch. **8** (1959), 199–257; English transl. in Amer. Math. Soc. Transl. (2) **25** (1963).

42. J. S. Bradley and B. D. Sleeman, *K-positive ordinary differential operators of the third order*, J. Math. Anal. Appl. **70** (1979), 249–257.

43. H. Brézis, *Quelques propriétés des opérateurs monotones et des semigroupes non linéaires*, Nonlinear Operators and the Calculus of Variations (Summer School, Brussels, 1975), Lecture Notes in Math., vol. 543, Springer-Verlag, 1976, pp. 56–82.

44. Haïm Brézis and Ivar Ekeland, *Un principe variationnel associé à certaines équations paraboliques. Le cas dépendant du temps*, C. R. Acad. Sci. Paris Sér. A-B **282** (1976), A1197–A1198.

45. Felix E. Browder, *Remarks on nonlinear functional equations*. I, Proc. Nat. Acad. Sci. U.S.A. **51** (1964), 985–989.

46. ____, *Remarks on nonlinear functional equations*. II, III, Illinois J. Math. **9** (1965), 608–616, 617–622.

47. ____, *Existence and uniqueness theorems for solutions of nonlinear boundary value problems*, Applications of Nonlinear Partial Differential Equations in Mathematical Physics, Proc. Sympos. Appl. Math., vol. 17, Amer. Math. Soc., Providence, R. I., 1965, pp. 24–49.

48. ____, *Nonlinear monotone and accretive operators in Banach spaces*, Proc. Nat. Acad. Sci. U.S.A. **61** (1968), 388–393.

49. V. I. Burenkov, *The density of infinitely differentiable functions in spaces of functions specified on an arbitrary open set*, Theory of Cubature Formulas and Applications of Functional Analysis to Problems of Mathematical Physics (Materials, School-Conf., Tashkent, 1974; S. L. Sobolev, editor), Inst. Mat., Sibirsk. Otdel Akad. Nauk SSSR, Novosibirsk, 1975, pp. 9–22. (Russian)

50. Alfred Carasso, *A least squares procedure for the wave equation*, Math. Comp. **28** (1974), 757–767.

51. ____, *Error bounds in the final value problem for the heat equation*, SIAM J. Math. Anal. **7** (1976), 195–199.

52. E. Conjura and W. V. Petryshyn, *Extension of nonlinear densely defined operators, rates of convergence for the error and the residual, and application to differential equations*, J. Math. Anal. Appl. **64** (1978), 651–694.

53. Philip Cooperman, *An extension of the method of Trefftz for finding local bounds on the solutions of boundary value problems, and on their derivatives*, Quart. Appl. Math. **10** (1953), 359–373.

54. E. T. Copson, *Partial differential equations and the calculus of variations*, Proc. Roy. Soc. Edinburgh **46** (1925/26), 126–135.

55. R. W. Cottle et al. (editors), *Variational inequalities and complementarity problems* (Proc. Internat. School, Erice, 1978), Wiley, 1980.

56. Richard Courant, *Remarks about the Rayleigh-Ritz method*, Boundary Problems in Differential Equations (Proc. Sympos., Madison, Wisc., 1959; R. E. Langer, editor), Univ. of Wisconsin Press, Madison, Wisc., 1960, pp. 273–277.

57. R. Courant and D. Hilbert, *Methoden der mathematischen Physik.* Vols. I (2nd ed.), II, Springer-Verlag, 1931, 1937; English transl. of Vol. I, Interscience, 1953; of Vol. II (drastically revised), Interscience, 1962.

58. James W. Daniel, *Applications and methods for the minimization of functionals*, Nonlinear Functional Analysis and Applications (Proc. Sem., Madison, Wisc., 1970; L. B. Rall, editor), Academic Press, 1971, pp. 399–424.

59. ____, *The approximate minimization of functionals*, Prentice-Hall, Englewood Cliffs, N. J., 1971.

60. David R. Davis, *The inverse problem of the calculus of variations in higher space*, Trans. Amer. Math. Soc. **30** (1928), 710–736.

61. Paul Dedecker, *Sur un problème inverse du calcul des variations*, Acad. Roy. Belgique Bull. Cl. Sci. (5) **36** (1950), 63–70.

62. V. F. Dem'yanov and A. M. Rubinov, *Approximate methods in optimization problems*, Izdat. Leningrad. Univ., Leningrad, 1968; English transl., Amer. Elsevier, New York, 1970.

63. A. A. Dezin, *Existence and uniqueness theorems for boundary value problems for partial differential equations in function spaces*, Uspekhi Mat. Nauk **14** (1959), no. 3(87), 21–73; English transl. in Amer. Mat. Soc. Transl. (2) **42** (1964).

64. ____, *General questions in the theory of boundary value problems*, "Nauka", Moscow, 1980. (Russian)

65. A. A. Dezin and V. P. Mikhaĭlov, *On boundary value problems for linear differential operators*, Proc. Fourth All-Union Math. Congr. (Leningrad, 1961), vol. 2, "Nauka", Leningrad, 1964, pp. 499–501. (Russian)

66. J. B. Diaz, *Upper and lower bounds for quadratic functionals*, Proc. Sympos. Spectral Theory and Differential Problems, Dept. of Math., Oklahoma Agric. and Mech. Coll., Stillwater, Okla., 1951, pp. 279–289.

67. ____, *Upper and lower bounds for quadratic functionals*, Collectanea Math. **4** (1951), 3–49.

68. ____, *Upper and lower bounds for quadratic integrals, and at a point, for solutions of linear boundary value problems*, Boundary Problems in

Differential Equations (Proc. Sympos., Madison, Wisc., 1959; R. E. Langer, editor), Univ. of Wisconsin Press, Madison, Wisc., 1960, pp. 47–83.

69. J. B. Diaz and H. J. Greenberg, *Upper and lower bounds for the solution of the first boundary value problem of elasticity*, Quart. Appl. Math. **6** (1948), 326–331.

70. ____, *Upper and lower bounds for the solution of the first biharmonic boundary value problem*, J. Math. and Phys. **27** (1948), 193–201.

71. V. P. Didenko, *Generalized solvability of boundary value problems for systems of differential equations of mixed type*, Differentsial'nye Uravneniya **8** (1972), 24–29; English transl. in Differential Equations **8** (1972).

72. ____, *Generalized solvability of the Tricomi problem*, Ukrain. Mat. Zh. **25** (1973), 14–24; English transl. in Ukrainian Math. J. **25** (1973).

73. ____, *Finding strong solutions by variational methods*, Proc. Sci. Conf. Computational Mathematics in Modern Scientific and Technological Progress (Kanev, 1974), Inst. Kibernet. Akad. Nauk Ukrain. SSR and Kiev. Gos. Univ., Kiev, 1974, pp. 154–159. (Russian)

74. ____, *A variational problem for equations of mixed type*, Differentsial'nye Uravneniya **13** (1977), 44–49; English transl. in Differential Equations **13** (1977).

75. ____, *A variational method of solving boundary value problems whose operator is not symmetric*, Dokl. Akad. Nauk SSSR **240** (1978), 1277–1280; English transl. in Soviet Math. Dokl. **19** (1978).

76. V. P. Didenko and A. A. Popova, *Inequalities with a negative norm and variational problems for asymmetric differential operators*, Preprint 81–29, Inst. Kibernet. Akad. Nauk Ukrain. SSR, Kiev, 1981. (Russian) MR **83j**:35009.

77. Dj. S. Djukić and T. M. Alancković, *Error bounds via a new extremum variational principle, mean square residual and weighted mean square residual*, J. Math. Anal. Appl. **75** (1980), 203–218.

78. V. V. Dodonov, V. I. Man'ko, and V. D. Skarzhinsky [Skarzhinskiĭ], *The inverse problem of the variational calculus and the nonuniqueness of the quantization of classical systems*, Hadronic J. **4** (1980/81), 1734–1804.

79. Jesse Douglas, *Solution of the inverse problem of the calculus of variations*, Trans. Amer. Math. Soc. **50** (1941), 71–128.

80. A. D. Dzhabrailov, *On some function spaces. Direct and inverse imbedding theorems*, Dokl. Akad. Nauk SSSR **159** (1964), 254–257; English transl. in Soviet Math. Dokl. **5** (1964).

81. ____, *On the theory of "imbedding theorems"*, Trudy Mat. Inst. Steklov. **89** (1967), 80–118; English transl. in Proc. Steklov Inst. Math. **89** (1967).

82. A. V. Dzhishkariani, *The Bubnov-Galerkin method*, Zh. Vychisl. Mat. i Mat. Fiz. **7** (1967), 1398–1402; English transl. in USSR Comput. Math. and Math. Phys. **7** (1967).

83. ____, *On the question of the stability of approximate methods of variational type*, Zh. Vychisl. Mat. i Mat. Fiz. **11** (1971), 569–579; English transl. in USSR Comput. Math. and Math. Phys. **11** (1971).

84. T. D. Dzhuraev, *Boundary value problems for equations of mixed and mixed-composite types*, "Fan", Tashkent, 1979. (Russian)

85. Arne Engvist, *On fundamental solutions supported by a convex cone*, Ark. Mat. **12** (1974), 1–40.

86. V. T. Erofeenko, *The connection between the fundamental solutions in cylindrical and spherical coordinates (with the same origin) for certain equations of mathematical physics*, Differentsial'nye Uravneniya **9** (1973), 1310–1317; English transl. in Differential Equations **9** (1973).

87. M. A. Evgrafov, *Estimation of the fundamental solution of an elliptic equation with constant coefficients*, Preprint No. 11, Keldysh Inst. Appl. Math. Acad. Sci. USSR, Moscow, 1980. (Russian) MR **81d**:35015.

88. I. I. Fedik, V. N. Mikhaĭlov, and V. I. Kozhuklovskiĭ, *Formalization of a variational method for solving boundary value problems in complicated two-dimensional domains*, Dokl. Akad. Nauk Ukrain. SSR Ser. A **1982**, no. 10, 51–54. (Russian)

89. M. Filar, *Construction of the fundamental solution of the equation* $\Delta^p u(x) + ku(x) = 0$, Prace Mat. **10** (1966), 131–140.

90. A. F. Filippov, *Uniqueness of generalized solutions*, Sibirsk Mat. Zh. **10** (1969), 217–222; English transl. in Siberian Math. J. **10** (1969).

91. V. M. Filippov, *A variational method for solving boundary value problems of mathematical physics, and function spaces*, Differentsial'nye Uravneniya **15** (1979), 2056–2065; English transl. in Differential Equations **15** (1979).

92. ____, *Strongly B-positive operators*, Differential Equations and Inverse Problems of Dynamics (R. G. Muklarlyamov, editor), Univ. Druzhby Narodov, Moscow, 1983, pp. 95–99. (Russian)

93. ____, *The relation of the least squares method to symmetrization of differential operators*, Differential Equations and Functional Analysis (V. N. Maslennikova, editor), Univ. Druzhby Narodov, Moscow, 1984, pp. 148–153. (Russian)

94. ____, *A variational method for solving boundary value problems for the wave equation*, Differentsial'nye Uravneniya **20** (1984), 1961–1968; English transl. in Differential Equations **20** (1984).

95. ____, *On some consequences of the isometry of the spaces* $W_{2(x,y)}^{(+\alpha,-\beta)}(\Omega)$ *to the spaces* $W_2^k(\Omega)$, Differential Equations and Functional Analysis (V. N. Maslennikova, editor), Univ. Druzhby Narodov, Moscow, 1984, pp. 141–147. (Russian)

96. ____, *A variational method for solving a hyperbolic equation with boundary conditions on the entire boundary of the domain*, Differential Equations and Functional Analysis (V. N. Maslennikova, editor), Univ. Druzhby Narodov, Moscow, 1983, pp. 114–119. (Russian)

97. ____, *On a direct variational method for solving an elliptic equation with a nonsymmetric and nonpositive operator*, Numerical Methods in Problems of Mathematical Physics, Univ. Druzhby Narodov, Moscow, 1985, pp. 61–66. (Russian)

98. ____, *On classes of operators in the variational method for solving nonlinear equations*, Numerical Methods in Problems of Mathematical Physics, Univ. Druzhby Narodov, Moscow, 1985, pp. 46–55. (Russian)

99. ____, *A direct variational method for studying boundary value problems for an integrodifferential hyperbolic equation*, Preprint, Univ. Druzhby Narodov, Moscow, 1985=Manuscript No. 5444-85, deposited at VINITI, 1985. (Russian) R.Zh.Mat. **1985**, 11Б793.

100. V. M. Filippov and A. N. Skorokhodov, *A quadratic functional for the heat equation*, Differentsial'nye Uravneniya **13** (1977), 1113–1123; English transl. in Differential Equations **13** (1977).

101. ____, *The principle of a minimum of a quadratic functional for a boundary value problem of heat conduction*, Differentsial'nye Uravneniya **13** (1977), 1434–1445; English transl. in Differential Equations **13** (1977).

102. Bruce A. Finlayson, *The method of weighted residuals and variational principles*, Academic Press, 1972.

103. ____, *Variational principles for heat transfer*, Numerical Properties and Methodologies in Heat Transfer (Proc. Second Nat. Sympos., College Park, Md., 1981; T. M. Shih, editor), Hemisphere, Washington, D. C., 1983, pp. 17–31.

104. G. B. Folland, *A fundamental solution for a subelliptic operator*, Bull. Amer. Math. Soc. **79** (1973), 373–376.

105. Kurt O. Friedrichs, *Ein Verfahren der Variationsrechnung das Minimum eines Integrals als das Maximum eines anderen Ausdruckes darzustellen*, Nachr. Ges. Wiss. Göttingen Math.-Phys. Kl. **1929**, 13–20.

106. ____, *Spektraltheorie halbbeschränkter Operatoren und Anwendung auf die Spektralzerlegung von Differentialoperatoren. I, II*, Math. Ann. **109** (1934), 465–487, 685–713.

107. ____, *The identity of weak and strong extensions of differential operators*, Trans. Amer. Math. Soc. **55** (1944), 132–151.

108. ____, *Symmetric positive linear differential equations*, Comm. Pure Appl. Math. **11** (1958), 333–418.

109. Bent Fuglede and Laurent Schwartz, *Un nouveau théorème sur les distributions*, C. R. Acad. Sci. Paris Sér. A-B. **263** (1966), A899–A901.

110. Tsutomu Fujino, *Variational principles of linear differential equations*, Mitsubishi Tech. Bull. No. 77 (1972), 1–25.

111. Hiroshi Fújita, *Contribution to the theory of upper and lower bounds in boundary value problems*, J. Phys. Soc. Japan **10** (1955), 1–8.

112. Paul Funk, *Variationsrechnung und ihre Anwendung in Physik und Technik*, Springer-Verlag, 1962.

113. D. H. Gage et al., *The non-existence of a general thermokinetic variational principle*, Non-equilibrium Thermodynamics, Variational Techniques and Stability (Proc. Sympos., Chicago, Ill., 1965; R. J. Donnelly et al., editors), Univ. of Chicago Press, Chicago, Ill., 1966, pp. 283–286.

114. Herbert Gajewski and Arno Langenbach, *Zur Konstruktion von Minimalfolgen für das Funktional des ebenen elastisch-plastischen Spannungszustandes*, Math. Nachr. **30** (1965), 165–180.

115. V. S. Gamidov, *Variational principles for B-positive-definite operators*, Seventh School on Theory of Operators in Function Spaces, Abstracts of Reports, Minsk, 1982, pp. 40–41. (Russian)

116. V. S. Gamidov and A. A. Popova, *A variational problem for B-positive-definite operators*, Manuscript No. 1276-80, deposited at VINITI, 1980. (Russian) R.Zh.Mat. **1980**, 7Б259.

117. I. M. Gel'fand and L. A. Dikiĭ, *Asymptotic behavior of the resolvent of Sturm-Liouville equations and the algebra of the Korteweg-deVries equations*, Uspekhi Mat. Nauk **30** (1975), no. 5(185), 67–100; English transl. in Russian Math. Surveys **30** (1975).

118. I. M. Gel'fand and I. Ya. Dorfman, *Hamiltonian operators and algebraic structures associated with them*, Funktsional. Anal. i Prilozhen. **13** (1979), no. 4, 13–30; English transl. in Functional Anal. Appl. **13** (1979).

119. I. M. Gel'fand and S. V. Fomin, *Calculus of variations*, Fizmatgiz, Moscow, 1961; English transl., Prentice-Hall, Englewood Cliffs, N. J., 1963.

120. I. M. Gel'fand and G. E. Shilov, *Generalized functions*. Vol. 2: *Spaces of test functions*, Fizmatgiz, Moscow, 1958; English transls., Academic Press, 1968; Gordon and Breach, 1968.

121. P. Glansdorff, *Sur une forme norwelle en cascade du terme aux limites de la variation d'une intégrale multiple*, Acad. Roy. Belgique Bull. Cl. Sci. (5) **38** (1952), 136–153.

122. R. Glowinski, J.-L. Lions, and R. Trémolières, *Analyse numérique des inéquations variationnelles*. Vols. 1, 2, Dunod, Paris, 1976.

123. E. A. Gorin, *Partially hypoelliptic partial differential equations with constant coefficients*, Sibirsk. Mat. Zh. **3** (1962), 500–526. (Russian)

124. L. M. Graves (editor), *Calculus of variations and its applications*, Proc. Sympos. Appl. Math., vol. 8, Amer. Mat. Soc., Providence, R. I., 1958.

125. H. J. Greenberg, *The determination of upper and lower bounds for the solution of the Dirichlet problem*, J. Math. and Phys. **27** (1948), 161–182.

126. V. V. Grushin, *Relation between the local and global properties of solutions of hypoelliptic equations with constant coefficients*, Mat. Sb. **66(108)** (1965), 525–550; English transl. in Amer. Math. Soc. Transl. (2) **67** (1968).

127. K. A. Gubaĭdullin, *Some boundary value problems for an equation of composite type*, Volzh. Mat. Sb. Vyp. 5 (1966), 104–113. (Russian)

128. F. Guil Guerrero and L. Martinez Alonzo, *Generalized variational derivatives in field theory*, J. Phys. A **13** (1980), 689–700.

129. M. E. Gurtin, *Variational principles for linear initial-value problems*, Quart. Appl. Math. **22** (1964/65), 252–256.

130. A. F. Guseĭnov, *On boundary value problems for quasilinear equations*, Trudy Inst. Fiz. i Mat. Akad. Nauk Azerbaĭdzhan. SSR Ser. Mat. 7 (1955), 129–162. (Russian)

131. István Gyarmati, *Non-equilibrium thermodynamics; field theory and variational principles*, Springer-Verlag, 1970.

132. P. Havas, *The range of application of the Lagrange formalism*, Nuovo Cimento (10) **5** (1957), Suppl., 363–388.

133. H. von Helmholtz, *Über die physikalische Bedeutung des Princips der kleinsten Wirkung*, J. Reine Angew. Math. **100** (1887), 133–166, 213–222.

134. Ismail Herrera and Jacobo Bielak, *A simplified version of Gurtin's variational principles*, Arch. Rational Mech. Anal. **53** (1974), 131–149.

135. Joseph Hersch, *Une transformation variationnelle apparentée à celle de Friedrichs, conduisant à la méthode des problèmes auxiliaires unidimensionnels*, L'Enseignement Math. (2) **11** (1965), 159–169.

136. Heinrich Kertz, *Die Prinzipien der Meckanik in neuem Zusammenhangedargestellt*, Barth, Leipzig, 1894; English transl., Macmillan, 1899; reprint, Dover, New York, 1956.

137. David Hilbert, *Über das Dirichlet'sche Princip*, Jber. Deutsch. Math.-Verein. **8** (1900), 184–188.

138. Arthur Hirsch, *Über eine charakteristische Eigenschaft der Differentialgleichungen der Variationsrechnung*, Math. Ann. **49** (1897), 49–72.

139. Ivan Hlaváček, *Variational principles for parabolic equations*, Apl. Mat. **14** (1969), 278–297.

140. Lars Hörmander, *On the theory of general partial differential operators*, Acta Math. **94** (1955), 161–248.

141. Gregory Walter Horndeski, *Differential operators associated with the Euler-Lagrange operator*, Tensor (N.S.) **28** (1974), 303–318.

142. V. A. Il'in, *On the solvability of mixed problems for hyperbolic and parabolic equations*, Uspekhi Mat. Nauk **15** (1960), no. 2(92), 97–154; English transl. in Russian Math. Surveys **15** (1960).

143. V. P. Il'in, *Some inequalities in function spaces and their application to the study of the convergence of variational processes*, Trudy Mat. Inst. Steklov. **53** (1959), 64–127; English transl. in Amer. Math. Soc. Transl. (2) **81** (1969).

144. B. Frank Jones, Jr., *A fundamental solutions for the heat equation which is supported in a strip*, J. Math. Anal. Appl. **60** (1977), 314–324.

145. P. B. Kagan, *On the solvability of a mixed boundary value problem for a parabolic equation*, Trudy Novokuznetsk. Gos. Ped. Inst. **4** (1962), 52–55. (Russian)

146. L. I. Kamynin and V. N. Maslennikova, *Solution of the first boundary value problem for a quasilinear parabolic equation in a noncylindrical domain*, Mat. Sb. **57**(99) (1962), 241–264. (Russian)

147. L. V. Kantorovich, *Functional analysis and applied mathematics*, Uspekhi Mat. Nauk **3** (1948), no. 6(28), 89–185; English transl., Report No. 1509, Nat. Bur. Standards, Washington, D. C., 1952.

148. S. Kaplan, *An analogy between the variational principles of reactor theory and those of classical mechanics*, Nuclear Sci. and Engrg. **23** (1965), 234–237.

149. ____, *Canonical and involutory transformations of variational problems involving higher derivatives*, J. Math. Anal. Appl. **22** (1968), 45–53.

150. S. Kaplan and James A. Davis, *Canonical involutory transformations of the variational problems of transport theory*, Nuclear Sci. and Engrg. **28** (1967), 166–176.

151. Tosio Kato, *On some approximate methods concerning the operators T^*T*, Math. Ann. **126** (1953), 253–262.

152. ____, *Demicontinuity, hemicontinuity and monotonicity*, Bull. Amer. Math. Soc. **70** (1964), 548–550.

153. ____, *Perturbation theory for linear operators*, Springer-Verlag, 1966.

154. G. G. Kazaryan, *The first boundary value problem for a general nonregular equation with constant coefficients*, Dokl. Akad. Nauk SSSR **251** (1980), 22–24; English transl. in Soviet Math. Dokl. **21** (1980).

155. ____, *A variational problem for a nonregular equation and uniqueness of the classical solution*, Differential'nye Uravneniya **18** (1982), 1907–1917; English transl. in Differential Equations **18** (1982).

156. P. Keast, *On the solutions of the wave equation in one space dimension under derivative boundary conditions*, SIAM J. Appl. Math. **17** (1969), 223–230.

157. Wilhelm Kecs, *On the fundamental solution of the Cauchy problem for a class of partial differential equations with constant coefficients*, An. Univ. Bucureşti Ser. Şti. Nat. Mat.-Mec. **17** (1968), no. 2, 59–68. (Romanian)

158. A. A. Kerefov, *Nonlocal boundary value problems for parabolic equations*, Differentsial'nye Uravneniya **15** (1979), 74–78; English transl. in Differential Equations **15** (1979).

159. Michael Kerner, *Die Differentiale in der allgemeinen Analysis*, Ann. of Math. (2) **34** (1933), 546–572.

160. D. F. Kharazov, *Symmetrizable linear operators which depend meromorphically on a parameter, and their applications*. I, II, Izv. Vyssh. Uchebn. Zaved. Mat. **1967**, no. 8(63), 82–93; no. 11(66), 90–97. (Russian)

161. Yu. S. Kolesov, *Periodic solutions of quasilinear parabolic equations of the second order*, Trudy Moskov. Mat. Obshch. **21** (1970), 103–134; English transl. in Trans. Moscow. Math. Soc. **21** (1970).

162. A. N. Kolmogorov and S. V. Fomin, *Elements of the theory of functions and of functional analysis*, 4th ed., "Nauka", Moscow, 1976; English transl. of 1st ed., Vols. I, II, Graylock Press, Albany, N. Y., 1957, 1961.

163. Vadim Komkov, *Application of Rall's theorem to classical elastodynamics*, J. Math. Anal. Appl. **14** (1966), 511–521.

164. ____, *Another look at the dual variational principles*, J. Math. Anal. Appl. **63** (1978), 319–323.

165. V. I. Kondrashov, *On the theory of boundary value problems and eigenvalue problems in domains with a degenerate contour for variational and differential equations*, Doctoral Dissertation, Steklov Inst. Math. Acad. Sci. USSR, Moscow, 1948. (Russian)

166. Yu. I. Kovach, *The Goursat problem for an n-wave equation*, Dokl. i Soobshch. Uzhgorod. Gos. Univ. Ser. Fiz.-Mat. Nauk No. 5 (1962), 98–101. (Russian)

167. L. A. Kozdoba, *Methods of solving nonlinear problems of heat conduction*, "Nauka", Moscow, 1975. (Russian)

168. M. Krawtchouk [M. Kravchuk], *Sur la résolution des équations linéaires différentielles et intégrales parla méthode des moments.* Vol. I, Vseukraïn. Akad. Nauk Prirod.-Tekhn. Vīddīl, Kiev, 1932. (Ukrainian; 25-page French summary).

169. M. G. Kreĭn, *On the theory of weighted integral equations,* Bul. Akad. Shtiintse RSS Moldoven, **1965**, no. 7, 40–46. (Russian)

170. S. G. Kreĭn et al., *Functional analysis,* "Nauka", Moscow, 1964; English transl., Noordhoff, 1972.

171. Nicolas Kryloff [N. V. Krylov], *Sur une méthode d'intégration approchée conterant comme cas particulier la méthode de W. Ritz, ainsi que celle des moindres carrés,* C. R. Acad. Sci. Paris **182** (1926), 676–678.

172. L. D. Kudryavtsev, *Direct and inverse imbedding theorems,* Trudy Mat. Inst. Steklov. **55** (1959), English transl., Amer. Math. Soc., Providence, R. I., 1974.

173. ____, *On the solution by the variational method of elliptic equations which degenerate on the boundary of the region,* Dokl. Akad. Nauk SSSR **108** (1956), 16–19. (Russian)

174. ____, *Onan integral inequality,* Nauchn. Dokl. Vyssh. Shkoly Fiz.-Mat. Nauki **1959**, no. 3, 25–32. (Russian)

175. ____, *A course in mathematical analysis.* Vols. I, II, "Vysshaya Shkola", Moscow, 1981. (Russian)

176. W. Kundt and E. T. Newman, *Hyperbolic differential equations in two dimensions,* J. Mathematical Phys. **9** (1968), 2193–2210.

177. B. A. Kupershmidt, *Lagrangian formalism in the calculus of variations,* Funktsional. Anal. i Prilozhen. **10** (1976), no. 2, 77–78; English transl. in Functional Anal. Appl. **10** (1976).

178. V. D. Kupradze, *Approximate solution of problems of mathematical physics,* Uspekhi Mat. Nauk **22** (1967), no. 2(134), 59–107; English transl. in Russian Math. Surveys **22** (1967).

179. L. P. Kuptsov, *The fundamental solution of a class of elliptic-parabolic equations of second order,* Differentsial'nye Uravneniya **8** (1972), 1649–1660; English transl. in Differential Equations **8** (1972).

180. O. A. Ladyzhenskaya and N. N. Ural'tseva, *Linear and quasilinear elliptic equations,* "Nauka", Moscow, 1964; English transl., Academic Press, 1968.

181. Arno Langenbach, *Variationsmethoden in der nichtlinearen Elastizitäts- und Plastizitätstheorie,* Wiss. Z. Humboldt-Univ. Berlin Math.-Nat. Reihe **9** (1959/60), 146–164.

182. ____, *Die Regularisierung nichtlinearer Gleichungen,* Math. Nachr. **24** (1962), 33–51.

183. ____, *Über Gleichungen mit Potentialoperatoren und Minimalfolgen nichtquadratischer Funktionale*, Math. Nachr. **32** (1966), 9–24.

184. ____, *On the application of a variational principle to some nonlinear differential equations*, Dokl. Akad. Nauk SSSR **121** (1958), 214–217. (Russian)

185. ____, *On an application of the method of least squares to nonlinear equations*, Dokl. Akad. Nauk SSSR **143** (1962), 31–34; English transl. in Soviet Math. Dokl. **3** (1962).

186. M. M. Lavrent'ev and B. Imomnazarov, *Strongly positive operators*, Dokl. Akad. Nauk SSSR **238** (1979), 23–25; English transl. in Soviet Math. Dokl. **19** (1978).

187. Peter D. Lax, *Symmetrizable linear transformations*, Comm. Pure Appl. Math. **7** (1954), 633–647.

188. ____, *On Cauchy's problem for hyperbolic equations and the differentiability of solutions of elliptic equations*, Comm. Pure Appl. Math. **8** (1955), 615–633.

189. Henri Lebesgue, *Sur le problème de Dirichlet*, Rend. Circ. Mat. Palermo **24** (1907), 371–402.

190. Jean Leray, *Hyperbolic differential equations*, Inst. Adv. Study, Princeton, N. J., 1953.

191. Beppo Levi, *Sul principo di Dirichlet*, Rend. Circ. Mat. Palermo **22** (1906), 293–360.

192. J. L. Lions, *Sur quelques problèmes aux limites relatifs à des opérateurs différentiels elliptiques*, Bull. Soc. Math. France **83** (1955), 225–250.

193. ____, *Problèmes mixtes abstraits*, Proc. Internat. Congr. Math. (Edinburgh, 1958), Cambridge Univ. Press, 1960, pp. 389–397.

194. J. L. Lions and E. Magenes, *Problèmes aux limites non homogènes et applications*. Vol. 1, Dunod, Paris, 1968; English transl., Springer-Verlag, 1972.

195. P. I. Lizorkin and S. M. Nikol'skiĭ, *A classification of differentiable functions in some fundamental spaces with dominant mixed derivative*, Trudy Mat. Inst. Steklov. **77** (1965), 143–167; English transl. in Proc. Steklov Inst. Math. **77** (1965).

196. A. D. Lyashko, *On the Galerkin-Petrov method*, Kazan. Gos. Univ. Uchen. Zap. **117** (1957), kn. 2, 42–44. (Russian)

197. ____, *Convergence of Galerkin type methods*, Dokl. Akad. Nauk SSSR **120** (1958), 242–244. (Russian)

198. ____, *A generalization of Galerkin's method*, Izv. Vyssh. Uchebn. Zaved. Mat. **1958**, no. 4(5), 153–160; correction, **1959**, no. 2(9), 275. (Russian)

199. ___, *Convergence of methods analogous to Galerkin's*, Izv. Vyssh. Uchebn. Zaved. Mat. **1958**, no. 6(7), 176–179. (Russian)

200. ___, *Some versions of the Galerkin-Krylov method*, Dokl. Akad. Nauk SSSR **128** (1959), 468–470. (Russian)

201. ___, *An approximate solution of one-dimensional boundary value problems*, Izv. Vyssh. Uchebn. Zaved. Mat. **1962**, no. 2(27), 95–99. (Russian)

202. ___, *A variational method for nonlinear operator equations*, Kazan. Gos. Univ. Uchen. Zap. **125** (1965), kn. 2, 95–101. (Russian)

203. A. D. Lyashko and M. M. Karchevskiĭ, *A study of the method of straight lines for nonlinear elliptic equations*, Zh. Vychils. Mat. i Mat. Fiz. **7** (1967), 677–680; English transl. in USSR Comput. Math. and Math. Phys. **7** (1967).

204. ___, *Study of difference schemes for nonlinear equations by a variational method*, Izv. Vyssh. Uchebn. Zaved. Mat. **1967**, no. 3(58), 59–65; English transl. in Amer. Math. Soc. Transl. (2) **111** (1978).

205. F. Magri, *Variational formulation for every linear equation*, Internat. J. Engrg. Sci. **12** (1974), 537–549.

206. Franco Mandras, *Su una classe di equazioni ellittiche degeneri di tipo non variazionale in due variabili*, Boll. Un. Mat. Ital. B (5) **18** (1981), 605–618.

207. G. I. Marchuk and V. I. Agoshkov, *Introduction to projection-difference methods*, "Nauka", Moscow, 1981. (Russian)

208. Joseph Marty, *Valeurs singulièrs d'une équation de Fredholm*, C. R. Acad. Sci. Paris **150** (1910), 1499–1502.

209. A. E. Martynyuk, *On a generalization of a variational method*, Dokl. Akad. Nauk SSSR **117** (1957), 374–377. (Russian)

210. ___, *On Galerkin's method*, Kazan. Gos. Univ. Uchen. Zap. **117** (1957), kn. 2, 70–74. (Russian)

211. ___, *Variational methods in boundary value problems for weakly elliptic equations*, Dokl. Akad. Nauk SSSR **122** (1959), 1222–1225. (Russian)

212. ___, *Some new applications of Galerkin type methods*, Mat. Sb. **49(91)** (1959), 85–108. (Russian)

213. ___, *Applications of Galerkin's method and Galerkin moments to a certain partial differential equation of Vekua type*, Izv. Vyssh. Uchebn. Zaved. Mat. **1960**, no. 3(16), 188–204. (Russian)

214. ___, *Solution of a fundamental boundary value problem for certain linear partial differential equations of even order*, Dokl. Akad. Nauk SSSR **147** (1962), 1288–1291; English transl. in Soviet Math. Dokl. **3** (1962).

215. ____, *Some new criteria for convergence of the method of successive approximations*, Kazan. Gos. Univ. Uchen. Zap. **124** (1964), kn. 6, 183–188. (Russian)

216. ____, *Some approximate methods for solving nonlinear equations with unbounded operators*, Izv. Vyssh. Uchebn. Zaved. Mat. **1966**, no. 6(55), 85–94; addendum, ibid. **1967**, no. 8(63), 111. (Russian)

217. ____, *Some approximate methods of Galerkin type and combined type*, Izv. Vyssh. Uchebn. Zaved. Mat. **1967**, no. 10(65), 62–75. (Russian)

218. V. N. Maslennikova, *On mixed problems for a system of equations of mathematical physics*, Dokl. Akad. Nauk SSSR **102** (1955), 885–888. (Russian)

219. T. Matsuzawa, *Construction of fundamental solutions of hypoelliptic equations by the use of a probabilistic method*, Nagoya Math. J. **72** (1978), 103–126.

220. Krzyztof Maurin, *Methods of Hilbert spaces*, PWN, Warsaw, 1959; English transl., 1967.

221. V. P. Mikhailov, *The Dirichlet problem and the first mixed problem for a parabolic equation*, Dokl. Akad. Nauk SSSR **140** (1961), 303–306; English transl. in Soviet Math. Dokl. **2** (1961).

222. ____, *The Dirichlet problem for a parabolic equation. I*, Mat. Sb. **61(103)** (1963), 40–64. (Russian)

223. S. G. Mikhlin, *The problem of the minimum of a quadratic functional*, GITTL, Moscow, 1952; English transl., Holden-Day, San Francisco, Calif., 1965.

224. ____, *Two theorems on regularizers*, Dokl. Akad. Nauk SSSR **125** (1959), 737–739. (Russian)

225. ____, *Variational methods in mathematical physics*, 2nd rev. aug. ed., "Nauka", Moscow, 1970; English transl. of 1st ed., Pergamon Press, Oxford, and Macmillan, New York, 1964.

226. Clark B. Millikan, *On the study of the motions of viscous, incompressible fluids with particular reference to a variation principle*, Philos. Mag. and J. Sci. (7) **7** (1929), 641–662.

227. George J. Minty, *Integrability conditions for vector fields in Banach spaces*, Ph. D. Thesis, Univ. of Michigan, Ann Arbor, Mich., 1958.

228. ____, *Monotone (nonlinear) operators in Hilbert space*, Duke Math. J. **29** (1962), 341–346.

229. A. R. Mitchell and R. Wait, *The finite element method in partial differential equations*, Wiley, 1977.

230. B. L. Moiseiwitsch, *Variational principles*, Interscience, 1966.

231. Cathleen S. Morawetz, *A weak solution for a system of equations of elliptic-hyperbolic type*, Comm. Pure Appl. Math. **11** (1958), 315–331.

232. P. P. Mosolov, *A boundary value problem for hypoelliptic operators*, Mat. Sb. **55(97)** (1961), 307–328. (Russian)

233. ____, *Variational methods in nonstationary problems*, Izv. Akad. Nauk SSSR Ser. Mat. **34** (1970), 425–457; English transl. in Math. USSR Izv. **4** (1970).

234. M. Nagumo, *Lectures on the contemporary theory of partial differential equations*, Kyōritsu Shuppan, Tokyo, 1957 (Japanese); Russian transl., "Mir", Moscow, 1967.

235. M. Z. Nashed, *The convergence of the method of steepest descents for nonlinear equations with variational or quasi-variational operators*, J. Math. and Mech. **13** (1964), 765–794.

236. ____, *Differentiability and related properties of nonlinear operators: some aspects of the role of differentials in nonlinear functional analysis*, Nonlinear Functional Analysis and Applications (Proc. Sem., Madison, Wisc., 1970; L. B. Rall, editor), Academic Press, 1971, pp. 103–309.

237. J. von Neumann, *Über adjungierte Funktionaloperatoren*, Ann. of Math. (2) **33** (1932), 294–310.

238. S. M. Nikol'skiĭ, *On the question of solving the polyharmonic equation by a variational method*, Dokl. Akad. Nauk SSSR **88** (1953), 409–411. (Russian)

239. ____, *Properties of differentiable functions of several variables at the boundary of their domain of definition*, Dokl. Akad. Nauk SSSR **146** (1962), 542–545; English transl. in Soviet Math. Dokl. **3** (1962).

240. ____, *On stable boundary values of differentiable functions of several variables*, Mat. Sb. **61(103)** (1963), 224–252. (Russian)

241. ____, *A variational problem*, Mat. Sb. **62(104)** (1963), 53–75; English transl. in Amer. Math. Soc. Transl. (2) **51** (1966).

242. ____, *Approximation of functions of several variables and embedding theorems*, "Nauka", Moscow, 1969; English transl., Springer-Verlag, 1975.

243. B. Noble, *Complementary variational principles for boundary value problems. I*, V. S. Army Math. Res. Center Report No. 473, Univ. of Wisconsin, Madison, Wisc., 1964.

244. B. Noble and M. J. Sewell, *On dual extremum principles in applied mathematics*, J. Inst. Math. Appl. **9** (1972), 123–193.

245. Yu. I. Nyashin, *Variational formulation of the time-dependent problem of heat conduction*, Perm. Politekhn. Inst. Sb. Nauchn. Trudov Vyp. 152 (1974), 3–9. (Russian)

246. Yu. I. Nyashin, A. N. Skorokhodov, and I. N. Aran'ev, *A variational method for calculating a time-dependent temperature field*, Inzh.-Fiz. Zh. **26** (1974), 770–776; English transl. in J. Engrg. Phys. **26** (1974).

247. M. R. Osborne, *The numerical solution of a periodic parabolic problem subject to a nonlinear boundary condition. II*, Numer. Math. **12** (1968), 280–287.

248. V. P. Palamodov, *On the singularity of fundamental solutions of hypoelliptic equations*, Sibirsk. Mat. Zh. **4** (1963), 1365–1375; English transl. in Amer. Math. Soc. Transl. (2) **54** (1966).

249. B. P. Paneyakh, *On the existence of well-posed boundary value problems for systems of differential equations*, Uspekhi Mat. Nauk **18** (1963), no. 2(110), 175–176. (Russian)

250. W. V. Petryshyn, *Direct and iterative methods for the solution of linear operator equations in Hilbert space*, Trans. Amer. Math. Soc. **105** (1962), 136–175.

251. ____, *The generalized overrelaxation method for the approximate solution of operator equations in Hilbert space*, J. Soc. Indust. Appl. Math. **10** (1962), 675–690.

252. ____, *On a general iterative method for the approximate solution of linear operator equations*. Math. Comp. **17** (1963), 1–10.

253. ____, *On the eigenvalue problem $Tu - \lambda Su = 0$ with unbounded and nonsymmetric operators T and S*, Philos. Trans. Roy. Soc. London Ser. A **262** (1967/68), 413–458.

254. ____, *On the extrapolated Jacobi or simultaneous displacements method in the solution of matrix and operator equations*, Math. Comp. **19** (1965), 37–55.

255. ____, *On two variants of a method for the solution of linear equations with unbounded operators and their applications*, J. Math. and Phys. **44** (1965), 297–312.

256. ____, *On generalized inverses and on the uniform convergence of $(I - \beta K)^n$ with application to iterative methods*, J. Math. Anal. Appl. **18** (1967), 417–439.

257. ____, *On a class of K-p.d. and non-K-p.d. operators and operator equations*, J. Math. Anal. Appl. **10** (1965), 1–24.

258. ____, *On the extension and the solution of nonlinear operator equations*, Illinois J. Math. **10** (1966), 255–274.

259. ____, *On the inversion of matrices and linear operators*, Proc. Amer. Math. Soc. **16** (1965), 893–901.

260. ____, *On the iteration, projection and projection-iteration methods in the solution of nonlinear functional equations*, J. Math. Anal. Appl. **21** (1968), 575–607.

261. ____, *Remarks on the generalized overrelaxation and the extrapolated Jacobi methods for operator equations in Hilbert space*, J. Math. Anal. Appl. **29** (1970), 558–568.

262. ____, *On the approximation-solvability of equations involving A-proper and pseudo-A-proper mappings*, Bull. Amer. Math. Soc. **81** (1975), 223–312.

263. ____, *Variational solvability of quasilinear elliptic boundary value problems at resonance*, Nonlinear Anal. **5** (1981), 1095–1108.

264. T. H. H. Pian, *Finite element methods by variational principles with relaxed continuity requirement*, Variational Methods in Engineering (Proc. Internat. Sympos., Southhampton, 1973; C. Brebbia and H. Tottenham, editors), Vol. I, Southampton Univ. Press, Southhampton, 1973, pp. 3/1–3/25.

265. H. Poincaré, *Les idées de Hertz sur la mécanique*, Rev. Gén. Sci. Pures Appl. **8** (1897), 734–743.

266. G. C. Pomraning, *Complementary variational principles and their application to neutron transport problems*, J. Mathematical Phys. **8** (1967), 2096–2108.

267. ____, *Reciprocal and canonical forms of variational problems involving linear operators*, J. Math. and Phys. **47** (1968), 155–169.

268. A. A. Popova, *Application of the method of steepest descent to the solution of asymmetric operator equations*, Vychisl. i Prikl. Mat. (Kiev) Vyp. 42 (1980), 17–24. (Russian)

269. ____, *Variational methods for the approximate solution of boundary value problems in mathematical physics with a nonsymmetric operator*, Author's Summary of Candidate's Dissertation, Kiev State Univ., Kiev, 1981. (Russian)

270. I. Prigogine and P. Glansdorff, *Variational properties and fluctuation theory*, Physica **3** (1965), 1242–1256.

271. L. B. Rall, *Variational methods for nonlinear integral equations*, Nonlinear Integral Equations (Proc. Adv. Sem., Madison, Wisc., 1963; P. M. Anselone, editor), Univ. of Wisconsin Press, Madison, Wisc., 1964, pp. 155–189.

272. ____, *On complementary variational principles*, J. Math. Anal. Appl. **14** (1966), 174–184.

273. I. M. Rapoport, *The inverse problem of the calculus of variations*, Zh. Inst. Mat. Akad. Nauk Ukrain. SSR **1938**, no. 4, 105–122. (Russian)

274. ____, *Le problème inverse du calcul des variations*, C. R. (Dokl.) Acad. Sci. URSS **18** (1938), 131–135.

275. William T. Reid, *Symmetrizable completely continuous linear transformations in Hilbert space*, Duke Math. J. **18** (1951), 41–56.

276. Karel Pektorys, *On application of direct variational methods to the solution of parabolic boundary value problems of arbitrary order in the space variables*, Czechoslovak Math. J. **21(96)** (1971), 318–339.

277. ____, *Variational methods in mathematics, science and engineering*, Reidel, 1977.

278. Peter D. Robinson, *Complementary variational principles*, Nonlinear Functional Analysis and Applications (Proc. Sem., Madison, Wisc., 1970; L. B. Rall, editor), Academic Press, 1971, pp. 507–576.

279. Marcel N. Roşculet, *Contributions to the study of partial differential equations of parabolic type. IV*, Acad. R. P. Romîne Stud. Cerc. Mat. **13** (1962), 419–437. (Romanian; French summary)

280. Hanno Rund, *The reduction of certain boundary value problems to variational problems by means of transversality conditions*, Numer. Math. **15** (1970), 49–56.

281. Kyuichi Sakuma, *On the fundamental solutions of ultrahyperbolic operators*, Sûgaku **12** (1960/61), 107–111. (Japanese)

282. David A. Sánchez, *Calculus of variations for integrals depending on a convolution product*, Ann. Scuola Norm. Sup. Pisa (3) **18** (1964), 233–254.

283. ____, *On extremals for integrals depending on a convolution product*, J. Math. Anal. Appl. **11** (1965), 213–216.

284. Ranbir S. Sandhu and Karl S. Pister, *Variational principles for boundary value and initial-boundary value problems in continuum mechanics*, Internat. J. Solids and Structures 7 (1971), 639–654.

285. ____, *Variational methods in continuum mechanics*, Variational Methods in Engineering (Proc. Internat. Sympos., Southhampton, 1973; C. Brebbia and H. Tottenham, editors), Vol. I, Southampton Univ. Press, Southhampton, 1973, pp. 1/13–1/25.

286. Ruggero Maria Santilli, *Foundations of theoretical mechanics. Vol. I*, Springer-Verlag, 1978.

287. Leonard Sarason, *Differentiable solutions of symmetrizable and singular symmetric first order systems*, Arch. Rational Mech. Anal. **26** (1967), 357–384.

288. R. S. Schechter, *The variational method in engineering*, McGraw-Hill, 1967.

289. Laurent Schwartz, *Les travaux de L. Gårding sur les équations aux dérivés partielles elliptiques*, Sém. Bourbaki 1951/52, Exposé 67, Secretariat Math., Paris, 1952.

290. L. I. Sedov, *Application of a basic variational equation for the construction of models of continuous media*, Selected Questions of Modern Mechanics (S. S. Grigoryan Fiftieth Birthday Vol.), Izdat. Moskov. Gos. Univ., Moscow, 1981, pp. 11–64. (Russian) cf. also *Applying the basic variational equation for building models of matter and fields*, Proc. IUTAM-ISIMM Sympos.Modern Developments in Analytical Mech.(Torino, 1982). Vol. II, Atti Accad. Sci. Torino Cl. Sci. Fis. Mat. Nat. **117** (1983), Suppl. 2, 765–816.

291. M. J. Sewell, *On dual approximation principles and optimization in continuum mechanics*, Philos. Trans. Roy. Soc. London Ser. A **265** (1969/70), 319–351.

292. V. M. Shalov, *A generalization of Friedrichs space*, Dokl. Akad. Nauk SSSR **151** (1963), 292–294; English transl. in Soviet Math. Dokl. **4** (1963).

293. ____, *Solution of nonselfadjoint equations by the variational method*, Dokl. Akad. Nauk SSSR **151** (1963), 511–512; English transl. in Soviet Math. Dokl. **4** (1963).

294. ____, *The principle of a minimum of a quadratic functional for a hyperbolic equation*, Differentsial'nye Uravneniya **1** (1965), 1338–1365; English transl. in Differential Equations **1** (1965).

295. ____, *Equations of continuum mechanics*, Differentsial'nye Uravneniya **9** (1973), 912–921; English transl. in Differential Equations **9**(1973).

296. F. A. Shelkovnikov, *A generalized Cauchy formula*, Uspekhi Mat. Nauk **6** (1951), no. 3(43), 157–159. (Russian)

297. G. E. Shilov, *Local properties of solutions of partial differential equations with constant coefficients*, Uspekhi Mat. Nauk **14** (1959), no. 5(89), 3–44; English transl. in Amer. Math. Soc. Transl. (2) **42** (1964).

298. I. I. Shmulev, *Periodic solutions of the first boundary value problem for parabolic equations*, Mat. Sb. **66**(108) (1965), 398–410; English transl. in Amer. Math. Soc. Transl. (2) **79** (1969).

299. N. N. Shopolov, *A mixed problem for the heat equation with a nonlocal initial condition*, C. R. Acad. Bulgare Sci. **34** (1981), 935–936. (Russian)

300. J. P. O. Silberstein, *Symmetrisable operators*, II, III, J. Austral. Math. Soc. **4** (1964), 15–30, 31–48.

301. L. N. Slobodetskiĭ, *Generalized Sobolev spaces and their application to boundary value problems for partial differential equations*, Leningrad,

Gos. Ped. Inst. Uchen. Zap. **197** (1958), 54–112; English transl. in Amer. Math. Soc. Transl. (2) **57** (1966).

302. M. G. Slobodyanskiĭ, *Estimate of the error of the quantity sought for in the solution of linear problems by a variational method*, Dokl. Akad. Nauk SSSR **86** (1952), 243–246. (Russian)

303. ——, *Estimates of errors in approximate solutions of linear problems*, Prikl. Mat. Mekh. **17** (1953), 229–244. (Russian)

304. ——, *Approximate solution of some boundary value problems for elliptic differential equations, and error estimates*, Dokl. Akad. Nauk SSSR **89** (1953), 221–224. (Russian)

305. ——, *On the construction of an approximate solution in linear problems*, Prikl. Mat. Mekh. **19** (1955), 571–588. (Russian)

306. ——, *On transformation of the problem of the minimum of a functional to the problem of the maximum*, Dokl. Akad. Nauk SSSR **91** (1953), 733–736. (Russian)

307. M. M. Smirnov, *The first boundary value problem for a hyperbolic equation of the fourth order*, Vestnik Leningrad. Univ. **1958**, no. 19 (Ser. Mat. Mekh. Astr. vyp. 4), 55–57. (Russian)

308. W. Smit, *Complementary variational principles for the solution of a large class of operator equations*, Variational Methods in Engineering (Proc. Internat. Sympos., Southhampton, 1973; C. Brebbia and H. Tottenham, editors), Vol. I, Southampton Univ. Press, Southhampton, 1973, pp. 1/29–1/42.

309. Donald R. Smith and C. V. Smith, Jr., *When is Hamilton's principle an extremum principle?*, AIAA J. **12** (1974), 1573–1576.

310. P. Smith [Peter Smith], *Approximate operators and dual extremum principles for linear problems*, J. Inst. Math. Appl. **22** (1978), 457–465.

311. S. L. Sobolev, *Applications of functional analysis in mathematical physics*, Izdat. Leningrad. Gos. Univ., Leningrad, 1950; English transl., Amer. Math. Soc., Providence, R. I., 1963.

312. S. L. Sobolev and M. I. Vishik, *Some functional methods in the theory of partial differential equations*, Proc. Third All-Union Math. Congr. (Moscow, 1956), Vol. 3, Izdat. Akad. Nauk SSSR, Moscow, 1956, pp. 152–162. (Russian)

313. P. E. Sobolevskiĭ, *On equations with operators forming an acute angle*, Dokl. Akad. Nauk SSSR **116** (1957), 754–757. (Russian)

314. Ivar Stakgold, *Green's functions and boundary value problems*, Wiley, 1979.

315. J. L. Synge, *The method of the hypercircle in function-space for boundary-value problems*, Proc. Roy. Soc. London Ser. A 191 (1947), 447–467.

316. ____, *The method of the hypercircle in elasticity when body forces are present*, Quart. Appl. Math. **6** (1947), 15–19.

317. ____, *The hypercircle in mathematical physics: a method for the approximate solution of boundary value problems*, Cambridge Univ. Press, 1957.

318. Zofia Szmydt, *Fourier transformation and linear differential equations*, PWN, Warsaw, and Peidel, Dordrecht, 1977.

319. A. T. Taldykin, *Variational methods for solving equations*, Differentsial′nye Uravneniya **10** (1974), 1714–1720; English transl. in Differential Equations **10** (1974).

320. Enzo Tonti, *Condizioni iniziali neiprincipi variazionali*, Ist. Lombardo Accad. Sci. Lett. Rend. A **100** (1966), 982–988.

321. ____, *Variational formulation of nonlinear differential equations*. I, II, Acad. Roy. Belgique Bull. Cl. Sci. (5) **55** (1969), 137–165, 262–278.

322. ____, *On the variational formulation for linear initial value problems*, Ann. Mat. Pura Appl. (4) **95** (1973), 331–359.

323. ____, *A systematic approach to the search for variational principles*, Variational Methods in Engineering (Proc. Internat. Sympos., Southhampton, 1973; C. Brebbia and H. Tottenham, editors), Vol. I, Southampton Univ. Press, Southhampton, 1973, pp. 1/1–1/12.

324. ____, *A general solution of the inverse problem of the calculus of variations*, Proc. First Internat. Conf. Nonpotential Interactions and Their Lie-Admissible Treatment (Orléans, 1982), Part C, Hadronic J. 5(1981/82), 1404–1450.

325. E. Trefftz, *Ein Gegenstück zum Ritzschen Verfahren*, Proc. Second Internat. Congr. Appl. Mech. (Zürich, 1926), Orell Füssli Verlag, Zürich, 1927, pp. 131–138.

326. ____, *Konvergenz und Fehlerabschätzung beim Ritzschen Verfahren*, Math. Ann. **100** (1928), 503–521.

327. J. F. Treves [François Trèves], *Lectures on linear partial differential equations with constant coefficients*, Notas de Mat., no. 27, Inst. Mat. Pura Apl. Conselho Nac. Pesquisas, Rio de Janeiro, 1961.

328. Hans Triebel, *Interpolation theory, function spaces, differential operators*, VEB Deutscher Verlag Wiss., Berlin, 1977, and North-Holland, Amsterdam, 1978.

329. Tu Guizhang, *On formal variational calculus of higher dimensions*, J. Math. Anal. Appl. **94** (1983), 348–365.

330. M. M. Vaĭnberg, *Some questions of the differential calculus in linear spaces*, Uspekhi Mat. Nauk **7** (1952), no. 4(50), 55–102. (Russian)

331. ___, *Variational methods for the study of nonlinear operators*, GITTL, Moscow, 1956; English transl., Holden-Day, San Francisco, Calif., 1964.

332. ___, *On some new principles in the theory of nonlinear equations*, Uspekhi Mat. Nauk **15** (1960), no. 1(19), 243–244. (Russian)

333. ___, *Convergence of the method of steepest descent for nonlinear equations*, Sibirsk. Mat. Zh. **2** (1961), 201–220. (Russian)

334. M. M. Vaĭnberg and R. I. Kachurovskiĭ, *On the variational theory of nonlinear operators and equations*, Dokl. Akad. Nauk SSSR **129** (1959), 1199–1202. (Russian)

335. A. L. Vanderbauwhede, *Potential operators and the inverse problem of classical mechanics*, Hadronic J. **1** (1978), 1177–1197.

336. R. S. Varga, *Functional analysis and approximation theory in numerical analysis*, Conf. Board Math. Sci. Regional Conf. Ser. Appl. Math., vol. 3, SIAM, Philadelphia, Pa., 1971.

337. Waldemar Velte, *Direkte Methoden der Variationsrechnung*, Teubner, Stuttgart, 1976.

338. ___, *Komplementäre Extremalprobleme*, Methoden und Verfahren der Mathematischen Physik, vol. 15 (B. Brosowski and E. Martensen, editors), Bibliograpisches Inst., Mannheim, 1976, pp. 1–44.

339. M. I. Vishik, *Mixed boundary value problems for equations having a first derivative with respect to time, and an approximate method for solving them*, Dokl. Akad. Nauk SSSR **99** (1954), 189–192. (Russian)

340. V. S. Vladimirov, *Mathematical problems in the uniform-speed theory of transport*, Trudy Mat. Inst. Steklov. **61** (1961). (Russian)

341. L. R. Volevich and B. P. Paneyakh, *Some spaces of generalized functions and imbedding theorems*, Uspekhi Mat. Nauk **20** (1965), no. 1(121), 3–74; English transl. in Russian Math. Surveys **20** (1965).

342. Vito Volterra, *Leçons sur les fonctions des lignes*, Gauthier-Villars, Paris, 1913.

343. Eugene L. Wachspress, *A rational finite element basis*, Academic Press, 1975.

344. K. Washizu, *Bounds for solutions of boundary value problems in elasticity*, J. Math. and Phys. **32** (1953), 117–128.

345. H. F. Weinberger, *Upper and lower bounds for torsional rigidity*, J. Math. and Phys. **32** (1953), 54–62.

346. Jürgen Weyer, *Über maximale Monotonie von Operatoren des Typs L*Φ∘L*, Report 78-24, Math. Inst., Univ. Köln, 1978; published as Manuscripta Math. **28** (1979), 305–316.

347. S. Ya. Yakubov, *Hilbert-Schmidt theory for integral and integro-differential equations with nonsymmetric kernels*, Izv. Akad. Nauk Azerbaidzhan. SSR Ser. Fiz.-Mat. i Tekhn. Nauk **1962**, no. 1, 35–45. (Russian)

348. A. C. Zaanen, *Über vollstetige symmetrische und symmetrisierbare Operatoren*, Nieuw Arch. Wiskunde (2) **22** (1943), 57–80.

349. ____, *Normalisable transformations in Hilbert space and systems of linear integral equations*, Acta Math. **83** (1950), 197–248.

350. B. Zainea, *The generalized energy method*, An. Univ. Bucureşti Mat.-Mec. **21** (1972), 119–128. (Romanian)

351. V. I. Zaplstnyï [V. Ī. Zaplatnii], *Construction of the density of the Lagrangian function from a given fourth-order partial differential equation*, Dokl. Akad. Nauk Ukrain. SSR Ser. A **1980**, no. 1, 33–35.

352. Eduardo H. Zarantonello, *The closure of the numerical range contains the spectrum*, Bull. Amer. Math. Soc. **70** (1964), 781–787.

353. Stanislas [Stanisław] Zaremba, *Sur le principe de minimum*, Bull. Internat. Acad. Sci. Cracovie Cl. Sci. Math. Nat. **1909** (deuxième semestre), no. 7, 199–264.

354. ____, *Sur un problème toujours possible comprenant à titre de cas particuliers, le problème de Dirichlet et celui de Neumann*, J. Math. Pures Appl. (9) **6** (1927), 127–163.

355. O. C. Zienkiewicz, *The finite element method in engineering science*, 2nd rev. ed., McGraw-Hill, 1971; German transl., Carl Hanser Verlag, Munich and Vienna, 1975.

356. George Dincă and Daniel Mateescu, *On the structure of linear and K-positive definite operators*, Rev. Roumaine Math. Pures Appl. **27** (1982), 677–687.

ABCDEFGHIJ – 89